Power System Relaying

Power System Relaying

Fifth Edition

Stanley H. Horowitz
Retired Consulting Engineer
American Electric Power
BSEE City College of New York, USA

Arun G. Phadke
University Distinguished Research Professor
Virginia Technical University, USA

Charles F. Henville
Principal
Henville Consulting Inc.
Delta, BC, Canada

This Fifth edition first published 2023
© 2023 John Wiley & Sons Ltd

Edition History
Research Studies Press Limited (3e, 2008); John Wiley & Sons Ltd (4e, 2014)

The right of Stanley H. Horowitz, Arun G. Phadke and Charles F. Henville to be identified as the authors of this work has been asserted in accordance with law.

Registered Offices
John Wiley & Sons Ltd, The Atrium, Southern Gate, Chichester, West Sussex, PO19 8SQ, UK

Editorial Office
The Atrium, Southern Gate, Chichester, West Sussex, PO19 8SQ, UK

For details of our global editorial offices, customer services, and more information about Wiley products visit us at www.wiley.com.

Wiley also publishes its books in a variety of electronic formats and by print-on-demand. Some content that appears in standard print versions of this book may not be available in other formats.

Library of Congress Cataloging-in-Publication Data applied for
Names: Horowitz, Stanley H., 1925- author. | Phadke, Arun G. author. |
    Henville, Charles F., author. | John Wiley & Sons, publisher.
Title: Power system relaying / Stanley H. Horowitz, Arun G. Phadke, Charles
    F. Henville.
Description: Fifth edition. | Hoboken, NJ : Wiley, 2023. | Includes
    bibliographical references and index.
Identifiers: LCCN 2022027194 (print) | LCCN 2022027195 (ebook) | ISBN
    9781119838432 (cloth) | ISBN 9781119838449 (adobe pdf) | ISBN
    9781119838456 (epub)
Subjects: LCSH: Protective relays. | Electric power systems–Protection.
Classification: LCC TK2861 .H67 2023 (print) | LCC TK2861 (ebook) | DDC
    621.31/7–dc23/eng/20220817
LC record available at https://lccn.loc.gov/2022027194
LC ebook record available at https://lccn.loc.gov/2022027195

Cover Design: Wiley
Cover Image: © Arun Phadke

Set in 9.5/12.5pt STIXTwoText by Straive, Pondicherry, India

Contents

Preface to the Fifth Edition

We are pleased to offer this fifth edition of 'Power System Relaying,' a book welcomed by teachers and students in many countries around the world for the last 20 years. We are saddened by the fact that our leading author—Stanley Horowitz—was unable to participate in this revision. We certainly miss his sure hand and deep knowledge of the protection field to guide us in preparing this edition.

In this edition, we have taken note of the thoughtful comments of several reviewers and made changes to include recent references to IEEE and IEC standards as they address relaying issues. We have also updated many of the chapters with present-day practices and technologies and provided new references to additional and updated IEEE and IEC standards.

We have also added a new Chapter 16 with solutions to the problems which appear at the end of many chapters. In the past, the solutions were provided to instructors who requested them for use in their protection classes. We believe including the solutions in the book itself will make them more useful to the instructors as well as the students.

August 2022

Arun G. Phadke
Portland, Oregon, United States
Charles F. Henville
Vancouver, BC, Canada

The readers will note that we have a new member joining the team of authors: Charles Henville. Mr. Henville is a distinguished protection engineer with vast experience both in theory and practice of protection.
Welcome Charles.

Preface to the Fourth Edition

The third edition of our book, issued with corrections in 2009, continued to be used as a text book in several universities around the world. The success of our book and the positive feed-back we continue to receive from our colleagues are gratifying. Since that time, we have had a few other typos and errors pointed out to us, which we are correcting in this fourth edition.

However, the major change in this edition is the inclusion of two new chapters. Chapter 14 gives an account of the application of wide-area measurements (WAMS) in the field of protection. WAMS technology using global positioning satellites (GPS) for syn-chronized measurements of power system voltages and currents is finding many applica-tions in monitoring, protection, and control of power systems. Field installations of such systems are taking place in most countries around the world, and we believe that an account of what is possible with this technology, as discussed in Chapter 14, is timely and will give the student using this edition an appreciation of these exciting changes taking place in the field of power system protection. Chapter 14 also provides a list of relevant references for the reader who is interested in pursuing this technology in depth.

Another major development in the field of power system protection on the advent of renewable resources for generation of electricity is reported in the new Chapter 15. This also is a technology that is being deployed in most modern power systems around the world. As the penetration of renewable resources in the mix of generation increases, many chal-lenges are faced by power system engineers. In particular, how to handle the interconnec-tion of these resources with proper protection systems is a very important subject. We are very fortunate to have a distinguished expert, James Niemira of S&C Electric Company, Chicago, contributor of this chapter to our book. The chapter also includes a list of refer-ences for the interested reader. We believe this chapter will answer many of the questions asked by our students.

Finally, we would welcome continued correspondence from our readers who give us val-uable comments about what they like in the book, and what other material should be included in future editions. We wish to thank all the readers who let us have their views, and assure them that we greatly value their inputs.

April 24, 2013

Stanley H. Horowitz
Columbus
Arun G. Phadke
Blacksburg

Preface to the Third Edition

The second edition of our book, issued in 1995, continued to receive favorable response from our colleagues and is being used as a textbook by universities and in industry courses worldwide. The first edition presented the fundamental theory of protective relaying as applied to individual system components. This concept was continued throughout the second edition. In addition, the second edition added material on generating plant auxiliary systems, distribution protection concepts, and the application of electronic inductive and capacitive devices to regulate system voltage. The second edition also presented additional material covering monitoring power system performance and fault analysis. The application of synchronized sampling and advanced timing technologies using the global positioning satellite (GPS) system was explained.

This third edition takes the problem of power system protection an additional step forward by introducing power system phenomena which influence protective relays and for which protective schemes, applications, and settings must be considered and implemented. The consideration of power system stability and the associated application of relays to mitigate its harmful effects are presented in detail. New concepts such as undervoltage load shedding, adaptive relaying, hidden failures, and the Internet standard COMTRADE and its uses are presented. The history of notable blackouts, particularly as affected by relays, is presented to enable students to appreciate the impact that protection systems have on the overall system reliability.

As mentioned previously, we are gratified with the response that the first and second editions have received as both a textbook and a reference book. Recent changes in the electric power industry have resulted in power system protection assuming a vital role in maintaining power system reliability and security. It is the authors' hope that the additions embodied in this third edition will enable all electric power system engineers, designers, and operators to better integrate these concepts and to understand the complex interaction of relaying and system performance.

Stanley H. Horowitz
Columbus
Arun G. Phadke
Blacksburg

Preface to the Second Edition

The first edition, issued in 1992, has been used as a textbook by universities and in industry courses throughout the world. Although not intended as a reference book for practicing protection engineers, it has been widely used as one. As a result of this experience and of the dialog between the authors, teachers, students, and engineers using the first edition, it was decided to issue a second edition, incorporating material which would be of significant value. The theory and fundamentals of relaying constituted the major part of the first edition and it remains so in the second edition. In addition, the second edition includes concepts and practices that add another dimension to the study of power system protection.

A chapter has been added covering monitoring power system performance and fault analysis. Examples of oscillographic records introduce the student to the means by which disturbances can be analyzed and corrective action and maintenance initiated. The application of synchronized sampling for technologies such as the GPS satellite is explained. This chapter extends the basic performance of protective relays to include typical power system operating problems and analysis. A section covering power plant auxiliary systems has been added to the chapter on the protection of rotating machinery. Distribution protection concepts have been expanded to bridge the gap between the protection of distribution and transmission systems. The emerging technology of static VAR compensators to provide inductive and capacitive elements to regulate system voltage has been added to the chapter on bus protection. The subject index has been significantly revised to facilitate reference from both the equipment and the operating perspective.

We are gratified with the response that the first edition has received as a text and reference book. The authors thank the instructors and students whose comments generated many of the ideas included in this second edition. We hope that the book will continue to be beneficial and of interest to students, teachers, and power system engineers.

Stanley H. Horowitz
Columbus
Arun G. Phadke
Blacksburg

Preface to the First Edition

This book is primarily intended to be a textbook on protection, suitable for final-year undergraduate students wishing to specialize in the field of electric power engineering. It is assumed that the student is familiar with techniques of power system analysis, such as three-phase systems, symmetrical components, short-circuit calculations, load flow, and transients in power systems. The reader is also assumed to be familiar with calculus, matrix algebra, Laplace and Fourier transforms, and Fourier series. Typically, this is the background of a student who is taking power option courses at a US university. The book is also suitable for a first-year graduate course in power system engineering.

An important part of the book is the large number of examples and problems included in each chapter. Some of the problems are decidedly difficult. However, no problems are unrealistic, and, difficult or not, our aim is always to educate the reader, and help the student realize that many of the problems that will be faced in practice will require careful analysis, consideration, and some approximations.

The book is not a reference book, although we hope it may be of interest to practicing relay engineers as well. We offer derivations of several important results, which are normally taken for granted in many relaying textbooks. It is our belief that by studying the theory behind these results, students may gain an insight into the phenomena involved, and point themselves in the direction of newer solutions which may not have been considered. The emphasis throughout the book is on giving the reader an understanding of power system protection principles. The numerous practical details of relay system design are covered to a limited extent only, as required to support the underlying theory. Subjects which are the province of the specialist are left out. The engineer interested in such detail should consult the many excellent reference works on the subject, and the technical literature of various relay manufacturers.

The authors owe a great deal to published books and papers on the subject of power system protection. These works are referred to appropriate places in the text. We would like to single out the book by the late C. R. Mason, *The Art and Science of Protective Relaying*, for special praise. We, and many generations of power engineers, have learned relaying from this book. It is a model of clarity, and its treatment of the protection practices of that day is outstanding.

Our training as relay engineers has been enhanced by our association with the Power System Relaying Committee of the Institute of Electrical and Electronics Engineers (IEEE) and the Study Committee SC34 of the Conférence Internationale des Grands Réseaux

Électriques a Haute Tension (CIGRE). Much of our technical work has been under the auspices of these organizations. The activities of the two organizations, and our interaction with the international relaying community, have resulted in an appreciation of the differing practices throughout the world. We have tried to introduce an awareness of these differences in this book. Our long association with the American Electric Power (AEP) Service Corporation has helped sustain our interest in electric power engineering, and particularly in the field of protective relaying. We have learned much from our friends in AEP. AEP has a well-deserved reputation for pioneering in many phases of electric power engineering, and particularly in power system protection. We are fortunate to be a part of many important relaying research and development efforts conducted at AEP. We have tried to inject this experience of fundamental theory and practical implementation throughout this text. Our colleagues in the educational community have also been instrumental in getting us started on this project, and we hope they find this book useful. No doubt some errors remain, and we will be grateful if readers bring these errors to our attention.

Stanley H. Horowitz
Columbus
Arun G. Phadke
Blacksburg

1

Introduction to Protective Relaying

1.1 What is relaying?

In order to understand the function of protective relaying systems, one must be familiar with the nature and the modes of operation of an electric power system. Electric energy is one of the fundamental resources of modern industrial society. Electric power is available to the user instantly, at the correct voltage and frequency, and exactly in the amount that is needed. This remarkable performance is achieved through careful planning, design, installation, and operation of a very complex network of generators, transformers, and transmission and distribution lines. To the user of electricity, the power system appears to be in a steady state: imperturbable, constant, and infinite in capacity. Yet, the power system is subject to constant disturbances created by random load changes, by faults created by natural causes, and sometimes as a result of equipment or operator failure. In spite of these constant perturbations, the power system maintains its quasi-steady state because of two basic factors: the large size of the power system in relation to the size of individual loads or generators, and correct and quick remedial action taken by the protective relaying equipment.

Relaying is the branch of electric power engineering concerned with the principles of design and operation of equipment (called "relays" or "protective relays") that detects abnormal power system conditions and initiates corrective action as quickly as possible in order to return the power system to its normal state. The quickness of response is an essential element of protective relaying systems—response times of the order of a few milliseconds are often required. Consequently, human intervention in the protection system operation is not possible. The response must be automatic, quick, and should cause a minimum amount of disruption to the power system. As the principles of protective relaying are developed in this book, the reader will perceive that the entire subject is governed by these general requirements: correct diagnosis of trouble, quickness of response, and minimum disturbance to the power system. To accomplish these goals, we must examine all possible types of fault or abnormal conditions that may occur in the power system. We must analyze the required response to each of these events and design protective equipment that will provide such a response. We must further examine the possibility that protective relaying equipment itself may fail to operate correctly, and provide for a backup protective function. It should be clear that extensive and sophisticated equipment is needed to accomplish these tasks.

Power System Relaying, Fifth Edition. Stanley H. Horowitz, Arun G. Phadke and Charles F. Henville.
© 2023 John Wiley & Sons Ltd. Published 2023 by John Wiley & Sons Ltd.

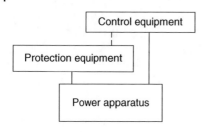

Figure 1.1 Three-layered structure of power systems.

1.2 Power system structural considerations

1.2.1 Multilayered structure of power systems

A power system is made up of interconnected equipment that can be said to belong to one of the three layers from the point of view of the functions performed. This is illustrated in Figure 1.1.

At the basic level is the power apparatus that generates, transforms, and distributes the electric power to the loads. Next, there is the layer of control equipment. This equipment helps to maintain the power system at its normal voltage and frequency, generates sufficient power to meet the load, and maintains optimum economy and security in the interconnected network. The control equipment is organized in a hierarchy of its own, consisting of local and central control functions. Finally, there is the protection equipment layer. The response time of protection functions is generally faster than that of the control functions. Protection acts to open- and closed-circuit breakers (CBs), thus changing the structure of the power system, whereas the control functions act continuously to adjust system variables, such as the voltages, currents, and power flow on the network. Oftentimes, the distinction between a control function and a protection function becomes blurred. This is becoming even more of a problem with the recent advent of computer-based protection systems in substations. For our purposes, we may arbitrarily define all functions that lead to operation of power switches or CBs to be the tasks of protective relays, while all actions that change the operating state (voltages, currents, and power flows) of the power system without changing its structure to be the domain of control functions.

1.2.2 Neutral grounding of power systems

Neutrals of power transformers and generators can be grounded in a variety of ways, depending upon the needs of the affected portion of the power system. As grounding practices affect fault current levels, they have a direct bearing upon relay system designs. In this section, we examine the types of grounding system in use in modern power systems and the reasons for each of the grounding choices. Influence of grounding practices on relay system design will be considered at appropriate places throughout the remainder of this book.

It is obvious that there is no ground fault current in a truly ungrounded system. This is the main reason for operating the power system ungrounded. As the vast majority of faults on a power system are ground faults, service interruptions due to faults on an ungrounded system are greatly reduced. However, as the number of distribution and transmission lines connected to the power system grows, the capacitive coupling of the feeder conductors with ground provides a path to ground, and a ground fault on such a system produces a capacitive fault current. This is illustrated in Figure 1.2a. The coupling capacitors to ground,

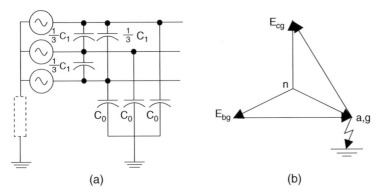

Figure 1.2 Neutral grounding impedance. (a) System diagram and (b) phasor diagram showing neutral shift on ground fault.

C_0, provide the return path for the fault current. The interphase capacitors $1/3C_1$ play no role in this fault. When the size of the capacitance becomes sufficiently large, the capacitive ground fault current becomes self-sustaining, and does not clear by itself. It then becomes necessary to open the CBs to clear the fault, and the relaying problem becomes one of detecting such low magnitudes of fault currents. In order to produce a sufficient fault current, a resistance is introduced between the neutral and the ground—inside the box shown by a dotted line in Figure 1.2a. One of the design considerations in selecting the grounding resistance is the thermal capacity of the resistance to handle a sustained ground fault.

Ungrounded systems produce good service continuity, but are subjected to high overvoltages on the unfaulted phases when a ground fault occurs. It is clear from the phasor diagram of Figure 1.2b that when a ground fault occurs on phase a, the steady-state voltages of phases b and c become $\sqrt{3}$ times their normal value. Transient overvoltages become correspondingly higher. This places additional stress on the insulation of all connected equipment. As the insulation level of lower-voltage systems is primarily influenced by lightning-induced phenomena, it is possible to accept the fault-induced overvoltages, as they are lower than the lightning-induced overvoltages. However, as the system voltages increase to higher than about 100 kV, the fault-induced overvoltages begin to assume a critical role in insulation design, especially of power transformers. At high voltages, it is therefore common to use solidly grounded neutrals (more precisely "effectively grounded"). The definition of effectively grounded requires that neutral grounding must be of sufficiently low impedance "to limit the buildup of voltages to levels below that, which may result in undue hazard to persons or to connected equipment" [1].

Effectively grounded systems have high ground fault currents, and each ground fault must be cleared by CBs. As high-voltage systems are generally heavily interconnected, with several alternative paths to load centers, operation of CBs for ground faults does not lead to a reduced service continuity.

In certain heavily meshed systems, particularly at 69 and 138 kV, the ground fault current could become excessive because of very low zero-sequence impedance at some buses. If ground fault current is beyond the capability of the CBs, it becomes necessary to insert

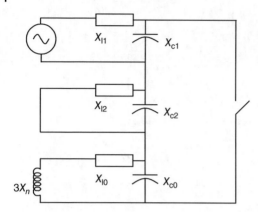

Figure 1.3 Symmetrical component representation for ground fault with grounding reactor.

an inductance in the neutral in order to limit the ground fault current to a safe value. As the network Thévenin impedance is primarily inductive, a neutral inductance is much more effective (than resistance) in reducing the fault current. Also, there is no significant power loss in the neutral reactor during ground faults.

In several lower-voltage networks, a very effective alternative to ungrounded operation can be found if the capacitive fault current causes ground faults to be self-sustaining. This is the use of a Petersen coil, also known as the ground fault neutralizer (GFN). Consider the symmetrical component representation of a ground fault on a power system that is grounded through a grounding reactance of X_n (Figure 1.3). If $3X_n$ is made equal to X_{c0} (the zero-sequence capacitive reactance of the connected network), the parallel resonant circuit formed by these two elements creates an open circuit in the fault path, and the ground fault current is once again zero. No CB operation is necessary upon the occurrence of such a fault, and service reliability is essentially the same as that of a truly ungrounded system. The overvoltages produced on the unfaulted conductors are comparable to those of ungrounded systems, and consequently GFN use is limited to system voltages below 100 kV. In practice, GFNs must be tuned to the entire connected zero-sequence capacitance on the network, and thus if some lines are out of service, the GFN reactance must be adjusted accordingly. Petersen coils have found much greater use in several European countries than in the United States.

1.3 Power system bus configurations

The manner in which the power apparatus is connected together in substations and switching stations, and the general layout of the power network, has a profound influence on protective relaying. It is therefore necessary to review the alternatives and the under-lying reasons for selecting a particular configuration. A radial system is a single-source arrangement with multiple loads, and is generally associated with a distribution system (defined as a system operating at voltages below 100 kV) or an industrial complex (Figure 1.4).

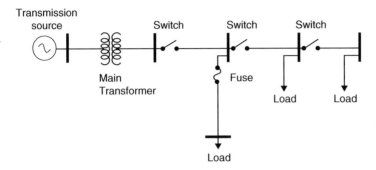

Figure 1.4 Radial power system.

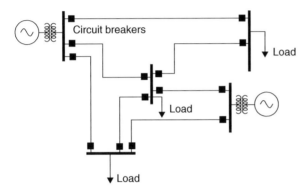

Figure 1.5 Network power system.

Such a system is most economical to build; but from the reliability point of view, the loss of the single source will result in the loss of service to all of the users. Opening main line reclosers or other sectionalizing devices for faults on the line sections will disconnect the loads downstream of the switching device. From the protection point of view, a radial system presents a less complex problem. The fault current can only flow in one direction, that is, away from the source and toward the fault. Since radial systems are generally electrically remote from generators, the fault current does not vary much with changes in generation capacity.

A network has multiple sources and multiple loops between the sources and the loads. Sub-transmission and transmission systems (generally defined as systems operating at voltages of 100–200 kV and above) are network systems (Figure 1.5).

In a network, the number of lines and their interconnections provide more flexibility in maintaining service to customers, and the impact of the loss of a single generator or transmission line on service reliability is minimal. Since sources of power exist on all sides of a fault, fault current contributions from each direction must be considered in designing the protection system. In addition, the magnitude of the fault current varies greatly with changes in system configuration and installed generation capacity. The situation is dramatically increased with the introduction of the smart grid discussed in Section 6.13 of Chapter 6.

Example 1.1 Consider the simple network shown in Figure 1.6. The load at bus 2 has secure service for the loss of a single power system element. Further, the fault current for a fault at bus 2 is $-j20.0$ pu when all lines are in service. If lines 2 and 3 go out of service, the fault current changes to $-j10.0$ pu. This is a significant change.

Now consider the distribution feeder with two intervening transformers connected to bus 2. All the loads on the feeder will lose their source of power if transformers 2–4 are lost. The fault current at bus 9 on the distribution feeder with system normal is $-j0.23$ pu, whereas the same fault when one of the two generators on the transmission system is lost is $-j0.229$ pu. This is an insignificant change. Of course, the reason for this is: with the impedances of the intervening transformers and transmission network, the distribution system sees the source as almost a constant impedance source, regardless of the changes taking place on the transmission network.

Substations are designed for reliability of service and flexibility in operation, and to allow for equipment maintenance with a minimum interruption of service. The most common bus arrangements in a substation are (i) single bus, single breaker, (ii) two buses, single breaker, (iii) two buses, two breakers, (iv) ring bus, and (v) breaker-and-a-half. These bus arrangements are illustrated in Figure 1.7.

A single-bus, single-breaker arrangement, shown in Figure 1.7a, is the simplest, and probably the least expensive to build. However, it is also the least flexible. To do maintenance work on the bus, a breaker, or a disconnect switch, de-energizing the associated transmission lines is necessary. A two-bus, single-breaker arrangement, shown in Figure 1.7b, allows the breakers to be maintained without de-energizing the associated line. For system flexibility, and particularly to prevent a bus fault from splitting the system too drastically, some of the lines are connected to bus 1 and some to bus 2 (the transfer bus). When maintaining a breaker, all of the lines that are normally connected to bus 2 are transferred to bus 1, the breaker to be maintained is bypassed by transferring its line to bus 2, and the bus tie breaker becomes the line breaker. Only one breaker can be maintained at a time. Note that the protective relaying associated with the buses and the line whose breaker is being maintained

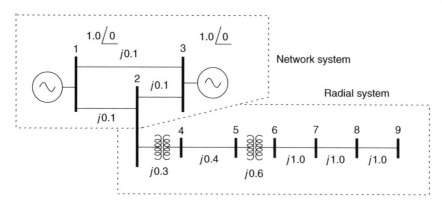

Figure 1.6 Power system for Example 1.1.

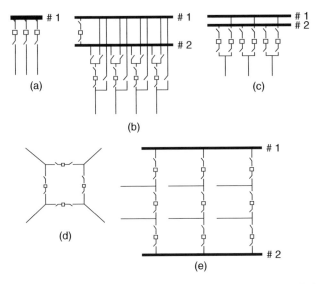

Figure 1.7 Substation bus arrangements: (a) single bus, single breaker; (b) two buses, one breaker; (c) two buses, two breakers; (d) ring bus; and (e) breaker-and-a-half.

must also be reconnected to accommodate this new configuration. This will be covered in greater detail as we discuss the specific protection schemes.

A two-bus, two-breaker arrangement is shown in Figure 1.7c. This allows any bus or breaker to be removed from service, and the lines can be kept in service through the companion bus or breaker. A line fault requires two breakers to trip to clear a fault. A bus fault must trip all of the breakers on the faulted bus, but does not affect the other bus or any of the lines. This station arrangement provides the greatest flexibility for system maintenance and operation; however, this is at a considerable expense: the total number of breakers in a station equals twice the number of the lines. A ring-bus arrangement shown in Figure 1.7d achieves similar flexibility while the ring is intact. When one breaker is being maintained, the ring is broken, and the remaining bus arrangement is no longer as flexible. Finally, the breaker-and-a-half scheme, shown in Figure 1.7e, is most commonly used in most extra high-voltage (EHV) transmission substations. It provides for the same flexibility as the two-bus, two-breaker arrangement at the cost of just one-and-a-half breakers per line on an average. This scheme also allows for future expansions in an orderly fashion.[1] In recent years, however, a new concept, popularly and commonly described as the "smart grid," has

---

1 The breaker-and-a-half bus configuration is the natural outgrowth of operating practices that developed as systems matured. Even in developing systems, the need to keep generating units in service was recognized as essential and it was common practice to connect the unit to the system through two CBs. Depending on the particular bus arrangement, the use of two breakers increased the availability of the unit despite line or bus faults or CB maintenance. Lines and transformers, however, were connected to the system through one CB per element. With one unit and several lines or transformers per station, there was a clear economic advantage to this arrangement. When the number of units in a station increased, the number of breakers increased twice as fast: one unit and two lines required four breakers, two units and two lines required six breakers, and so on. It is attractive to rearrange the bus design, so that the lines and transformers shared the unit breakers. This gave the same maintenance advantage to the lines, and when the number of units exceeded the number of other elements, reduced the number of breakers required.

entered the lexicon of bus configuration, introducing ideas and practices that are changing the fundamental design, operation, and performance of the "distribution" system. The fundamental basis of the "smart grid" transforms the previously held definition of a "distribution system," that is, a single-source, radial system to a transmission-like configuration with multiple generating sites, communication, operating, and protective equipment similar to high-voltage and extra-high-voltage transmission.

The impact of system and bus configurations on relaying practices will become clear in the chapters that follow.

1.4 The nature of relaying

We will now discuss certain attributes of relays that are inherent to the process of relaying, and can be discussed without reference to a particular relay. The function of protective relaying is to promptly remove from service any element of the power system that starts to operate in an abnormal manner. In general, relays do not prevent damage to equipment: they operate after some detectable damage has already occurred. Their purpose is to limit, to the extent possible, further damage to equipment, to minimize danger to people, to reduce stress on other equipment, and, above all, to remove the faulted equipment from the power system as quickly as possible, so that the integrity and stability of the remaining system are maintained. The control aspect of relaying systems also helps to return the power system to an acceptable configuration as soon as possible, so that service to customers can be restored.

1.4.1 Reliability, dependability, and security

Reliability is generally understood to measure the degree of certainty that a piece of equipment will perform as intended. Relays, in contrast with most other equipment, have two alternative ways in which they can be unreliable: they may fail to operate when they are expected to, or they may operate when they are not expected to. This leads to a two-pronged definition of reliability of relaying systems: a reliable relaying system must be dependable and secure [1]. Dependability is defined as the measure of the certainty that the relays will operate correctly for all the faults for which they are designed to operate. Security is defined as the measure of the certainty that the relays will not operate incorrectly for any fault.

Most protection systems are designed for high dependability. In other words, a fault is always cleared by some relay. As a relaying system becomes dependable, its tendency to become less secure increases. Thus, in present-day relaying system designs, there is a bias toward making them more dependable at the expense of some degree of security. Consequently, a majority of relay system misoperations are found to be the result of unwanted trips caused by insecure relay operations. This design philosophy correctly reflects the fact that a power system provides many alternative paths for power to flow from generators to loads. Loss of a power system element due to an unnecessary trip is therefore less objectionable than the presence of a sustained fault. This philosophy is no longer appropriate when the number of alternatives for power transfer is limited, as in a radial power system, or in a power system in an emergency operating state.

Example 1.2 Consider the fault F on the transmission line shown in Figure 1.8. In normal operation, this fault should be cleared by the two relays R_1 and R_2 through the CBs B_1 and B_2. If R_2 does not operate for this fault, it has become unreliable through a loss of dependability. If relay R_5 operates through breaker B_5 for the same fault, and before breaker B_2 clears the fault, it has become unreliable through a loss of security. Although we have designated the relays as single entities, in reality they are likely to be collections of several relays making up the total protection system at each location. Thus, although a single relay belonging to a protection system may lose security, its effect is to render the complete relaying system insecure, and hence unreliable.

1.4.2 Selectivity of relays and zones of protection

The property of security of relays, that is, the requirement that they not operate for faults for which they are not designed to operate, is defined in terms of regions of a power system—called zones of protection—for which a given relay or protective system is responsible. The relay will be considered to be secure if it responds only to faults within its zone of protection. Relays usually have inputs from several current transformers (CTs), and the zone of protection is bounded by these CTs. The CTs provide a window through which the associated relays "see" the power system inside the zone of protection. While the CTs provide the ability to detect a fault inside the zone of protection, the CBs provide the ability to isolate the fault by disconnecting all of the power equipment inside the zone. Thus, a zone boundary is usually defined by a CT and a CB. When the CT is part of the CB, it becomes a natural zone boundary. When the CT is not an integral part of the CB, special attention must be paid to the fault detection and fault interruption logic. The CT still defines the zone of protection, but communication channels must be used to implement the tripping function from appropriate remote locations where the CBs may be located. We return to this point later in Section 1.5 where CBs are discussed.

In order to cover all power equipment by protection systems, the zones of protection must meet the following requirements.

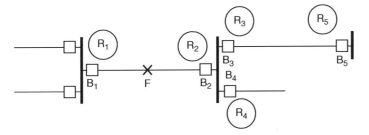

Figure 1.8 Reliability of protection system.

- All power system elements must be encompassed by at least one zone. Good relaying practice is to be sure that the more important elements are included in at least two zones.
- Zones of protection must overlap to prevent any system element from being unprotected. Without such an overlap, the boundary between two nonoverlapping zones may go unprotected. The region of overlap must be finite but small, so that the likelihood of a fault occurring inside the region of overlap is minimized. Such faults will cause the protection belonging to both zones to operate, thus removing a larger segment of the power system from service.

A zone of protection may be closed or open. When the zone is closed, all power apparatus entering the zone is monitored at the entry points of the zone. Such a zone of protection is also known as "differential," "unit," or "absolutely selective." Conversely, if the zone of protection is not unambiguously defined by the CTs, that is, the limit of the zone varies with the fault current, the zone is said to be "nonunit," "unrestricted," or "relatively selective." There is a certain degree of uncertainty about the location of the boundary of an open zone of protection. Generally, the nonpilot protection of distribution and transmission lines employs open zones of protection.

Example 1.3 Consider the fault at F_1 in Figure 1.9. This fault lies in a closed zone, and will cause CBs B_1 and B_2 to trip. The fault at F_2, being inside the overlap between the zones of protection of the transmission line and the bus, will cause CBs B_1, B_2, B_3, and B_4 to trip, although opening B_3 and B_4 is unnecessary. Both of these zones of protection are closed zones.

   Now consider the fault at F_3. This fault lies in two open zones. The fault should cause CB B_6 to trip. B_5 is the backup breaker for this fault, and will trip if for some reason B_6 fails to clear the fault.

1.4.3 Relay speed

It is, of course, desirable to remove a fault from the power system as quickly as possible. However, the relay must make its decision based upon voltage and current waveforms that are severely distorted due to transient phenomena which must follow the occurrence of a fault. The relay must separate the meaningful and significant information contained in these waveforms upon which a secure relaying decision must be based. These considerations demand that the relay takes a certain amount of time to arrive at a decision with

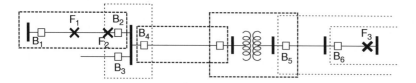

Figure 1.9 Closed and open zones of protection.

the necessary degree of certainty. The relationship between the relay response time and its degree of certainty is an inverse one [2], and this inverse-time operating characteristic of relays is one of the most basic properties of all protection systems.

Although the operating time of relays often varies between wide limits, relays are generally classified by their speed of operation as follows.

1) Instantaneous: These relays operate as soon as a secure decision is made. No intentional time delay is introduced to slow down the relay response.[2]
2) Time delay: An intentional time delay is inserted between the relay decision time and the initiation of the trip action.[3]
3) High speed: A relay that operates in less than a specified time. The specified time in present practice is 50 ms (three cycles on a 60 Hz system).
4) Ultrahigh speed: This term is not included in the relay standards, but is commonly considered to be in operation in 4 ms or less.

1.4.4 Primary and backup protection

A protection system may fail to operate and, as a result, fail to clear a fault. It is thus essential that provision be made to clear the fault by some alternative protection system or systems [3, 4]. These alternative protection system(s) are referred to as redundant (or dual primary), backup, or breaker failure protection systems. The main protection system(s) for a given zone of protection is (or are) called the primary protection system(s). It operates in the fastest time possible and removes the least amount of equipment from service. On EHV and high voltage systems, it is common to use redundant primary protection systems in case an element in one primary protection chain may fail to operate. Redundant protection will have adequate performance to ensure that power system performance requirements are met even if one protection system fails [5]. This duplication is therefore intended to cover the failure of the relays themselves. One may use relays from a different manufacturer, or relays based upon a different principle of operation, so that some inadequacy in the design of one of the primary relays is not repeated in the redundant system.

It is not always practical for every element of the protection chain to be redundant—on high-voltage and EHV systems, the transducers or the CBs are very expensive, and the cost of redundant equipment may not be justified. On lower-voltage systems, even the relays themselves may not be redundant. In such situations, only backup relaying is used. Backup relays are generally slower than the primary relays and may remove more system elements than may be necessary to clear a fault. Backup relaying may be installed locally, that is, in the same substation as the primary protection, or remotely. Remote backup relays are

---

2 There is no implication relative to the speed of operation of an instantaneous relay. It is a characteristic of its design. A plunger-type overcurrent relay will operate in one to three cycles depending on the operating current relative to its pickup setting. A 125 V DC hinged auxiliary relay, operating on a 125 V DC circuit, will operate in three to six cycles, whereas a 48 V DC tripping relay operating on the same circuit will operate in one cycle. All are classified as instantaneous.
3 The inserted time delay can be achieved by an R–C circuit, an induction disc, a dashpot, or other electrical or mechanical means. A short-time induction disc relay used for bus protection will operate in three to five cycles, a long-time induction disc relay used for motor protection will operate in several seconds, and bellows or geared timing relays used in control circuits can operate in minutes.

completely independent of the relays, transducers, batteries, and CBs of the protection system they are backing up. There are no common failures that can affect both sets of relays. However, complex system configurations may significantly affect the ability of remote backup relays to "see" all the faults for which backup is desired. In addition, remote backup relays may remove more loads in the system than can be allowed. Local backup relaying does not suffer from these deficiencies, but it does use common elements such as the transducers, batteries, and CBs, and can thus fail to operate for the same reasons as the primary protection.

Breaker failure relays are a subset of local backup relaying that is provided specifically to cover a failure of the CB. This can be accomplished in a variety of ways. The most common, and simplest, breaker failure relay system consists of a separate timer that is energized whenever protective relaying energizes the breaker trip coil and is de-energized when the fault current through the breaker disappears. If the fault current persists for longer than the timer setting, a trip signal is given to all local and remote breakers that are required to clear the fault. A separate set of relays may be installed to provide this breaker failure protection, in which case it may use independent transducers and batteries (also see Chapter 12 [Section 12.4]).

These ideas are illustrated by the following example, and will be further examined when specific relaying systems are considered in detail later.

Example 1.4 Consider the fault at location F in Figure 1.10. It is inside the zone of protection of transmission line AB. Primary relays R_1 and R_5 will clear this fault by acting through breakers B_1 and B_5. At station B, a duplicate primary relay R_2 may be installed to trip the breaker B_1 to cover the possibility that the relay R_1 may fail to trip. R_2 will operate in the same time as R_1 and may use the same or different elements of the protection chain. For instance, on EHV lines, it is usual to provide separate CTs, but use the same potential device with separate windings. The CBs are not duplicated, but the battery may be. On lower-voltage circuits, it is not uncommon to share all of the transducers and direct current (DC) circuits. The local backup relay R_3 is designed to operate at a slower speed than R_1 and R_2; it is probably set to see more of the system. It will first attempt to trip breaker B_1 and then its breaker failure relay will trip breakers B_5, B_6, B_7, and B_8. This is local backup relaying, often known as breaker failure protection, for CB B_1. Relays R_9, R_{10}, and R_4 constitute the

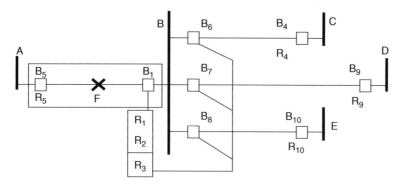

Figure 1.10 Duplicate primary, local backup, and remote backup protection.

remote backup protection for the primary protection R_1. No elements of the protection system associated with R_1 are shared by these protection systems, and hence no common modes of failure between R_1 and R_4 and R_9 and R_{10} are possible. These remote backup protections will be slower than R_1, R_2, or R_3, and also remove additional elements of the power system—namely lines BC, BD, and BE—from service, which would also de-energize any loads connected to these lines.

A similar set of backup relays is used for the system behind station A.

1.4.5 Single- and three-phase tripping and reclosing

The prevailing practice in the United States is to trip all three phases of the faulted power system element for all types of faults. In several European and Asian countries, it is a common practice to trip only the faulted phase for a phase-to-ground fault, and to trip all three phases for all multiphase faults on transmission lines. These differences in the tripping practice are the result of several fundamental differences in the design and operation of power systems, as discussed in Section 1.6.

As a large proportion of faults on a power system are of a temporary nature, the power system can be returned to its prefault state if the tripped CBs are reclosed as soon as possible. Reclosing can be manual. That is, it is initiated by an operator working from the switching device itself, from a control panel in the substation control house or from a remote system control center through a supervisory control and data acquisition (SCADA) system. Clearly, manual reclosing is too slow for the purpose of restoring the power system to its prefault state when the system is in danger of becoming unstable. Automatic reclosing of CBs is initiated by dedicated relays for each switching device, or it may be controlled from a substation or central reclosing computer. All reclosing operations should be supervised (i.e., controlled) by appropriate interlocks to prevent an unsafe, damaging, or undesirable reclosing operation. Some of the common interlocks for reclosing are the following.

1) Voltage check: Used when good operating practice demands that a certain piece of equipment be energized from a specific side. For example, it may be desirable to always energize a transformer from its high-voltage side. Thus, if a reclosing operation is likely to energize that transformer, it would be good to check that the CB on the low-voltage side is closed only if the transformer is already energized.
2) Synchronizing check: This check may be used when the reclosing operation is likely to energize a piece of equipment from both sides. In such a case, it may be desirable to check that the two sources that would be connected by the reclosing breaker are in synchronism and approximately in phase with each other. If the two systems are already in synchronism, it would be sufficient to check that the phase angle difference between the two sources is within certain specified limits. If the two systems are likely to be unsynchronized, and the closing of the CB is going to synchronize the two systems, it is necessary to monitor the phasors of the voltages on the two sides of the reclosing CB and close the breaker as the phasors approach each other.
3) Equipment check: This check is to ensure that some piece of equipment is not energized inadvertently.

These interlocks can be used either in the manual or in the automatic mode. It is the practice of some utilities, however, not to inhibit the manual reclose operation of CBs, on the assumption that the operator will make the necessary checks before reclosing the CB. In extreme situations, sometimes the only way to restore a power system is through operator intervention, and automatic interlocks may prevent or delay the restoration operation. On the other hand, if left to the operator during manual operation, there is the possibility that the operator may not make the necessary checks before reclosing.

Automatic reclosing can be high speed, or it may be delayed. The term high speed generally implies reclosing in times shorter than a second. Many utilities may initiate high-speed reclosing for some types of faults (such as ground faults), and not for others. Delayed reclosing usually operates in several seconds or even in minutes. The timing for the delayed reclosing is determined by specific conditions for which the delay is introduced.

1.5 Elements of a protection system

Although, in common usage, a protection system may mean only the relays, the actual protection system consists of many other subsystems that contribute to the detection and removal of faults. As shown in Figure 1.11, the major subsystems of the protection system are the transducers, relays, battery, CBs, and communications and/or human–machine interface (HMI).

The transducers, that is, the current and voltage transformers, constitute a major component of the protection system, and are considered in detail in Chapter 3. Relays are the logic elements that initiate the tripping and closing operations, and we will, of course, discuss relays and their performance in the rest of this book.

1.5.1 Battery and DC supply

Since the primary function of a protection system is to remove a fault, the ability to trip a CB through a relay must not be compromised during a fault, when the alternating current (AC) voltage available in the substation may not be of sufficient magnitude. For example, a close-in three-phase fault can result in zero AC voltage at the substation AC outlets. Tripping power, as well as the power required by the relays, cannot therefore be obtained from the AC system, and is usually provided by the station battery.

The battery is permanently connected through a charger to the station AC service, and, normally, during steady-state conditions, it floats on the charger. The charger is of a sufficient volt-ampere capacity to provide all steady-state loads powered by the battery. Usually,

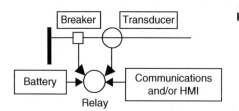

Figure 1.11 Elements of a protection system.

the battery is also rated to maintain adequate DC power for 8–12 h following a station black-out. Although the battery is probably the most reliable piece of equipment in a station, in EHV stations, it is not uncommon to have duplicate batteries, each connected to its own charger and complement of relays. Electromechanical relays are known to produce severe transients on the battery leads during operation, which may cause misoperation of other sensitive relays in the substation, or may even damage them. It is therefore common prac-tice, insofar as practical, to separate electromechanical and solid-state equipment by con-necting them to different batteries.

Positive and negative busses on DC supplies are usually floating with respect to ground. This is so that a ground fault on one or the other of the busses will not cause an outage of any part of the DC system. The fault will be alarmed and may be troubleshot while the con-nected protection systems remain mostly operational.

1.5.2 Circuit breakers

It would take too much space to describe various CB designs and their operating principles here. Indeed, several excellent references do just that [5, 6]. Instead, we will describe a few salient features about CBs, which are particularly significant from the point of view of relaying.

Knowledge of CB operation and performance is essential to an understanding of protec-tive relaying. It is the coordinated action of both that results in successful fault clearing. The CB isolates the fault by interrupting the current at or near a current zero. At the present time, an EHV CB can interrupt fault currents of the order of 10^5A at system voltages up to 800 kV. It can do this as quickly as the first current zero after the initiation of a fault, although it more often interrupts at the second or third current zero. As the CB contacts move to interrupt the fault current, there is a race between the establishment of the dielec-tric strength of the interrupting medium and the rate at which the recovery voltage reap-pears across the breaker contacts. If the recovery voltage wins the race, the arc reignites, and the breaker must wait for the next current zero when the contacts are farther apart.

CBs of several designs can be found in a power system. One of the first designs, and one that is still in common use, incorporates a tank of oil in which the breaker contacts and operating mechanism are immersed. The oil serves as the insulation between the tank, which is at the ground potential, and the main contacts, which are at line potential. The oil also acts as the cooling medium to quench the arc when the contacts open to interrupt load or fault current. An oil CB rated for 138 kV service is shown in Figure 1.12.

As transmission system voltages increased, it was not practical to build a tank large enough to provide the dielectric strength required in the interrupting chamber. In addition, better insulating materials, better arc-quenching systems, and faster operating require-ments resulted in a variety of CB characteristics: interrupting medium of oil, gas, air, or vacuum; insulating medium of oil, air, gas, or solid dielectric; and operating mechanisms using impulse coil, solenoid, spring–motor–pneumatic, or hydraulic. This broad selection of CB types and accompanying selection of ratings offer a high degree of flexibility. Each user has unique requirements and no design can be identified as the best or preferred design. One of the most important parameters to be considered in the specification of a CB is the interrupting medium. Oil does not require energy input from the operating mechanism

Figure 1.12 A 138 kV oil circuit breaker. *Source:* Courtesy of Appalachian Power Company.

to extinguish the arc. It gets that energy directly from the arc itself. Sulfur hexafluoride (SF$_6$), however, does require additional energy and must operate at high pressure or develop a blast of gas or air during the interruption phase. When environmental factors are considered, however, oil CBs produce high noise and ground shock during interruption, and for this reason may be rejected. They are also potential fire hazards or water-table pollutants. SF$_6$ CBs have essentially no emission, although the noise accompanying their operation may require special shielding and housing. And as with all engineering decisions, the cost of the CB must be an important consideration. At present, oil-filled CBs are the least expensive, and may be preferred if they are technically feasible, but this may change in the future. Figure 1.13a shows a live-tank SF$_6$ CB and Figure 1.13b shows a dead-tank SF$_6$ CB.

An important design change in CBs with a significant impact on protection systems was the introduction of the "live-tank" design [7]. By placing the contact enclosure at the same potential as the contacts themselves, the need for the insulation between the two was eliminated. However, the dead-tank (Figures 1.12 and 1.13b) designs incorporate CTs in the bushing pocket of the tank, thereby providing CTs on both sides of the contacts. This arrangement provides a very nice mechanism for providing overlapping zones of protection on the two sides of the CBs. In the live-tank design, since the entire equipment is at line potential, it is not possible to incorporate CTs that have their secondary windings essentially at the ground potential. It then becomes necessary to design the CTs with their own insulating system, as separate freestanding devices, a design that is quite expensive. With free-standing CTs, it is no longer economical to provide CTs on both sides of a CB, and one must make do with only one CT on one side of the breaker. Of course, a freestanding CT has multiple secondaries, and protection zone overlap is achieved using secondary windings on opposite sides of the zones of protection. This is illustrated in Figure 1.14a. A live-tank, air-blast CB and a freestanding CT rated at 800 kV are shown in Figure 1.15. The location of the primary winding and the protective assignments of the secondary winding of the CTs have a very significant implication for the protection being provided. The dead-tank CB can provide CTs on either side of the interrupting mechanism and allow the protection to easily determine the appropriate tripping scheme. The live-tank, air-blast CB, associated with the

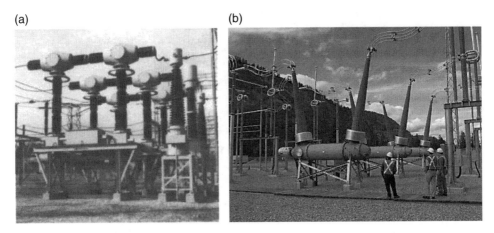

Figure 1.13 (a) Live-tank 345 kV SF$_6$ circuit breaker. *Source:* Courtesy of Appalachian Power Company. (b) Dead-tank 500 kV SF$_6$ circuit breaker.

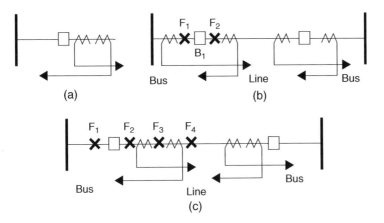

Figure 1.14 Zone overlap with different types of CTs and circuit breakers.

Figure 1.15 Live-tank, air-blast circuit breaker and a current transformer for 800 kV. *Source:* Courtesy of Appalachian Power Company.

higher voltages, introduces, with the CTs located on only one side of the tripping mechanism forces, a more complex tripping logic. This is illustrated in Example 1.5 in the following text. With advanced technology, however, using SF_6 for tripping and quenching the arc, the interruption of EHV faults within a dead tank, that is, a tank whose enclosure can be grounded, is possible and therefore the ability to provide grounded CTs on either side of the interrupter removes the difficulty discussed in the earlier text. This is illustrated in the following example.

Example 1.5   Consider the dead-tank CB shown in Figure 1.14b. The bushing CTs are on either side of the breaker, and the secondaries are connected to the bus and line protection, so that they overlap at the breaker. For a fault at F_1, both protective systems will operate. The bus differential relays will trip B_3 and all other breakers on the bus. This will clear the fault. The line protection will similarly trip breaker B_3, and the corresponding relays at the remote station will also trip their associated breakers. This is unnecessary, but unavoidable. If there are tapped loads on the line, they will be de-energized until the breakers reclose. For a fault at F_2, again both protective systems will operate. For this fault, tripping the other bus breakers is not necessary to clear the fault, but tripping the two ends of the line is necessary.

Now consider the live-tank design shown in Figure 1.14c. For a fault at F_1, only the bus protection sees the fault and correctly trips B_1 and all the other bus breakers to clear the fault. For a fault at F_2, however, tripping the bus breakers does not clear the fault, since it is still energized from the remote end, and the line relays do not operate. This is a blind spot in this configuration. Column protection or breaker failure protection will cover this area. For a fault at F_3 and F_4, the line relays will operate and the fault will be cleared from both ends. The fault at F_3 again results in unnecessary tripping of the bus breakers.

1.5.3   Communications and HMI

Protective relays will normally include adjustment features to enable a variety of different settings to alter characteristics and operating values and other functions. These settings will be adjustable through an HMI.

On electromechanical relays, there will only be a few adjustments available. These will be screw plugs to adjust coil taps, multiposition links, and a variety of levers and dials.

On digital relays, there may be hundreds or thousands of different setting adjustments available. Such numerous adjustments are made available because of the expanded functionality of multifunction digital relays as compared to the limited functionality of discrete electromechanical relays or solid-state analog relays. The functionality of relays is now so expanded into control and automation areas that they fall into the higher-level classification of intelligent electronic devices (IEDs). Digital relay setting adjustments will either be made through push buttons or a touch-screen interface, or (more commonly) through a computer connection. Modern digital relays usually have external software to facilitate setting adjustment. The adjustments can be made off line and then downloaded to or uploaded from the relay through a communications port.

Thus, digital relays will be equipped with a variety of communications ports for multiple purposes:

- To allow HMI for operation and setting adjustment of the relay.
- To facilitate data exchange with SCADA systems.
- To facilitate data exchange with engineering applications.
- To facilitate data exchange with other IEDs in the same substation or in remote substations or with field-installed devices.

Communications systems that facilitate peer to peer interaction need to meet the same performance and reliability specifications as the relays themselves. Further, there is often a need for interoperability between relays made by different manufacturers. Peer to peer communications between relays and other IEDs are standardized for interoperability by a suite of standards under the umbrella of Standard IEC 61850. The scope of these standards is too large to address in this book; however, the interested reader may explore further in Reference [8].

1.6 International practices

Although the fundamental protective and relay-operating concepts are similar throughout the world, there are very significant differences in their implementation. These differences arise through different traditions, operating philosophies, experiences, and national standards. Electric-power utilities in many countries are organs of the national government. In such cases, the specific relaying schemes employed by these utilities may reflect the national interest. For example, their preference may be for relays manufactured inside their respective countries. In some developing countries, the choice of relays may be influenced by the availability of low-cost, hard-currency loans or a transfer-of-technology agreement with the prospective vendor of the protective equipment. The evolutionary stage of the power system itself may have an influence on the protection philosophy. Thus, more mature power systems may opt for a more dependable protection system at the expense of some degradation of its (protection system's) security. A developing power network has fewer alternative paths for power transfer between the load and generation, and a highly secure protection system may be the desired objective. Long transmission lines are quite common in countries with large areas, for example, the United States or Russia. Many European and Asian countries have relatively short transmission lines, and, since the protection practice for long lines is significantly different from that for short lines, this may be reflected in the established relaying philosophy.

As mentioned in Section 1.4, reclosing practices also vary considerably among different countries. When one phase of a three-phase system is opened in response to a single-phase fault, the voltage and current in the two healthy phases tend to maintain the fault arc after the faulted phase is de-energized. Depending on the length of the line, system operating voltage, and load, compensating shunt reactors may be necessary to extinguish this "secondary" arc [9]. Where the transmission lines are short, such secondary arcs are not a problem, and no compensating reactors are needed. Thus, in countries with short transmission lines, single-phase tripping and reclosing may be a sound and viable operating strategy. By contrast, when transmission lines are long, the added cost of compensation may dictate that three-phase tripping and reclosing be used for all faults. The loss of synchronizing power flow created by three-phase tripping is partially mitigated by the use of high-speed reclosing. Also, use is made of high-speed relaying (three cycles or less) to reduce the impact of three-phase tripping and reclosing. Of course, there are exceptional situations that may dictate a practice that is out of the ordinary in a given country. Thus, in the United States, where high-speed tripping with three-phase tripping and reclosing is the general trend, exception may be made when a single transmission line is used to connect a remote generator to the power system. Three-phase tripping of such a line for a ground fault may cause the loss of the generator for too many faults, and single-phase tripping and reclosing may be the desirable alternative.

An important factor in the application of specific relay schemes is associated with the configuration of the lines and substations. Multiple circuit towers as found throughout Europe have different fault histories than single circuit lines, and therefore have different protection system needs. The same is true for double-bus, transfer-bus, or other breaker bypassing arrangements. In the United States, EHV stations are almost exclusively breaker-and-a-half or ring-bus configurations. This provision to do maintenance work on a breaker significantly affects the corresponding relaying schemes. The philosophy of installing several complete relay systems also affects the testing capabilities of all relays. In the United States, it is not the common practice to remove more than one phase or zone relay at a time for calibration or maintenance. In other countries, this may not be considered to be as important, and the testing facilities built in the relays may not be as selective.

The use of turnkey contracts to design and install complete substations also differs considerably between countries, being more prevalent in many European, South American, and certain Asian countries than in North America. This practice leads to a manufacturer or consulting engineering concern taking total project responsibility, as opposed to the North American practice where the utilities themselves serve as the general contractor. In the latter case, the effect is to reduce the variety of protection schemes and relay types in use.

1.7 Summary

In this chapter, we have examined some of the fundamentals of protective relaying philosophy. The concepts of reliability and its two components, dependability and security, have been introduced. Selectivity has been illustrated by closed and open zones of protection and local versus remote backup. The speed of relay operation has been defined. Three-phase tripping, the prevailing practice in the United States, has been compared to the more prevalent European practice of single-phase tripping. We have discussed various reclosing and interlocking practices and the underlying reasons for a given choice. We have also given a brief account of various types of CBs and their impact on the protection system design.

Problems

1.1   Write a computer program to calculate the three-phase fault current for a fault at F in Figure 1.16, with the network normal, and with one line at a time removed from service. The positive-sequence impedance data are given in the accompanying table. Use the commonly made assumption that all prefault resistance values are $(1.0 + j0.0)$ pu, and neglect all resistance values. Calculate the contribution to the fault flowing through the CB B_1, and the voltage at that bus. For each calculated case, consider the two possibilities: CB B_2 closed or open. The latter is known as the "stub-end" fault.

System data for Figure 1.16

From	To	Positive-sequence impedance
1	2	$0.0 + j0.1$
2	6	$0.05 + j0.15$
2	5	$0.04 + j0.2$
2	4	$0.01 + j0.1$
3	5	$0.015 + j0.15$
3	6	$0.01 + j0.19$
4	5	$0.01 + j0.19$
4	6	$0.03 + j0.1$
6	7	$0.0 + j0.08$

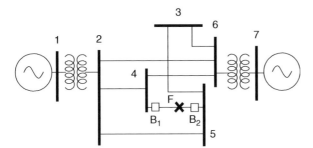

Figure 1.16 Problem 1.1.

1.2 Using the usual assumptions about the positive- and negative-sequence impedances of the network elements, what are the currents at breaker B_1 for b–c fault for each of the faults in Problem 1.1? What is the voltage between phases b and c for each case?

1.3 For the radial power system shown in Figure 1.17, calculate the line-to-ground fault current flowing in each of the CBs for faults at each of the buses. The system data are given in the accompanying table. Also determine the corresponding faulted phase voltage, assuming that the generator is ideal, with a terminal voltage of 1.0 pu.

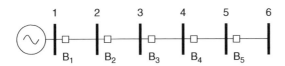

Figure 1.17 Problem 1.3.

System data for Figure 1.17

From	To	Positive-sequence impedance	Zero-sequence impedance
1	2	$0.01 + j0.05$	$0.02 + j0.13$
2	3	$0.003 + j0.04$	$0.01 + j0.16$
3	4	$0.008 + j0.04$	$0.04 + j0.15$
4	5	$0.01 + j0.05$	$0.03 + j0.15$
5	6	$0.003 + j0.02$	$0.01 + j0.06$

1.4 In a single-loop distribution system shown in Figure 1.18, determine the fault currents flowing in CBs B_1, B_2, and B_3 for a "b–c" fault at F. What are the corresponding phase-to-phase voltages at those locations? Consider the generator to be of infinite short-circuit capacity, and with a voltage of 1.0 pu. Consider two alternatives: (i) both transformers T_1 and T_2 in service and (ii) one of the two transformers out of service. The system data are given in the accompanying table.

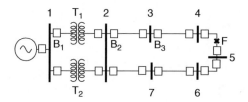

Figure 1.18 Problem 1.4.

System data for Figure 1.18

From	To	Positive-sequence impedance
1	2	$0.0 + j0.01$ (T_1)
		$0.0 + j0.01$ (T_2)
2	3	$0.0 + j0.08$
3	4	$0.02 + j0.05$
4	5	$0.01 + j0.03$
5	6	$0.0 + j0.06$
6	7	$0.01 + j0.09$
2	7	$0.01 + j0.09$

1.5 In the double-bus arrangement shown in Figure 1.19, CB B_1 must be taken out of service for repair. Starting with all equipment in service, make a list of operations required to take the CB out of service, and to return it to service. Repeat for all the bus arrangements shown in Figure 1.7. Remember that disconnect switches are generally not designed to break or make load current.

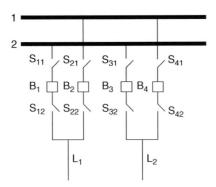

Figure 1.19 Problem 1.5.

1.6 Consider the various bus arrangements shown in Figure 1.7. Assume that each of the device types, bus, disconnect switch, CB, and transmission line may develop a fault and is removed from service. Prepare a table listing (single) faults that would cause loss of load connected to the remote end of one of the transmission lines in each of those configurations. What conclusions can you draw from such a table?

1.7 For the system shown in Figure 1.20, the fault at F produces these differing responses at various times: (i) $R_1 B_1$ and $R_2 B_2$ operate; (ii) $R_1 B_1$, R_2, $R_3 B_3$, and $R_4 B_4$ operate; (iii) $R_1 B_1$, $R_2 B_2$, and $R_5 B_5$ operate; (iv) $R_1 B_1$, $R_5 B_5$, and $R_6 B_6$ operate. Analyze each of these responses for fault F and discuss the possible sequence of events that may have led to these operations. Classify each response as being correct, incorrect, appropriate, or inappropriate. Note that "correct–incorrect" classification refers to relay operation, whereas "appropriate–inappropriate" classification refers to the desirability of that particular response from the point of view of the power system. Also determine whether there was a loss of dependability or a loss of security in each of these cases.

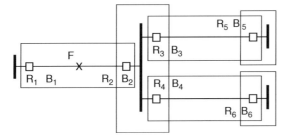

Figure 1.20 Problem 1.7.

1.8 In the systems shown in Figure 1.21a,b, it is desired to achieve overlap between the zones of protection for the bus and the transmission line. Show how this may be achieved through the connection of CTs to the appropriate protection systems.

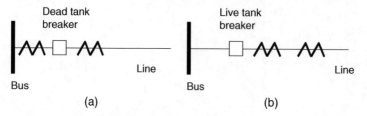

Figure 1.21 Problem 1.8.

1.9 In the part of the network shown in Figure 1.22, the minimum and maximum operating times for each relay are 0.8 and 2.0 cycles (of the fundamental power system frequency), and each CB has minimum and maximum operating times of 2.0 and 5.0 cycles. Assume that a safety margin of 3.0 cycles between any primary protection and backup protection is desirable. P_2 is the local backup for P_1, and P_3 is the remote backup. Draw a timing diagram to indicate the various times at which the associated relays and breakers must operate to provide a secure (coordinated) backup coverage for fault F.

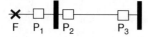

Figure 1.22 Problem 1.9.

1.10 For the system shown in Figure 1.23, the following CBs are known to operate: (i) B_1 and B_2; (ii) B_3, B_4, B_1, B_5, and B_7; (iii) B_7 and B_8; (iv) B_1, B_3, B_5, and B_7. Assuming that all primary protection has worked correctly, where is the fault located in each of these cases?

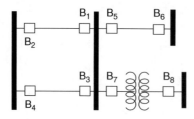

Figure 1.23 Problem 1.10.

References

1 ANSI/IEEE 100 (2000)). *IEEE Standard Dictionary of Electrical and Electronic Terms.* Standards Association of IEEE, New York.

2 Thorp, J.S., Phadke, A.G., Horowitz, S.H., and Beehler, J.E. (1979). Limits to impedance relaying. *IEEE Trans. PAS* **98** (1): 246–260.

3 Kennedy, L.F. and McConnell, A.J. (1957). An appraisal of remote and local backup relaying. *AIEE Trans.* **76** (1): 735–741.

4 IEEE (2022). *IEEE standard C37.120/D11.8 Draft Guide for Protection System Redundancy for Power System Reliability.* Piscataway, NJ: Standards Association of IEEE.

5 Lythall, R.T. (1953). *The J. & P. Switchgear Book.* London: Johnson & Phillips.

6 Willheim, R. and Waters, M. (1956). *Neutral Grounding in High-Voltage Transmission.* Elsevier.

7 Shores, R.B., Beatty, J.W., and Seeley, H.T. (1959). A line of 115 kV through 460 kV air blast breakers. *AIEE Trans. PAS* **58**: 673–691.

8 Mackiewicz, R.E. Overview of IEC 61850 and benefits. In: *2006 IEEE PES Power Systems Conference and Exposition*, 623–630. https://doi.org/10.1109/PSCE.2006.296392.

9 Shperling, B.R., Fakheri, A.J., Shih, C., and Ware, B.J. (1981). Analysis of single phase switching field tests on the AEP 765 kV system. *IEEE Trans. PAS* **100**: 1729–1735.

2

Relay Operating Principles

2.1 Introduction

Since the purpose of power system protection is to detect faults or abnormal operating conditions, relays must be able to evaluate a wide variety of parameters to establish that corrective action is required. The most common parameters that reflect the presence of a fault are the voltages and currents at the terminals of the protected apparatus or at the appropriate zone boundaries. Occasionally, the relay inputs may also include states—open or closed—of some contacts or switches. A specific relay, or a protection system, must use the appropriate inputs, process the input signals, and determine that a problem exists, and then initiate some action. In general, a relay can be designed to respond to any observable parameter or effect. The fundamental problem in power system protection is to define the quantities that can differentiate between normal and abnormal conditions. This problem of being able to distinguish between normal and abnormal conditions is compounded by the fact that "normal" in the present sense means that the disturbance is outside the zone of protection. This aspect—which is of the greatest significance in designing a secure relaying system—dominates the design of all protection systems. For example, consider the relay shown in Figure 2.1. If one were to use the magnitude of a fault current to determine whether some action should be taken, it is clear that a fault on the inside (fault F_1), or on the outside (fault F_2), of the zone of protection is electrically the same fault, and it would be impossible to tell the two faults apart based upon the current magnitude alone. Much ingenuity is needed to design relays and protection systems that would be reliable under all the variations to which they are subjected throughout their life.

Whether, and how, a relaying goal is met is dictated by the power system and the transient phenomena it generates following a disturbance. Once it is clear that a relaying task can be performed, the job of designing the hardware to perform the task can be initiated. The field of relaying is almost 100 years old. Ideas on how relaying should be done have evolved over this long period, and the limitations of the relaying process are well understood. As time has gone on, the hardware technology used in building the relays has gone through several major changes: relays began as electromechanical devices, then progressed to solid-state hardware in the late 1950s, and more recently they are being implemented on microcomputers. We will now examine—in general terms—the functional operating principles of relays and certain of their design aspects.

Power System Relaying, Fifth Edition. Stanley H. Horowitz, Arun G. Phadke and Charles F. Henville.
© 2023 John Wiley & Sons Ltd. Published 2023 by John Wiley & Sons Ltd.

Figure 2.1 Problem of relay selectivity for faults at a zone boundary.

2.2 Detection of faults

In general, as faults (short circuits) occur, currents increase in magnitude, and voltages go down. Besides these magnitude changes of the alternating current (AC) quantities, other changes may occur in one or more of the following parameters: phase angles of current and voltage phasors, harmonic components, active and reactive power, frequency of the power system, and so on. Relay operating principles may be based upon detecting these changes, and identifying the changes with the possibility that a fault may exist inside its assigned zone of protection. We will divide relays into categories based upon which of these input quantities a particular relay responds.

2.2.1 Level detection

This is the simplest of all relay operating principles. As indicated in the earlier text, fault current magnitudes are almost always greater than the normal load currents that exist in a power system. Consider the motor connected to a 4 kV power system as shown in Figure 2.2. The full-load current for the motor is 245 A. Allowing for an emergency overload capability of 25%, a current of $1.25 \times 245 = 306$ A or lower should correspond to normal operation. Any current above a set level (chosen to be above 306 A by a safety margin in the present example) may be taken to mean that a fault, or some other abnormal condition, exists inside the zone of protection of the motor. The relay should be designed to operate and trip the circuit breaker for all currents above the setting, or, if desired, the relay may be connected to sound an alarm, so that an operator can intervene and trip the circuit breaker manually or take other appropriate action.

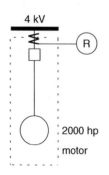

Figure 2.2 Overcurrent protection of a motor.

The level above which the relay operates is known as the pickup setting of the relay. For all currents above the pickup, the relay operates, and for currents smaller than the pickup value, the relay takes no action. It is of course possible to arrange the relay to operate for values smaller than the pickup value, and take no action for values above the pickup. An undervoltage relay is an example of such a relay.

The operating characteristics of an overcurrent relay can be presented as a plot of the operating time of the relay versus the current in the relay. It is best to normalize the current as a ratio of the actual current to the pickup setting. The operating time for (normalized) currents less than 1.0 is infinite, while for values greater than 1.0 the relay operates. The actual time for operation will

Figure 2.3 Characteristic of a level detector relay.

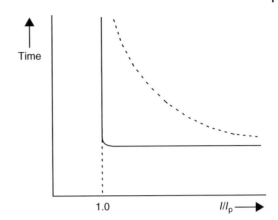

depend upon the design of the relay, and is discussed further in later chapters. The ideal level detector relay would have a characteristic as shown by the solid line in Figure 2.3. In practice, the relay characteristic has a less abrupt transition, as shown by the dotted line.

2.2.2 Magnitude comparison

This operating principle is based upon the comparison of one or more operating quantities with each other. For example, a current balance relay may compare the current in one circuit with the current in another circuit, which should have equal or proportional magnitudes under normal operating conditions.

The relay will operate when the current division in the two circuits varies by a given tolerance. Figure 2.4 shows two identical parallel lines that are connected to the same bus at either end. One could use a magnitude comparison relay that compares the magnitudes of the two line currents I_A and I_B. If $|I_A|$ is greater than $|I_B| + \varepsilon$ (where ε is a suitable tolerance), and line B is not open, the relay would declare a fault on line A and trip it. Similar logic would be used to trip line B if its current exceeds that in line A, when the latter is not open. Another instance in which this relay can be used is when the windings of a machine have two identical parallel subwindings per phase.

2.2.3 Differential comparison

Differential comparison is one of the most sensitive and effective methods of providing protection against faults. The concept of differential comparison is quite simple, and can be best understood by referring to the generator winding shown in Figure 2.5. As the winding is

Figure 2.4 Magnitude comparison relaying for two parallel transmission lines.

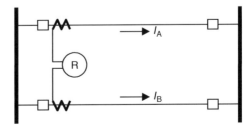

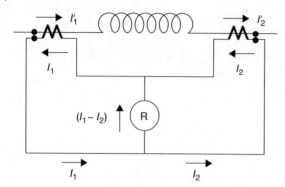

electrically continuous, current entering one end, I_1, must equal the current leaving the other end, I_2. One could use a magnitude comparison relay described in the earlier text to test for a fault on the protected winding. When a fault occurs between the two ends, the two currents are no longer equal. Alternatively, one could form an algebraic sum of the two currents entering the protected winding, that is, $(I_1 - I_2)$, and use a level detector relay to detect the presence of a fault. In either case, the protection is termed a differential protection. In general, the differential protection principle is capable of detecting very small magnitudes of fault currents. Its only drawback is that it requires currents from the extremities of a zone of protection, which restricts its application to power apparatus, such as transformers, generators, motors, buses, capacitors, and reactors. We discuss specific applications of differential relaying in later chapters.

2.2.4 Phase angle comparison

This type of relay compares the relative phase angle between two AC quantities. Phase angle comparison is commonly used to determine the direction of a current with respect to a reference quantity. For instance, the normal power flow in a given direction will result in the phase angle between the voltage and the current varying around its power factor angle, say approximately $\pm30°$. When the power flows in the opposite direction, this angle will become $(180° \pm 30°)$. Similarly, for a fault in the forward or reverse direction, the phase angle of the current with respect to the voltage will be $-\phi$ and $(180° - \phi)$, respectively, where ϕ, the impedance angle of the fault circuit, is close to 90° for power transmission networks. These relationships are explained for two transmission lines in Figure 2.6. This difference in phase relationships created by a fault is exploited by making relays that respond to phase angle differences between two input quantities—such as the fault voltage and the fault current in the present example.

2.2.5 Distance measurement

As discussed in the earlier text, the most positive and reliable type of protection compares the current entering the circuit with the current leaving it [1]. On transmission lines and feeders, the length, voltage, and configuration of the line may make this principle uneconomical. Instead of comparing the local line current with the far-end line current, the relay

Figure 2.6   Phase angle comparison for a fault on a transmission line.

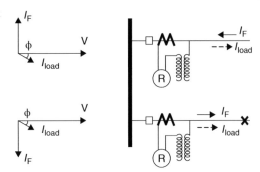

compares the local current with the local voltage. This, in effect, is a measurement of the impedance of the line as seen from the relay terminal. An impedance relay relies on the fact that the length of the line (i.e., its distance) for a given conductor diameter and spacing determines its impedance.

2.2.6   Pilot relaying

Certain relaying principles are based upon the information obtained by the relay from a remote location. The information is usually—although not always—in the form of contact status (open or closed). The information is sent over a communication channel using power line carrier, microwave, or telephone circuits. We consider pilot relaying in greater detail in Chapter 6.

2.2.7   Harmonic content

Currents and voltages in a power system usually have a sinusoidal waveform of the fundamental power system frequency. There are, however, deviations from a pure sinusoid, such as the third harmonic voltages and currents produced by the generators that are present during normal system operation. Other harmonics occur during abnormal system conditions, such as the odd harmonics associated with transformer saturation, or transient components caused by the energization of transformers. These abnormal conditions can be detected by sensing the harmonic content through filters in electromechanical or solid-state relays, or by calculation in digital relays. Once it is determined that an abnormal condition exists, a decision can be made whether some control action is required.

2.2.8   Frequency sensing

Normal power system operation is at 50 or 60 Hz, depending upon the country. Any deviation from these values indicates that a problem exists or is imminent. Frequency can be measured by filter circuits, by counting zero crossings of waveforms in a unit of time, or by special sampling and digital computer techniques [2]. Frequency-sensing relays may be used to take corrective actions that will bring the system frequency back to normal.

The various input quantities described in the earlier text, upon which fault detection is based, may be used either singly or in any combination, to calculate power, power factor, directionality, impedance, and so on, and can in turn be used as relay-actuating quantities.

Some relays are also designed to respond to mechanical devices such as fluid-level detectors, pressure sensors, or temperature sensors, and so on. Relays may be constructed from electromechanical elements such as solenoids, hinged armatures, induction discs, solid-state elements such as diodes, silicon-controlled rectifiers (SCRs), transistors, or magnetic or operational amplifiers, or digital computers using analog-to-digital converters and microprocessors. It will be seen that, because the electromechanical relays were developed early on in the development of protection systems, the description of all relay characteristics is often in terms of electromechanical relays. The construction of a relay does not inherently change the protection concept, although there are advantages and disadvantages associated with each type. We will examine the various hardware options for relays in the following section.

2.3 Relay designs

It is beyond the scope of this book to cover relay designs in any depth. Our interest is to achieve a general understanding of relay design and construction to assist us in realizing their capabilities and limitations. The following discussion covers a very small sample of the possible designs and is intended only to indicate how parameters required for fault detection and protection can be utilized by a relay. Specific details can be obtained from manufacturers' literature. Several excellent books [1, 3, 4] also provide valuable insights into relay design and related considerations.

2.3.1 Fuses

Before examining the operating principles of relays, we should introduce the fuse, which is the oldest and simplest of all protective devices. The fuse is a level detector, and is both the sensor and the interrupting device. It is installed in series with the equipment being protected and operates by melting a fusible element in response to the current flow. The melting time is inversely proportional to the magnitude of the current flowing in the fuse. It is inherently a one-shot device, since the fusible link is destroyed in the process of interrupting the current flow. There can be mechanical arrangements to provide multiple shots as discussed in the following text. Fuses may only be able to interrupt currents up to their maximum short-circuit rating, or they may have the ability to limit the magnitude of the short-circuit current by interrupting the flow before it reaches its maximum value. This current-limiting action is a very important characteristic that has application in many industrial and low-voltage installations. This is discussed in more detail in Chapter 4.

The study of fuses and their application is a complex and extensive discipline that is beyond the scope of this book. However, historically and technically, fuses form the background of protective relaying, particularly for radial feeders such as distribution lines or auxiliary systems of power plants. A review of fuse characteristics and performance may be found in the technical literature [5], and is discussed further in Chapter 4. The two major disadvantages of fuses are the following:

1) The single-shot feature referred to in the earlier text requires that a blown fuse be replaced before service can be restored. This means a delay and the need to have the

correct spare fuses and qualified maintenance personnel who must go and replace the fuses in the field. It is possible to provide a multiple-shot feature by installing a number of fuses in parallel and provide a mechanical triggering mechanism, so that the blowing of one fuse automatically transfers another in its place.

2) In a three-phase circuit, a single-phase-to-ground fault will cause one fuse to blow, de-energizing only one phase, permitting the connected equipment—such as motors—to stay connected with the remaining phases, with subsequent excessive heating and vibration because of the unbalanced voltage supply.

To overcome these disadvantages, protective relays were developed as logic elements that are divorced from the circuit interruption function. Relays are devices requiring low-level inputs (voltages, currents, or contacts). They derive their inputs from transducers, such as current or voltage transformers, and switch contacts. They are fault-detecting devices only and require an associated interrupting device—a circuit breaker—to clear the fault. Segregating the fault detection function from the interruption function was a most significant advance, as it gave the relay designer an ability to design a protection system that matched the needs of the power system. This separation of protection design from power system design was further aided by standardization of input devices, which is discussed in detail in Chapter 3.

2.4 Electromechanical relays

The early relay designs utilized actuating forces that were produced by electromagnetic interaction between currents and fluxes, much as in a motor. Some relays were also based upon the forces created by expansion of metals caused by a temperature rise due to a flow of current. In electromechanical relays, the actuating forces were created by a combination of the input signals, stored energy in springs, and dashpots. The plunger-type relays are usually driven by a single actuating quantity, while the induction-type relays may be activated by single or multiple inputs. There are a large number of legacy in-service electromechanical relays, since they are very durable and have a life span of decades. However, modern relays use digital technology.

2.4.1 Plunger-type relays

Consider a round-moving plunger placed inside a stationary electromagnet, as shown in Figure 2.7. With no current in the coil, the plunger is held partially outside the coil by the force $\mathfrak{F}_s$ produced by a spring. Let x be the position of the plunger tip inside the upper opening of the coil, as shown in the figure. When the coil is energized by a current i, and saturation phenomena are neglected, the energy $W(\lambda, i)$ and the co-energy $W'(i, x)$ stored in the magnetic field are given by Fitzgerald et al. [6]:

$$W(\lambda, i) = W'(i, x) = \frac{1}{2} Li^2, L = \frac{\mu_0 \pi d^2 N^2}{4(x + gd/4a)} \tag{2.1}$$

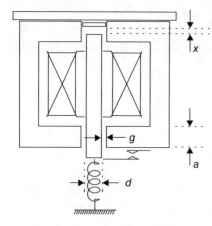

Figure 2.7 Plunger-type relay.

where λ is the flux linkage of the coil and L is the inductance of the coil. The force that tries to pull the plunger inside the coil is given by

$$\mathfrak{I}_m = \frac{\partial}{\partial x} W'(i, x) = K \frac{i^2}{(x + gd/4a)^2},$$

(2.2)

where K is a constant depending upon the constants of the electromagnetic circuit and a is the height of the pole-piece as shown in Figure 2.7. The plunger moves when $\mathfrak{I}_m$ exceeds $\mathfrak{I}_s$. If the current is sinusoidal with a root mean square (rms) value of I, the average force is proportional to I^2, and the value of the current (I_p) at which the plunger just begins to move—known as the "pickup" setting of the relay—is given by

$$I_P = \left\{ \sqrt{\mathfrak{I}_s/K} \right\} (x_0 + gd/4a),$$

(2.3)

where x_0 is the displacement of the plunger when no current is flowing in the coil. The operating time of the relay depends upon the mass of the plunger, and can be made to suit a particular need. The general shape of the relay characteristic, that is, its operating time plotted as a function of the current through the coil, is as shown in Figure 2.8. The plunger travels some distance, from x_0 to x_1, before it closes its contacts and hits a stop.

The energizing current must drop below a value I_d, known as the "dropout" current, before the plunger can return to its original position x_0. The dropout current is given by

$$I_d = \left\{ \sqrt{\mathfrak{I}_s/K} \right\} (x_1 + gd/4a).$$

(2.4)

As x_1 is smaller than x_0, the dropout current is always smaller than the pickup current. This is a very important—and common—feature of relays, and has significant implications as far as application of relays is concerned. We will return to a discussion of this point a little later.

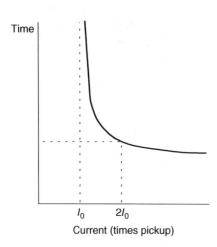

Figure 2.8 Operating time versus current of a plunger relay.

Example 2.1 Consider a plunger-type relay with a pickup current of 5 A rms. The pole face has a height of 1.5 cm, while the spring holds the plunger 1 cm out of the coil when the current is below the pickup value. The air gap g is 0.2 cm, and $gd/4a = 0.05$. Let the spring force be a constant, with a value of 0.001 N, and let the mass of the plunger be 0.005 kg. Let the travel of the plunger be 3 mm before it hits a stop and closes its contacts. The ratio of the dropout current to the pickup current is

$$I_d/I_p = \frac{(0.05 + 0.7)}{(0.05 + 1.0)} = 0.714.$$

For a normalized current of magnitude I (i.e., actual current divided by the pickup current), the accelerating force on the plunger is

$$\mathfrak{J} = \mathfrak{J}_m - \mathfrak{J}_s = K \frac{(I \times I_P)^2}{(x + gd/4a)^2} - \mathfrak{J}_s.$$

Substituting for $\mathfrak{J}_s$ from Equation (2.3) and using centimeters for all linear dimensions give

$$\mathfrak{J} = \mathfrak{J}_s \left\{ \frac{(x_0 + gd/4a)^2}{(x + gd/4a)^2} I^2 - 1 \right\} = 0.001 \times \left\{ \frac{(1.05)^2}{(0.05 + x)^2} I^2 - 1 \right\}.$$

The equation of motion for the plunger is

$$m\ddot{x} = \mathfrak{J},$$

where m is the mass of the plunger, and the force acts to reduce the displacement x:

$$0.005\ddot{x} = -0.001 \times \left\{ \frac{(1.05)^2}{(0.05 + x)^2} I^2 - 1 \right\}.$$

This equation can be integrated twice to provide the operating time of the relay, that is, the time it takes the plunger to travel from x_0 to x_1. However, because of the nature of the dependence of the force on x, the integrals in question are elliptic integrals, and must be evaluated numerically for given displacements. We may calculate an approximate result using a constant force equal to the average, taken over its travel from x_0 to x_1. For x equal to 1 cm, the force is $0.001(I^2 - 1)$N; for x equal to 0.7 cm, the force is $0.001(1.96I^2 - 1)$N. Therefore, the average force is $0.001(1.48I^2 - 1)$N. Using this expression for the force, the approximate equation of motion for the plunger is

$$0.005\ddot{x} = -0.001 \times \left(1.48I^2 - 1 \right),$$

and integrating this twice, the operating time (in seconds) of the relay is

$$t = \sqrt{\frac{10(x_0 - x_1)}{(1.48I^2 - 1)}} = \sqrt{\frac{0.3}{(1.48I^2 - 1)}}\ \text{s}.$$

The abovementioned formula is not accurate at $I = 1.0$ (i.e., at the pickup setting) because of the approximations made in the force expression. However, it does show the inverse-time behavior of the relay for larger values of the current. At or near the pickup values, the characteristic is asymptotic to the pickup setting, as shown in Figure 2.8.

Notice that the relay characteristic shown in Figure 2.8 has the ratio of the actual current to the pickup current as its abscissa. This method of normalization is quite common in defining the operating characteristic of relays. Most relays also have several taps available on the winding of the actuating coil, so that the pickup current can be adjusted over a wide range. For example, a plunger-type overcurrent relay may be available with tap settings of, for example, 1.0, 2.0, 3.0, 4.0, 5.0, 6.0, 8.0, and 10.0 A. Alternatively, the pickup can be controlled by adjusting the plunger within the coil, that is, the value of x in Figure 2.7. Plunger-type relays will operate on direct current (DC) as well as on AC currents. Hinged armature relays, also known as clapper-type relays, have similar characteristics, but have a smaller ratio of the dropout-to-pickup currents.

2.4.2 Induction-type relays

These relays are based upon the principle of operation of a single-phase AC motor. As such, they cannot be used for DC currents. Two variants of these relays are fairly standard: one with an induction disc and the other with an induction cup. In both cases, the moving element (disc or cup) is equivalent to the rotor of the induction motor. However, in contrast to the induction motor, the iron associated with the rotor in the relay is stationary. The moving element acts as a carrier of rotor currents, while the magnetic circuit is completed through stationary magnetic elements. The general constructions of the two types of relay are shown in Figures 2.9 and 2.10.

Induction-type relays require two sources of alternating magnetic flux in which the moving element may turn. The two fluxes must have a phase difference between them; otherwise, no operating torque is produced. Shading rings mounted on pole faces may be used to provide one of the two fluxes to produce motor action. In addition to these two sources of magnetic flux, other sources of magnetic flux—such as permanent magnets—may be used to provide special damping characteristics. Let us assume that the two currents in the coils of the relay, i_1 and i_2, are sinusoidal:

$$i_1(t) = I_{m1} \cos \omega t, \tag{2.5}$$

$$i_2(t) = I_{m2} \cos(\omega t + \theta). \tag{2.6}$$

If L_m is the mutual inductance between each of the coils and the rotor, each current produces a flux linkage with the rotor given by

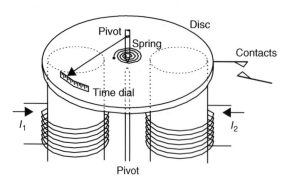

Figure 2.9 Principle of construction of an induction disc relay. Shaded poles and damping magnets are omitted for clarity.

Figure 2.10 Moving cup
induction relay.

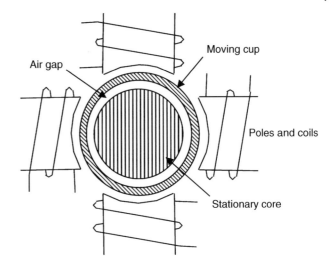

$$\lambda_1(t) = L_m I_{m1} \cos \omega t, \tag{2.7}$$

and

$$\lambda_2(t) = L_m I_{m2} \cos(\omega t + \theta) \tag{2.8}$$

Each of these flux linkages in turn induces a voltage in the rotor, and since the rotor is a metallic structure with low self-inductance [3], a rotor current in phase with the induced voltages flows in the rotor. Assuming the equivalent rotor resistance to be R_r, the induced rotor currents are given by

$$i_{r1}(t) = \frac{1}{R_r}\frac{d\lambda_1}{dt} = -\frac{\omega L_m I_{m1}}{R_r} \sin \omega t, \tag{2.9}$$

$$i_{r2}(t) = \frac{1}{R_r}\frac{d\lambda_2}{dt} = -\frac{\omega L_m I_{m2}}{R_r} \sin(\omega t + \theta). \tag{2.10}$$

Each of the rotor currents interacts with the flux produced by the other coil, producing a force. The two forces are in opposite directions with respect to each other, and the net force, or, what amounts to the same thing, the net torque τ, is given by

$$\tau \alpha [\lambda_1 \, i_{r2} - \lambda_2 \, i_{r1}]. \tag{2.11}$$

Substituting for the flux linkages and the rotor currents from Equations (2.7)–(2.10) and absorbing all the constants in a new constant K, we can write

$$\tau = K I_{m1} I_{m2} [\cos \omega t \, \sin(\omega t + \theta) - \cos(\omega t + \theta) \, \sin \omega t], \tag{2.12}$$

or, using a trigonometric identity,

$$\tau = K I_{m1} I_{m2} \sin \theta. \tag{2.13}$$

The direction of the torque is from the coil with the leading current to the one with the lagging current [3].

Note that the net torque is constant in this case, and does not change with time, nor as the disc (or the cup) turns. If the phase angle between the two coil currents is zero, there is no

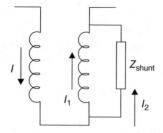

Figure 2.11 Phase shift for torque production.

torque produced. This is in keeping with the theory of the single-phase induction motor [6]. The induction-type relay is a very versatile device. By an appropriate choice of the source of the two coil currents, this relay could be made to take on the characteristic of a level detector, a directional relay, or a ratio relay. For example, using the same current to flow through the two coils, one could make a level detector, provided one arranged to produce a phase shift between the current carried by one of the coils and the original current. This is quite easily done by placing in parallel with one of the coils a shunt with an impedance angle that is different from that of the coil. This is illustrated in Figure 2.11.

The current in the first coil I and the current in the second coil I_1 have a phase difference between them, and the relay will produce a torque. Since both coil currents are proportional to the current I, the net torque produced by this relay is

$$\tau = K_1 I^2, \tag{2.14}$$

where the constant K_1 has been suitably modified to include the term $\sin\theta$, which is a constant for the relay. A spring keeps the disc from turning. When the torque produced by the current (the pickup current of the relay) just exceeds the spring torque τ_S, the disc begins to turn. After turning an angle ϕ (another constant of the relay design), the relay closes its contacts. As the torque does not depend upon the angular position of the rotor, the current at which the spring overcomes the magnetic torque and returns the relay to open position (the dropout current of the relay) is practically the same as the pickup current.

Example 2.2 Consider an induction disc relay designed to perform as an overcurrent relay. The spring torque τ_S is 0.001 Nm, and the pickup current of the relay is 10 A. The constant of proportionality K_1 is given by

$$K_1 = \frac{0.001}{10^2} = 10^{-5}.$$

For a normalized current (times pickup) I, the magnetic torque is given by

$$\tau_m = 10^{-5} \times (10I)^2 = 10^{-3}I^2.$$

The accelerating torque on the disc is the difference between the magnetic torque and the spring torque:

$$\tau = \tau_m - \tau_S = 10^{-3}\left(I^2 - 1\right).$$

If the moment of inertia of the disc is 10^{-4} kg·m^2, the equation of motion of the disc is

$$10^{-4}\ddot{\theta} = 10^{-3}\left(I^2 - 1\right),$$

where θ is the angle of rotation of the disc. θ starts at 0 and it stops at ϕ, where the relay closes its contacts. Let φ be 2′ or 0.035 rad. Integrating the equation of motion twice gives

$$\theta = 5\left(I^2 - 1\right)t^2,$$

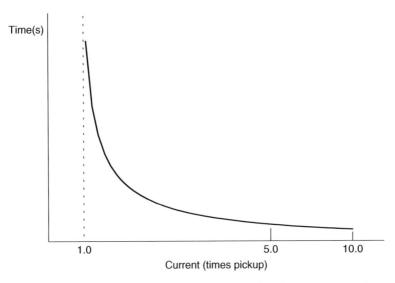

Time(s)

1.0   5.0   10.0

Current (times pickup)

Figure 2.12 Inverse time characteristic of an induction disc overcurrent relay.

and the operating time (in seconds) of the relay is

$$T = \sqrt{\frac{0.035}{5(I^2 - 1)}}\ \text{s.}$$

The relay operating characteristic is plotted in Figure 2.12, and it is seen to be an inverse-time relationship.

Induction disc-type or cup-type relays may be energized from voltage sources to produce undervoltage or overvoltage relays. Also, by providing one of the coils with a current source and the other coil with a voltage source, the relay may be made to respond to a product of current and voltage inputs. It should be remembered that the phase angle between the currents in the current coil and the voltage coil appears in the torque equation. The current in the voltage coil generally lags the voltage by an angle equal to the impedance angle of the voltage coil, while the current coil carries the actual input current. If the angle between the voltage and the current in the voltage coil is $-\phi$, the torque is proportional to $VI \sin(\theta + \phi)$. In general, all of these combinations of energizing quantities may be applied to several coils, and a composite torque expression for the accelerating torque can be obtained:

$$\tau = \tau_\text{m} - \tau_\text{S} = K_1 I^2 + K_2 V^2 + K_3 VI \sin(\theta + \phi) - \tau_\text{S}. \tag{2.15}$$

By selecting appropriate values for the various constants in Equation (2.15), a very wide range of relay characteristics can be obtained. This is explained in Example 2.3.

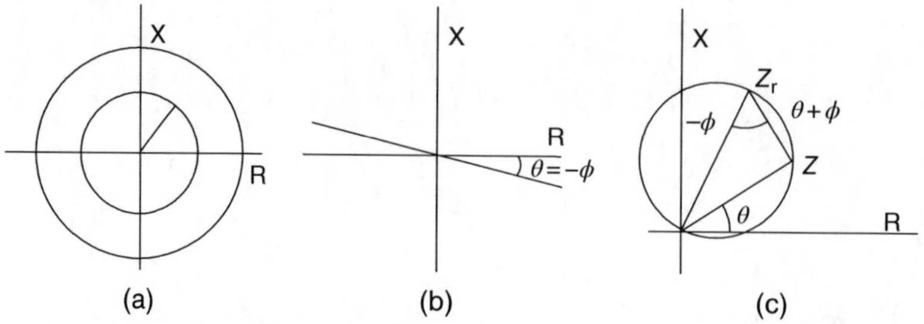

Figure 2.13 Characteristics obtained from the universal relay equation: (a) impedance relay; (b) directional relay; and (c) mho relay.

Example 2.3 Consider the choice of K_3 equal to zero in Equation (2.15), and that the control spring torque τ_S is negligible. When the relay is on the verge of operation (i.e., at its balance point) $\tau = 0$, and, if the voltage-induced torque is arranged to be in the opposite direction to that produced by the current (i.e., replacing K_2 by $-K_2$), then

$$|Z| = \frac{V}{I} = \sqrt{\frac{K_1}{K_2}},$$

which is the equation of a circle in the R–X plane, as shown in Figure 2.13a. This is known as an impedance or ohm relay. The torque is greater than this pickup value when the ratio of voltage to current (or impedance) lies inside the operative circle. By adding a current-carrying coil on the structure carrying a current proportional to the voltage, the torque equation at the balance point is

$$0 = K_1 I^2 - K_2 (V + K_4 I)^2,$$

or, in the R–X plane,

$$|Z + K_4| = \sqrt{\frac{K_1}{K_2}},$$

which is an equation of a circle with its center offset by a constant.

Now consider the choice of K_1, K_2, and τ_S equal to zero in Equation (2.15). These choices produce a balance point equation,

$$VI \sin(\theta + \varphi) = 0,$$

or, dividing through by I^2, and assuming that I is not zero, the balance point equation is

$$Z \sin(\theta + \varphi) = 0.$$

This is the equation of a straight line in the R–X plane, passing through the origin, and at an angle of φ to the R axis, as shown in Figure 2.13b. This is the characteristic of a directional relay. Finally, by setting K_1 and τ_S equal to zero, and reversing the sign of the torque produced by the VI term, the torque equation at the balance point becomes

$$0 = K_2 V^2 - K_3 VI \sin(\theta + \varphi),$$

or, dividing through by I^2—and assuming that I is not zero—the balance point equation is

$$|Z| = \frac{K_3}{K_2} \sin(\theta + \varphi).$$

This is the equation of a circle passing through the origin in the R–X plane, with a diameter of K_3/K_2. The diameter passing through the origin makes an angle of maximum torque of $-\varphi$ with the X axis, as shown in Figure 2.13c. This is known as an admittance or a mho relay characteristic.

2.5 Solid-state relays

The expansion and growing complexity of modern power systems have brought a need for protective relays with a higher level of performance and more sophisticated characteristics. This has been made possible by the development of semiconductors and other associated components that can be utilized in relay designs, generally referred to as solid-state or static relays. All of the functions and characteristics available with electromechanical relays can be performed by solid-state devices, either as discrete components or as integrated circuits. Solid-state relays use low-power components with rather limited capability to tolerate extremes of temperature and humidity, or overvoltages and overcurrents. This introduces concerns about the survivability of solid-state relays in the hostile substation environment. Indeed, early designs of solid-state relay were plagued by a number of failures attributable to the harsh environment in which they were placed. Solid-state relays also require independent power supplies, since springs and driving torques from the input quantities are not present. These issues introduce design and reliability concerns, which do not exist to the same degree in electromechanical relays. However, there are performance, and perhaps economic, advantages associated with the flexibility and reduced size of solid-state devices. In general, solid-state relays are more accurate. Their settings are more repeatable and hold to closer tolerances. Their characteristics can be shaped by adjusting logic elements as opposed to the fixed characteristics of induction discs or cups. This is a significant advantage where relay settings are difficult, because of unusual power system configurations or heavy loads. Solid-state relays are not affected by vibration or dust, and often require less mounting space and need not be mounted in a particular orientation. Solid-state relays are designed, assembled, and tested as a system. This puts the overall responsibility for proper operation of the relays on the manufacturer. In many cases, especially when special equipment or expertise for assembly and wiring is required, this results in more reliable equipment at a lower cost.

Solid-state relay circuits may be divided into two categories: analog circuits that are either fault-sensing or measuring circuits, and digital logic circuits for operation on logical variables. There is a great variety of circuit arrangements that would produce a desired relaying characteristic. It is impossible, and perhaps unnecessary, to go over relay circuit design

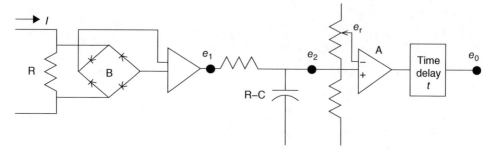

Figure 2.14 Possible circuit configuration for a solid-state instantaneous overcurrent relay.

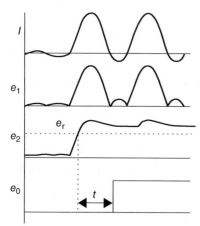

Figure 2.15 Waveforms of a solid-state instantaneous overcurrent relay.

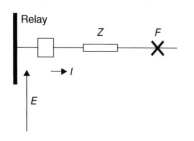

Figure 2.16 Distance protection of a transmission line.

practices that are currently in use. We will describe some examples of circuits that can provide desired relay characteristics.

2.5.1 Solid-state instantaneous overcurrent relays

Consider the circuit shown in Figure 2.14. The input current I is passed through the resistive shunt R, full-wave rectified by the bridge rectifier B, filtered to remove the ripple by the R–C filter, and applied to a high-gain summing amplifier A. The other input of the summing amplifier is supplied with an adjustable reference voltage e_r. When the input on the positive input of the summing amplifier exceeds the reference setting, the amplifier output goes high, and this step change is delayed by a time-delay circuit, in order to provide immunity against spurious transient signals in the input circuit. Waveforms at various points in this circuit are shown in Figure 2.15 for an assumed input fault current of a magnitude above the pickup setting e_r of the relay. By making the time-delay circuit adjustable, and by making the amount of delay depend upon the magnitude of the input current, a time-delay overcurrent relay characteristic can be obtained.

2.5.2 Solid-state distance (mho) relays

It was shown in Example 2.3 that a mho characteristic is defined by the equation $Z = (K_3/K_2)\sin(\theta + \varphi)$, where (K_3/K_2) is a constant for a relay design [7]. Using the symbol Z_r for (K_3/K_2), the performance equation of the mho relay becomes $Z = Z_r \sin(\theta + \varphi)$. Multiplying both sides by the relay input current I and replacing IZ by E (Figure 2.16), the voltage at the relay location, the performance equation is

Figure 2.17 Phasor diagram for a mho distance relay.

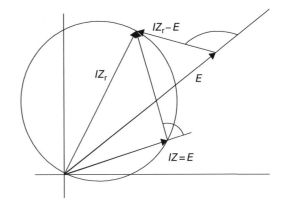

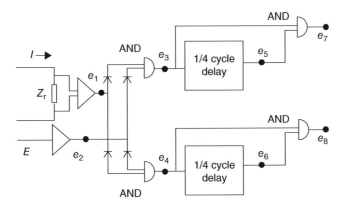

Figure 2.18 Possible circuit configuration for a solid-state distance relay.

$$E - I Z_\mathrm{r} \sin\left(\theta + \varphi\right) = 0 \tag{2.16}$$

The mho characteristic may be visualized as the boundary of the circle, with all points inside the circle leading to a trip and all points outside the circle producing a no-trip— or a block—signal. The points external to the circle are such that the phase angle between the phasor E and the phasor $(IZ_\mathrm{r} - E)$ is greater than 90°, while for all the points inside the circle, the angle between those two phasors is less than 90° (Figure 2.17). Conversely, if the angle between $(E - I Z_\mathrm{r})$ and E is greater than 90°, the fault is inside the zone of the relay; if this angle is smaller than 90°, the fault is outside the zone.

An analog circuit may be designed to measure the angle between the two input wave-forms corresponding to those two phasors. For example, consider the circuit shown in Figure 2.18. The relay input current is passed through a shunt with an impedance Z_r. This is known as a replica impedance. The negative of this signal and the relay input voltage signal are fed to high-gain amplifiers, which serve to produce rectangular pulses with zero-crossing points of the original sinusoidal waveform retained in the output, as shown in Figure 2.19.

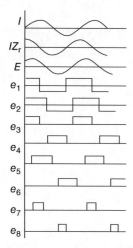

Figure 2.19 Waveforms in the circuit of Figure 2.18.

The positive and negative portions of these square waves are isolated by two half-wave bridges, and supplied to a logic AND gate. Assuming steady-state sine wave current and voltage inputs, the outputs of the two AND gates are at logic level 1 for the duration equal to the phase angle between the phasors $-I Z_r$ and E. If the angle is greater than 90°, that is, if the duration of the outputs of these two AND gates is greater than 4.16 ms (for a 60 Hz power system), the relay should operate. This condition may be tested using an edge-triggered 4.16 ms timer, and by checking if the input and output of the timer are ever at logic 1 level simultaneously. By using an AND gate on the input and the output signals of the timer, a logic 1 output of this AND gate would indicate an internal fault, while for external faults, this output would remain at logic 0. Waveforms at various points of this circuit are also shown in Figure 2.19. Similar comparisons may also be made on the negative half-cycles of the signals.

This is a very simplified analysis of the circuit, and issues of transient components in the input signals, response to noise pulses in the signals, and other practical matters have not been discussed. A great many more features must be included in the relay design for this to be a practical relay. However, our discussion should give some idea as to how a given relay characteristic may be produced in a solid-state relay.

As is evident from the previous discussion, a substantial part of a solid-state relay design includes logic circuits commonly found in digital circuit design. Many of the logic circuit elements, and the symbols used to represent them, are included in an Institute of Electrical and Electronics Engineers (IEEE) standard [8].

2.6 Computer relays

The observation has often been made that a relay is an analog computer. It accepts inputs, processes them electromechanically, or electronically, to develop a torque, or a logic output representing a system quantity, and makes a decision resulting in a contact closure or output signal. With the advent of rugged, high-performance microprocessors, it is obvious that a digital computer can perform the same function. Since the usual relay inputs consist of power system voltages and currents, it is necessary to obtain a digital representation of these parameters. This is done by sampling the analog signals, and using an appropriate computer algorithm to create suitable digital representations of the signals. This is done by a digital filter algorithm. The functional blocks shown in Figure 2.20 represent a possible configuration for a digital relay [2].

The current and voltage signals from the power system are processed by signal conditioners consisting of analog circuits, such as transducers, surge suppression circuits, and antialiasing filters, before being sampled and converted to digital form by the analog-to-digital converter. The sampling clock provides pulses at sampling frequency. Typical sampling frequencies in use in modern digital relays vary between 8 and 32 times the

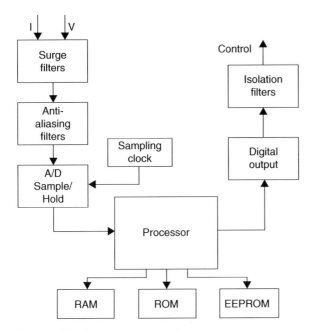

Figure 2.20 Major subsystems of a computer relay.

fundamental power system frequency. The analog input signals are generally frozen by a sample-and-hold circuit, in order to achieve simultaneous sampling of all signals regardless of the data conversion speed of the analog-to-digital converter. The relaying algorithm processes the sampled data to produce a digital output. The algorithm is, of course, the core of the digital relay, and a great many algorithms have been developed and published in the literature. It is not our purpose to examine or evaluate any algorithms. The book by Phadke and Thorp [2] contains an exhaustive treatment of the subject, and the interested reader may wish to consult it for additional information about computer-relaying algorithms.

In the early stages of their development, computer relays were designed to replace existing protection functions, such as transmission line, transformer, or bus protection. Some relays used microprocessors to make the relaying decision from digitized analog signals, others continued to use analog concepts to make the relaying decision and digital techniques for the necessary logic and auxiliary functions. In all cases, however, a major advantage of the digital relay was its ability to diagnose itself, a capability that could only be obtained in an analog relay—if at all—with great effort, cost, and complexity. In addition, the digital relay provides a communication capability that allows it to warn system operators when it is not functioning properly, permits remote diagnostics and possible correction, and provides local and remote readout of its settings and operations.

As digital relay investigations continued, and confidence mounted, another dimension was added to the reliability of the protective system. The ability to adapt itself, in real time, to changing system conditions is an inherent feature in the software-dominated digital relay [9]. These changes can be initiated by local inputs or by signals sent from a central

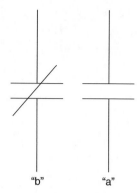

"b"   "a"

Figure 2.21 Conventions for contact status.

computer. The self-diagnostic capability, the hierarchical nature, the data-sharing abilities of microprocessors, and the ability to adapt settings and other characteristics to actual system conditions in real time make digital relays the preferred present-day protective system. The problem of mixing analog and digital devoiced within a common overall protection system, and the lack of standardization between manufacturers is being addressed and is discussed in Section 13.7 of Chapter 13.

2.7 Other relay design considerations

2.7.1 Contact definition

In an electromechanical relay, the operating mechanism is directed to physically move a contact structure to close or open its contact. A relay may operate and either open or close the contacts depending on the circumstances. Most relays have a spring or use gravity to make the contact assume a given state when the relay is completely de-energized. A contact that is closed under this condition, often referred to as its condition "on-the-shelf," is said to be a "normally closed" or a "b" contact. If the contact is open "on-the-shelf," it is referred to as a "normally open" or an "a" contact. It is important to note that the word "normally" does not refer to its condition in normal operation. An auxiliary relay with "a" and "b" contacts, if de-energized in service, would have its contacts as described; if the relay, however, is normally energized in service, the contact description would be the opposite. For example, a fail-safe relay that stays energized when the power is on and drops out with loss of power would have its "a" contact closed in service. It is also conventional to show the contacts on schematic (known as elementary diagrams) or wiring diagrams in the on-the-shelf condition regardless of the operation of the relay in the circuit. These contact definitions are illustrated in Figure 2.21.

2.7.2 Targets

Protective relays are invariably provided with some indication that shows whether or not the relay operated. In electromechanical relays, this indication is a target, that is, a brightly colored flag that becomes visible upon operation of the relay. These targets can be electrical or mechanical. An electrical target is usually preferred, because it is activated by the trip current, and shows that the trip current actually flowed. There are some instances, however, when a mechanical target is useful. For example, in a trip circuit with several logic elements in series, such as separate overcurrent and directional contacts, a mechanical target will show which element operated, even if all did not operate and no trip occurred.

Solid-state and digital relays use more complex targeting schemes that allow one to trace the tripping sequence more completely. For example, the logic elements associated with phase or ground fault detection, timing elements, and the tripping sequence are all capable of being brought to indicating lights, which are used as targets.

2.7.3 Seal-in circuit

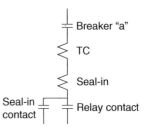

Figure 2.22 Principle of a seal-in relay circuit (TC, trip coil).

Electromechanical relay contacts are designed as a part of the overall relay design, which places restrictions on the size and mass of the contacts. They are not designed to interrupt the breaker trip relay coil current. In order to protect the relay contacts against damage, some electromechanical relays are provided with a holding mechanism. This is a small electromagnet whose coil is in series with the relay contacts and whose contact is in parallel with them. The electromagnet is energized, closing its contacts in parallel with the relay contact as soon as the trip coil (TC) is energized, and drops out when the circuit breaker opens. This allows the circuit breaker "a" switch to de-energize the TC and the holding coil. Figure 2.22 shows the details of a seal-in circuit.

2.7.4 Operating time

Operating time is a very important feature that can be used to amend any basic relaying function in order to reach a specific goal. Examples of using time delays in protection system design can be found in many of the following chapters. Time delay can be an integral part of a protective device, or may be produced by a timer. For example, the operating time of a fuse or an overcurrent relay is an inverse function of the operating current, that is, the greater the current, the shorter the operating time. The time delay is an integral part of the fuse or overcurrent relay and varies with the magnitude of the operating quantity. A clock or a pneumatic timer may be used as an auxiliary relay, and will operate in its set time regardless of the operating quantities that actuate the main relay. Figure 2.23 shows the operating characteristics of these two timing applications.

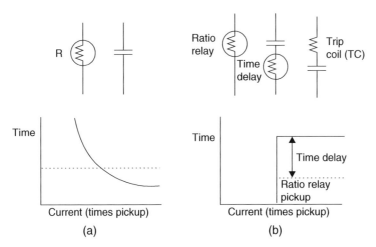

Figure 2.23 Time-delay relays: (a) integral time-delay relay; (b) external time-delay relay added to the circuit.

2.7.5 Ratio of pickup to reset

The pickup and dropout (also known as the reset) currents of a relay have already been mentioned. A characteristic that affects some relay applications is the relatively large difference between the pickup and dropout values. For instance, a plunger-type relay will shorten its air gap as it picks up, permitting a smaller magnitude of coil current to keep it picked up than it took to pick it up. If the relay trips a circuit breaker, the coil current drops to zero, so there is no problem. If, however, a low reset relay is used in conjunction with other relays in such a way that the coil current does not go to zero, the application should be carefully examined. When the reset value is a low percentage of the pickup value, there is a possibility that an abnormal condition might cause the relay to pick up, but a return to normal condition might not reset the relay.

2.8 Control circuits: a beginning

We will now begin a discussion of the station-battery-powered control circuits that are used to perform the actual tripping and reclosing functions in a substation. The output contacts of relays, circuit breakers, and other auxiliary relays, as well as the circuit breaker TCs and timers and reclosers, are connected to the battery terminals. Recall the discussion of "a" and "b" contacts in Section 2.7. The status of contacts in a control circuit is always shown in their "on-the-shelf" state. The battery terminals are shown as a positive DC bus and a negative DC bus. The DC circuits are usually isolated from ground. A test lamp circuit is arranged in a balanced configuration, as shown in Figure 2.24, with both lamps burning at half of full brilliance. If either of the battery buses is accidentally grounded, the indicating light connected to the grounded bus is extinguished, and the other light burns at full brilliance. If the accidental ground has some resistance, the lamp's intensity will be proportional to the fault resistance, that is, the lamp associated with the faulted bus will be less brilliant than the lamp associated with the unfaulted bus. A series alarm relay can be added (devices

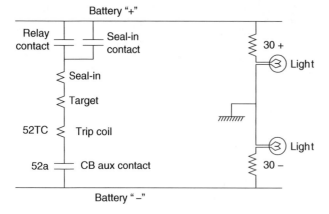

Figure 2.24 Control circuit showing various contacts and test circuit.

30+ and 30− in Figure 2.24), which can actuate an audible alarm in the station or transmit the alarm to a control center; so maintenance personnel can inspect the wiring and eliminate the accidental ground. It is extremely important to eliminate the first accidental ground even though it does not create a fault on the battery. Should a second fault occur, it would short-circuit the battery and produce a catastrophic failure.

Now consider the breaker TC connection in the control circuit. The TC is usually connected in series with a circuit breaker auxiliary "a" contact to the negative DC bus. The breaker is normally closed when the associated power equipment is energized, which ensures that the TC is normally connected to the negative bus. As there is relay contact corrosion due to electrolytic action at the positive terminals, this practice avoids any problems in the TC terminal connections [3]. The seal-in coil and the target coil are in series with the TC, and the entire chain is connected to the positive DC bus through the contacts of the tripping relay. As mentioned in Section 2.7, the contacts of the seal-in relay bypass the tripping relay output contacts to protect it from accidental opening while carrying the TC current. The breaker TC current is interrupted by the circuit breaker auxiliary "a" contact.

We will develop the connections of the DC control circuit throughout the rest of the book, as we introduce other relaying functions and auxiliary relays. It should be apparent that the diagram for a fully developed relay system will soon become quite complicated. In order to simplify the labeling on the control circuit, most relays and control devices found in substations are given a standard designation number. Thus, a circuit breaker device function number is 52; an instantaneous overcurrent relay is 50, while a differential relay is 87. Other standard device numbers can be found in the literature [4, 10] (see Appendix A). However, we will continue to use the device names in our DC control circuit development for the sake of clarity.

2.9 Summary

In this chapter, we have very briefly examined the general operating and design considerations of some typical protective relays. Our purpose is not to cover any relay design or relay construction beyond that which would give us an understanding of its operating principles; so we will better understand the specific applications as they are discussed in later chapters. We have identified the primary fault-detecting parameters of level detection, magnitude and phase angle comparison, and the differential principle. We have examined fundamental relay construction of electromechanical, solid-state, and digital relays. Finally, we have introduced DC control circuit diagrams as an aid in understanding how the various elements of a protection system combine to provide the desired protection function.

Problems

2.1 Consider the transmission line connected to a generator as shown in Figure 2.25. The impedance data for the generator and the line are given in the figure. A relay to be located at terminal A is to detect all faults on the transmission line.

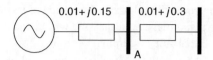

0.01+ *j*0.15   0.01+ *j*0.3

A

Figure 2.25 System diagram for Problem 2.1.

Assume a prefault voltage of 1.0 pu, and allow for a possible steady-state overvoltage of 1.2 pu during normal operation. Determine the pickup settings for an overcurrent relay and an undervoltage relay to be used as fault detectors for this circuit. Allow a sufficient margin between the normal conditions and the pickup settings to accommodate any inaccuracies in relay performance. Assume the maximum load current to be 1.0 pu.

2.2 A phase angle detector is to be installed at location A in the power system shown in Figure 2.26 to detect faults between the buses A and B. Determine the appropriate voltage and current for this detection function for the case of a b-to-c fault. Show the phase relationship for a fault between the buses A and B, as well as for a fault behind the relay at bus A. Assume reasonable values of various reactances and voltages, and construct a phasor diagram to illustrate the cases of the two types of fault. What voltage and current would be appropriate to detect a forward fault between phase and ground? Also, construct a phasor diagram showing the phase relationship between the phase a current and the current in the transformer neutral for the case of the phase-to-ground fault. Can the neutral current be used as a reference phasor for direction detection?

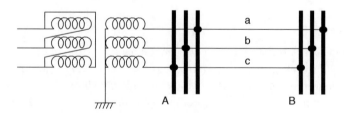

a

b

c

A   B

Figure 2.26 System diagram for Problem 2.2.

2.3 Derive Equation (2.2) and determine the value of the constant K for the relay structure shown in Figure 2.7. You may assume that the steel core of the electromagnet has infinite permeability. Derive the inductance for the coil, and then use Equation (2.1) to derive the required result.

2.4 Write a computer program to integrate the equation of motion of the plunger in Example 2.1 without making the approximation of a constant force on the plunger. You may use Euler's integration formula, or any other method of numerical integration. Plot the result as a curve showing the operating time of the relay as a function of the current. It should produce an asymptotic shape for currents near the pickup value.

2.5 A plunger-type relay is as shown in Figure 2.7. The inductance of the coil when the plunger is at its de-energized position is 2.0 H, while it is 2.05 H when the plunger has completed its travel of 2 mm. The height of the pole-piece a is 1 cm. What is the ratio of its dropout-to-pickup current?

2.6 Starting from Equation (2.12), derive Equation (2.13). Under what conditions is the torque produced by the relay equal to zero?

2.7 Prove that $Z \sin(\theta + \varphi) = 0$ is an equation of a straight line in the R–X plane. What is the slope of this line? Recall that this equation is obtained by dividing the torque equation by I. Thus, the straight-line characteristic is not valid for $I = 0$. If the spring constant is not negligible, show that the performance equation describes a circle through the origin, with the diameter of the circle depending upon the magnitude of the voltage [3]. Even for rather small values of the voltage, the diameter of the circle is so large that a straight-line approximation is still valid.

2.8 Show that the equation $|Z + K_4| = \sqrt{(K_1 + K_2)}$ in Example 2.3 describes a circle in the R–X plane. Where is the origin of this circle? What is the radius of the circle? This characteristic is known as the offset impedance relay.

2.9 Derive the equation of the mho characteristic given in Example 2.3. Show that the characteristic is a circle passing through the origin, and that its diameter is independent of the voltage or current. What is the length of the diameter? What angle does the diameter passing through the origin make with the R axis?

2.10 Consider the power system shown in Figure 2.27. The pu impedances of the two line sections and the generator are shown in the figure. Concentrating on three-phase faults only, assume that the relay at bus A is set to pickup for a fault at bus C. Assuming that the pickup setting is equal to one-third of the fault current, what is the pickup setting of this relay? If the dropout-to-pickup ratio for the relay is 0.1, what is the maximum load current at bus B for which the relay will drop out after the fault at bus C is cleared by the protection at B? Recall that after the protection at B clears

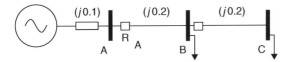

Figure 2.27 System diagram for Problem 2.10.

the fault at C, the current seen by the relay at A is equal to the load connected to bus B, and if the relay does not drop out at this current level, the relay at A will trip its breaker on the load current.

2.11 Show that Equation (2.16) is the equation of a circle in the voltage phasor plane. Show that on the boundary of the circle, the angle between E and $(E - I Z_r)$ is 90°. What is the diameter of this circle, and what angle does the diameter passing through the origin make with the horizontal axis?

References

1 van Cortlandt Warrington, A.R. (1962). *Protective Relays, the Theory and Practice*, vol. **1**. London; New York: Chapman and Hall; Wiley.

2 Phadke, A.G. and Thorp, J.S. (1988). *Computer Relaying for Power Systems*. London; New York: Research Studies Press; Wiley.

3 Mason, C.R. (1956). *The Art and Science of Protective Relaying*. New York: Wiley.

4 Westinghouse (1976). *Applied Protective Relaying*. Newark, NJ: Westinghouse Electric Corporation.

5 IEEE (1986). ANSI/IEEE STD 242–1986 *IEEE Recommended Practices for Protection and Coordination of Industrial and Commercial Power Systems*, IEEE Buff Book.

6 Fitzgerald, A.E., Kingsley, C., and Umans, S.D. Jr. (1983). *Electric Machinery*, 4e. New York: McGraw-Hill.

7 General Electric, *Static Phase Distance Relays, Type SLY, Instructions*, publication GEK-26487B, Power Systems Management Department, General Electric, Philadelphia, PA.

8 IEEE (1991). *Graphic Symbols for Logic Functions* (Includes IEEE Std 91A-1991 Supplement, and IEEE Std 91-1984), IEEE Std 91a-1991 & IEEE Std 91–1984 0–1. https://doi.org/10.1109/IEEESTD.1991.81068.

9 Horowitz, S.H., Phadke, A.G., and Thorp, J.S. (1988). Adaptive transmission system relaying. *IEEE Trans. Power Deliv.* **3** (4): 1436–1445.

10 IEEE (2008). *Standard Electrical Power System Device Function Numbers, Acronyms, and Contact Designations*, IEEE Std C37.2-2008 (Revision of IEEE Std C37.2-1996) 1–48. https://doi.org/10.1109/IEEESTD.2008.4639522.

3

Current and Voltage Transformers

3.1 Introduction

The function of current transformer (CT) and voltage transformer (VT) (collectively known as transducers) is to transform power system currents and voltages to lower magnitudes and to provide galvanic isolation between the power network and the relays and other instruments connected to the transducer secondary windings. The ratings of the secondary windings of transducers have been standardized, so that a degree of interchangeability among different manufacturers' relays and meters can be achieved. In the United States and several other countries, CT secondary windings are rated for 5 A, while in Europe, a second standard of 1 A secondary is also in use. VT secondary windings are rated at 120 V for phase-to-phase voltage connections, or, equivalently, at 69.3 V for phase-to-neutral connections. These are nominal ratings, and the transducers must be designed to tolerate higher values for abnormal system conditions. Thus, CTs are designed to withstand fault currents (which may be as high as 50 times the load current) for a few seconds, while VTs are required to withstand power system dynamic overvoltages (of the order of 20% above the normal value) almost indefinitely, since these types of overvoltage phenomena may last for long durations. CTs are magnetically coupled, multiwinding transformers, while the VTs, in addition to the magnetically coupled VT, may include a capacitive voltage divider for higher system voltages. In the latter case, the device is known as a coupling capacitor voltage transformer (CCVT), and when the transformer primary winding is directly connected to the power system, it is known as a VT. CTs and VTs may be freestanding devices, or they may be built inside the bushing of some power apparatus (such as a circuit breaker or a power transformer) with a grounded tank. Newer types of transducers using electronic and fiber-optic components are described briefly later in this chapter.

The function of transducers is to provide current and voltage signals to the relays (and meters) that are faithful reproductions of the corresponding primary quantities. Although modern transducers do so quite well in most cases, one must be aware of the errors of transformation introduced by the transducers, so that the performance of the relays in the presence of such errors can be assessed.

Power System Relaying, Fifth Edition. Stanley H. Horowitz, Arun G. Phadke and Charles F. Henville.

3.2 Steady-state performance of current transformers

CTs are single-primary, single-secondary, magnetically coupled transformers, and, as such, their performance can be calculated from an equivalent circuit commonly used in the analysis of transformers [1]. Some CTs are used for metering, and consequently their performance is of interest during normal loading conditions. Metering transformers may have very significant errors during fault conditions, when the currents may be several times their normal value for a very short time. Since metering functions are not required during faults, this is not significant. CTs used for relaying are designed to have small errors during faulted conditions, while their performance during normal steady-state operation, when the relay is not required to operate, may not be as accurate. In spite of this difference, all (measuring or relaying) CT performance may be calculated with the same equivalent circuit [2, 3]. The different values of equivalent circuit parameters are responsible for the difference in performance between the various types of CTs.

Consider the equivalent circuit shown in Figure 3.1a. Since the primary winding of a CT is connected in series with the power network, its primary current I'_1 is dictated by the network. Consequently, the leakage impedance of the primary winding Z'_{x1} has no effect on the performance of the transformer, and may be omitted. Referring all quantities to the secondary winding, the simplified equivalent circuit of Figure 3.1b is obtained. Using the turns ratio (l:n) of the ideal transformer of Figure 3.1a, one can write

$$I_1 = \frac{I'_1}{n},\tag{3.1}$$

$$Z_m = n^2 Z'_m.\tag{3.2}$$

The load impedance Z_b includes the impedance of all the relays and meters connected in the secondary winding, as well as that of the leads connecting the secondary winding terminals of the CT located in the substation yard to the protection equipment, which is located in the control house of the substation. Often, the lead impedance is a significant part of the total load impedance.

The load impedance Z_b is also known as the burden on the CT, and is described as a burden of Z_b Ω, or as a burden of $I^2 Z_b$ volt-amperes. If 5 A is the rated secondary current at which the burden is specified, the burden would be $25 Z_b$ volt-amperes. Referring to the phasor diagram in Figure 3.2, the voltage E_m across the magnetizing impedance Z_m is given by

$$E_m = E_b + Z_{x2} I_2,\tag{3.3}$$

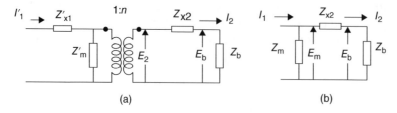

(a)   (b)

Figure 3.1 (a) and (b) CT equivalent circuit and its simplification.

Figure 3.2 CT phasor diagram.

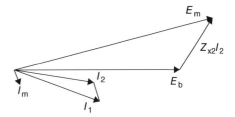

and the magnetizing current I_m is given by

$$I_m = \frac{E_m}{Z_m}. \tag{3.4}$$

The primary current I_1 (referred to the secondary winding) is given by

$$I_1 = I_2 + I_m. \tag{3.5}$$

For small values of the burden impedance, E_b and E_m are also small, and consequently I_m is small. The per unit current transformation error defined by

$$\varepsilon = \frac{I_1 - I_2}{I_1} = \frac{I_m}{I_1} \tag{3.6}$$

is, therefore, small for small values of Z_b. In other words, CTs work at their best when they are connected to very low-impedance burdens. In the limiting case of zero-burden impedance (and a small Z_{x2}), $I_1 = I_2$, and the CT error is zero.

More often, the CT error is presented in terms of a ratio correction factor R instead of the per unit error E discussed in the earlier text. The ratio correction factor R is defined as the constant by which the nameplate turns ratio n of a CT must be multiplied to obtain the effective turns ratio. It follows from Equations (3.5) and (3.6), and the definition of R, that

$$R = \frac{1}{1 - \varepsilon}. \tag{3.7}$$

Although ε and R are complex numbers, it is sometimes necessary to use the error and the ratio correction factor as real numbers equal to their respective magnitudes. This is approximate, but not excessively so.

Example 3.1 Consider a CT with a turns ratio of 500:5, a secondary leakage impedance of $(0.01 + j0.1)$ Ω, and a resistive burden of 2.0 Ω. If the magnetizing impedance is $(4.0 + j15)$ Ω, then for a primary current (referred to the secondary) of I_1,

$$E_m = \frac{I_1(0.0.1 + j0.1 + 2.0)(4.0 + j15.0)}{(0.01 + j0.1 + 2.0 + 4.0 + j15.0)} = I_1 \times 1.922\angle 9.62°$$

and

$$I_m = \frac{I_1 \times 1.922\angle 9.62°}{(4.0 + j15.0)} = I_1 \times 0.1238\angle -65.45°.$$

Thus, if the burden impedance and the magnetizing impedance of the CT are constant, the per unit CT error,

$$\varepsilon = \frac{I_{\mathrm{m}}}{I_1} = 0.1238\angle - 65.45°,$$

is constant, regardless of the magnitude of the primary current. However, the error does depend upon the magnitude and phase angle of the burden impedance. Thus, in this example, for a burden of 1.0 Ω, $\in$ is $0.064\angle - 66°$, and for an inductive burden of $j2$ Ω, $\in$ is $0.12\angle12.92°$. The corresponding ratio correction factor can be found for each of these burdens:

$$R = \frac{1}{(1.0 - 0.1238\angle - 65.45°)} = 1.0468\angle - 6.79° \quad \text{for } Z_{\mathrm{b}} = 2\,\Omega,$$

$$R = 1.025\angle - 3.44° \quad \text{for } Z_{\mathrm{b}} = 1\,\Omega,$$

and

$$R = 1.13\angle 1.73° \quad \text{for } Z_{\mathrm{b}} = j2\,\Omega.$$

Since the magnetizing branch of a practical transformer is nonlinear, Z_{m} is not constant, and the actual excitation characteristic of the transformer must be taken into account in determining the factor R for a given situation. The magnetizing characteristic of a typical CT is shown in Figure 3.3. This being a plot of the root mean square (rms) magnetizing current versus the rms secondary voltage, I_{m} for each E_{m} must be obtained

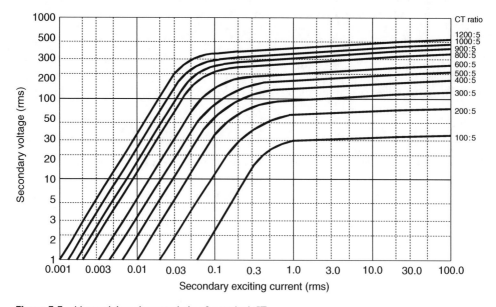

Figure 3.3 Magnetizing characteristic of a typical CT.

from this curve, and then used in Equations (3.5)–(3.7) to calculate the ratio correction factor. The procedure is illustrated by Example 3.2 and one of the problems at the end of this chapter. As can be seen from these problems, the calculation of CT performance with nonlinear magnetizing characteristic is a fairly complicated procedure. A much simpler and approximate procedure is available with the help of the standard class designations described next.

Note that Figure 3.3 contains a family of characteristics. Each CT may be provided with several taps, which can be used to obtain a turns ratio that is most convenient in a given application. The turns ratios for CTs have also been standardized [4], and some of the standard ratios are given in Table 3.1. Turns ratios other than those provided by the standard may be obtained on special order. However, this is quite expensive, and rarely can be justified for relaying applications.

When the slopes of the curves in Figure 3.3 transition from the steeper portion to the flatter portion, the CT passes through the saturation point. When the core of the CT is saturated, the magnetic properties of the steel are no longer relevant and the CT becomes effectively an air-core transformer. The region of transition into the saturated state is called the "knee point" region. The knee point of a CT is defined in Reference [2] as the point at which the tangent of the curve is sloped at 45° to the abscissa.

Note that in some CTs the knee point voltage can be higher than a few thousand volts. The multiple secondary turns in a CT step down the current, but will step up the voltage if open-circuited when energized. Since open-circuited CT secondaries can produce dangerously high voltages, secondary circuit design pays special attention to preventing accidental open circuits. Wiring terminals may be ring-type that cannot come off unless the securing screw is completely removed. Alternatively, connection terminals may use double-screw clamps. CT secondary test and isolation facilities also include special shorting features to always maintain a secondary short circuit if the normal low-impedance burden is open circuited or removed for test purposes.

Table 3.1 Standard current transformer multiratios (MR represents multiratio CTs).

600:5 MR	1200:5 MR	2000:5 MR	3000:5 MR
50:5	100:5	300:5	300:5
100:5	200:5	400:5	500:5
150:5	300:5	500:5	800:5
200:5	400:5	800:5	1000:5
250:5	500:5	1100:5	1200:5
300:5	600:5	1200:5	1500:5
400:5	800:5	1500:5	2000:5
450:5	900:5	1600:5	2200:5
500:5	1000:5	2000:5	2500:5
600:5	1200:5		3000:5

Example 3.2 Consider a CT with a turns ratio of 600:5, and the magnetizing characteristic corresponding to this ratio in Figure 3.3. It is required to calculate the current in its secondary winding for a primary current of 5000 A, if the total burden impedance is $(9 + j2)\,\Omega$ and the secondary leakage impedance is negligible. The impedance angle of the magnetizing branch is $60°$. Since the magnetizing branch is nonlinear, we may consider the equivalent circuit to be made up of a linear part consisting of a current source of $5000 \times 5/600 = 41.66$ A in parallel with the burden, and connected across the nonlinear impedance Z_m, as shown in Figure 3.4. The corresponding Thévenin equivalent consists of a voltage source of $41.66 \times (9 + j2) = 384.1\angle12.53°$ volts, in series with the burden. Since the impedance angle of Z_m is known to be $60°$, the magnetizing current I_m and the secondary voltage E_2 can be expressed in terms of the magnitude of Z_m with the Thévenin voltage as the reference phasor:

$$I_m = \frac{384.1}{[|Z_m| \times (0.5 + j0.866) + (9.0 + j2.0)]},$$

$$E_2 = I_m Z_m.$$

These two equations may be solved to produce values of E_2 and I_m in terms of $|Z_m|$ as the parameter (Table 3.2). Plotting the curve of these values on Figure 3.3, it is found to intersect the magnetizing characteristic at $I_m = 17$ A, $E_2 = 260$ V. Finally, reworking the equations to find the phase angles of the currents, the various currents are $I_1 = 41.66\angle0°$ (in that case, $E_{th} = 384.1\angle12.53°$), $I_m = 17\angle - 29.96°$, and $I_z = 28.24\angle17.51°$. The error $\in$ is therefore $0.408\angle - 29.96°$, and the ratio correction factor $R = 1.47\angle - 17.51°$.

Clearly, this CT is in severe saturation at this current and at the burden chosen. In practice, it must be used with much smaller burdens to provide reasonable accuracies under faulted conditions.

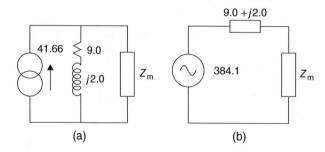

(a) (b)

Figure 3.4 (a) and (b) Calculation of CT performance in Example 3.2.

Table 3.2 Values of secondary voltage E_2 and magnetizing current I_m in terms of magnetizing impedance Z_m as the parameter.

$|Z_m|$	$|I_m|$	$|E_2|$
∞	0	384.1
100	3.61	361.0
10	21.82	218.2

3.2.1 Standard class designation

The equivalent circuit method of calculating the performance of a CT depends upon the availability of the magnetizing characteristic. When this is not readily available, an approximate assessment of the CT performance may be made through its standard class designation [2, 4] as defined by the American National Standards Institute (ANSI) and the Institute of Electrical and Electronics Engineers (IEEE). The ANSI/IEEE class designation of a CT is defined in one of the three classes "C" or "T" or "X."

In the case of "C" or "T" class, the letter is followed by a terminal voltage; for example, C400 or T300. At rated current, the upper limit on the error is 3%. Under fault conditions when the current is 20 times rated, the upper limit on the error is 10%, when the voltage at its secondary terminals is less than the rated terminal voltage. In cases where the CT secondary windings are rated at 5 A secondary (typical in North America), 20 times rated corresponds to a secondary current of 100 A. The C400 CT, for example, will have an error of less than or equal to 10% at a secondary current of 100 A for burden impedances that produce 400 V or less at its secondary terminals. If the magnetizing impedance is assumed to be linear, the error made will be approximately proportional to the developed voltage. The letter "C" in the class designation means that the transformer design is such that the CT performance can be calculated from the excitation curve, whereas the letter "T" signifies some uncertainties in the transformer design, and the performance of the CT must be determined by testing the CT.

In the case of "X" class CTs, at rated current, the maximum error current is limited to 1%. At higher currents, such as 20 times rated, similar to "C" class, the performance can also be calculated from the excitation curve. However, three specific secondary excitation requirements are specified:

Ek: the minimum knee point voltage.
Ik: the maximum exciting current at Ek.
Rct: the maximum allowable secondary winding resistance corrected to 75 °C.

The specified secondary excitation requirements can then be used to calculate the transient performance of the CT as described in Section 3.3 of this chapter.

The International Electrotechnical Commission (IEC) also has performance standards for CT accuracy [5], defining steady-state accuracy for protection applications using the class designation "P." Their additional performance classes similar to ANSI/IEEE Class X are as follows:

TPS for low-leakage-flux design CTs.
TPX for closed-core CT for specified transient duty cycle.
TPY for gapped-core CT with low remanence for specified transient duty cycle.
TPZ linear CT with no magnetic core.

Example 3.3 Consider a 600:5 turns ratio CT of the class C400. The C400 CT will provide 100 A in the secondary with no more than 10% error at 400 V secondary terminal voltage. Thus, the magnitude of the magnetizing impedance is approximately $400/(0.1 \times 100) = 40\,\Omega$. With a primary current of 5000 A, the nominal secondary current will be $5000 \times 5/600 = 41.66$ A. With a maximum error of 10%, this will allow a magnetizing current of about 4 A. At this magnetizing current, it may have a maximum secondary terminal voltage of $4.16 \times 400/10$, or 167 V. Since the primary current is 41.66 A, the maximum burden impedance that will produce 167 V at the secondary is $167/(41.66-4.16) = 4.45\,\Omega$. All burdens of a smaller magnitude will produce smaller errors.

In the abovementioned calculation, we have assumed that the magnetizing current is in phase with the secondary current and the current in the burden. Considering all other approximations made in this procedure (such as the assumed linearity of the magnetizing characteristic), this approximation is justifiable. We should remember that calculations such as these give us a limit for safe operation; oftentimes the errors will be much under those limits, and the approximations will not matter.

3.2.2 Polarity markings on CT windings

Polarity markings of transformer windings are a means of describing the relative directions in which the two windings are wound on the transformer core. The terminals identified by solid marks indicate the starting ends of the two windings, meaning that if these are considered to be the starting points, and we trace the two windings along the transformer core, both windings will go around the core in the same sense (i.e., counterclockwise or clockwise). In a transformer, if one of the winding currents is considered to be flowing into the marked terminal, the current in the other winding should be considered to be leaving its marked terminal. The two currents will then be (approximately) in phase with each other. Similarly, the voltages of the two windings, when measured from the unmarked terminal to the marked terminal, will be (approximately) in phase with each other. This convention for polarity marking is also used for CTs. An alternative way is to label the primary winding terminals H_1 and H_2 (or P_1 and P_2), and the secondary winding terminals X_1 and X_2 (or S_1 and S_2). In North America, the H and X notations are usually used. H_1 and X_1 may then be assumed to have the polarity mark on them. Both of these conventions are shown in Figure 3.5.

Since CT secondary windings are connected in quite complex networks in the overall protection systems for three-phase apparatus, it is extremely important that the meaning of the polarity marking be clearly understood. A current I_1 in the primary winding of the CT will produce a current I_2 in its secondary winding, where the magnitudes of I_1 and I_2 are in inverse proportion to the turns ratio (neglecting the magnetizing current for the moment), and their phase angles will be as indicated by the polarity markings. An easy way to remember this is to think of H_1 being the same terminal as X_1. The continuity of the current is then reflected by the polarity markings. It is good to think of the CT secondary winding as a constant current source of I_2 as determined by I_1. If I_1 is zero, I_2 also must be zero, and the secondary winding of such a CT may be considered to be open-circuited. Some of the problems at the end of this chapter reinforce these ideas, especially in the context of wye or delta connections of the CTs.

Polarity marks on CT secondaries are important when the phase angle of the secondary current must be compared with the phase angle of other secondary currents or voltages. CT inputs on relays will also include polarity marks, so that they can be properly

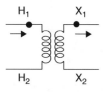

Figure 3.5 Polarity markings of a CT.

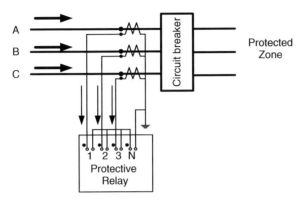

Figure 3.6 Polarity marks on a protection relay connect to polarity marks on CTs.

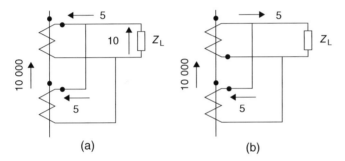

Figure 3.7 (a) and (b) CT connections for Example 3.4.

connected to instrument transformers to "look" in the correct direction. In Figure 3.6, note how the current entering the protected zone is in phase with the current entering the polarity marks on the protection relay.

Example 3.4   Consider the CTs shown in Figure 3.7a. If the primary current is 1000 A, and the two CT ratios are 1000:5 and 1000:5, respectively, the current in the burden impedance Z_L is 10 A. If the CT secondaries are connected as shown in Figure 3.6b, the burden current becomes zero. In reality, because of CT errors, the burden current in the first case will be less than 10 A, and in the second case it will be small, but not equal to zero. If the magnetizing characteristics of the two CTs are taken into account, the current in the burden must be calculated iteratively. One of the problems at the end of this chapter will illustrate this case.

3.3   Transient performance of current transformers

The performance of CTs when they are carrying the load current is not of concern as far as relaying needs are concerned. When faults occur, the current magnitudes could be much larger, the fault current may have substantial amounts of direct current (DC) components, and there may be remanence in the CT core. All of these factors may lead to saturation of the CT

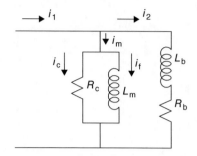

Figure 3.8 CT equivalent circuit for transient analysis.

core and cause significant distortion of the secondary current waveform. Although saturation of the core is a nonlinear phenomenon, we will first establish a relationship between the primary fault current of a CT, its burden, and the flux in its core when the core of the CT has a linear magnetizing characteristic. After this is done, we will qualitatively examine the effect of the nonlinearity on the performance of the CT. Consider the CT equivalent circuit shown in Figure 3.1b, with the total impedance in the secondary circuit—that is, the sum of the secondary leakage impedance, CT secondary winding resistance, lead impedance, and load impedance—given by $Z_b = (R_b + j\omega L_b)$. In Laplace domain, $Z_b = (R_b + s L_b)$. Assume further that the magnetizing impedance Z_m is a parallel combination of the core loss resistance R_c and the magnetizing inductance L_m (Figure 3.8). The primary current $i_1(t)$ (as reflected in the secondary), containing an exponentially decaying DC offset, is given by

$$i_1(t) = I_{max} \left[\cos(\omega t - \theta) - \varepsilon^{-t/T} \cos\theta \right] \text{ for } t > 0$$
$$= 0 \qquad\qquad\qquad\qquad\qquad \text{for } t < 0,$$

(3.8)

where I_{max} is the peak value of the sinusoidal steady-state fault current, T is the time constant of the primary fault circuit, and θ is the angle on the voltage wave where the fault occurs; it has been assumed that there was no primary current before the inception of the fault. In Laplace domain [6], the primary current is given by

$$i_1(s) = I_{max} \cos\theta \left(\frac{s}{s^2 + \omega^2} + \frac{T}{1 + sT} \right) + I_{max} \sin\theta \left(\frac{\omega}{s^2 + \omega^2} \right).$$

(3.9)

Also,

$$v_2(s) = R_c i_c = s L_m i_f = i_2(R_b + s L_b),$$

(3.10)

and the flux linkages λ of the core are given by $L_m i_f$. Furthermore,

$$i_2 = i_1 - (i_f - i_c).$$

(3.11)

Equations (3.10) and (3.11) can be solved for λ and v_2 in terms of i_1. A somewhat simple result may be resistive burden. Setting L_b equal obtained for the case of a purely to zero:

$$\lambda = \frac{R_c R_b}{R_c + R_b} \frac{1}{s + 1/\tau} i_1,$$

(3.12)

$$v_2 = \frac{R_c R_b}{R_c + R_b} \frac{1}{s + 1/\tau} I_1,$$

(3.13)

where

$$\tau = \frac{R_c L_m + R_b L_m}{R_b R_c}.$$

(3.14)

Substituting for i_i from Equation (3.9), and taking the inverse Laplace transform of Equations (3.12) and (3.13), the time behavior of the core flux linkages, and the secondary current, can be derived. The time domain expressions for λ and i_2 are

$$\lambda = I_{max} \cos\theta \frac{R_c R_b}{R_c + R_b} \left\{ \varepsilon^{-t/\tau} \left[-\frac{\tau T}{\tau - T} + \tau \left(\sin\varphi \cos\varphi \tan\theta - \cos^2\varphi \right) \right] \right. $$
$$\left. + \varepsilon^{-t/T} \left(\frac{\tau T}{\tau - T} \right) + \tau \frac{\cos\varphi}{\cos\theta} \cos\left(\omega t - \theta - \varphi\right) \right\} \tag{3.15}$$

and

$$i_2 = \frac{1}{B_b} \frac{d\lambda}{dt}$$
$$= I_{max} \cos\theta \frac{R_c}{R_c + R_b} \left\{ \varepsilon^{-t/\tau} \left[-\frac{T}{\tau - T} + \left(\sin\varphi \cos\varphi \tan\theta - \cos^2\varphi \right) \right] \right. \tag{3.16}$$
$$\left. - \varepsilon^{-t/T} \left(\frac{\tau}{\tau - T} \right) - \omega\tau \frac{\cos\varphi}{\cos\theta} \sin\left(\omega t - \theta - \varphi\right) \right\}$$

where $\tan\varphi = \omega\tau$.

Example 3.5 Consider the case of a purely resistive burden of 0.5 Ω being supplied by a CT with a core loss resistance of 100 Ω, and a magnetizing inductance of 0.005 H. Let the primary current with a steady-state value of 100 A be fully offset. Let the primary fault circuit time constant be 0.1 s. For this case,

$$i_1 = 141.4 \times \varepsilon^{-10t} - 141.4 \cos\left(\omega t\right),$$
$$\theta = \pi, R_c = 100, R_b = 0.5, T = 0.1 \text{ s}.$$

Hence,

$$\tau = \frac{(100 + 0.5) \times 0.005}{100 \times 0.5} = 0.01005,$$
$$\omega\tau = 377 \times 0.01005 = 3.789,$$

and

$$\varphi = \tan^{-1}(3.789) = 75.21° = 1.3127 \text{ rad}.$$

Substituting these values in Equations (3.15) and (3.16) gives

$$\lambda = -0.7399\varepsilon^{-99.5t} + 0.786\varepsilon^{-10t} - 0.1804 \cos\left(\omega t - 1.3127\right),$$
$$i_2 = 147.24\varepsilon^{-99.5t} - 15.726\varepsilon^{-10t} + 136.02 \sin\left(\omega t\right) - 1.3127.$$

These expressions for i_1, i_2, and λ have been plotted in Figure 3.9. When the burden is inductive, L_b cannot be neglected, and the expressions for i_2 and λ are far more complicated. Essentially, additional time constants are introduced in their expressions. It is usual to solve such circuits by one of the several available time-domain simulation programs such as mentioned in Reference [7].

The important fact to be noted in Figure 3.9 is the time behavior of the flux linkages λ. The DC component in the fault current causes the flux linkages to increase considerably above

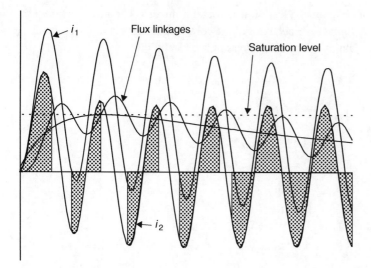

Figure 3.9 Primary and secondary currents and core flux linkages of a CT.

their steady-state peak. Now consider the effect of saturation. The dotted line in Figure 3.9 represents the flux level at which the transformer core goes into saturation. As an approximation, assume that in the saturated region the magnetizing curve is horizontal, that is, the incremental core inductance is zero. Thus, for the duration that λ is above the dotted line in Figure 3.9, it is held constant at the saturation level, and the magnetizing inductance L_m in the equivalent circuit of Figure 3.8 becomes zero. As this short-circuits the load impedance, the secondary current for this period also becomes zero. This is represented by the non-shaded i_2 curve in Figure 3.9. It should be noted that the flux linkages will return to zero DC offset in time, so that the CT will get out of saturation after some time, depending upon the circuit parameters. It should also be clear that any remnant flux in the CT core will also affect the time to saturate. Additional details of CT performance with different types of burden, and the expected time to saturate, can be found in the literature [2, 7, 8].

The significance of the abovementioned discussion is that the secondary current of a CT may not represent the primary current faithfully if the CT goes into saturation, and hence relays that depend upon the secondary current are likely to misoperate during this period. The possibility of CT saturation must be taken into account when designing a relaying system.

3.4 Special connections of current transformers

3.4.1 Auxiliary current transformers

Auxiliary CTs are used in many relaying applications for providing galvanic separation between the main CT secondary and some other circuit. They are also used to provide an adjustment to the overall current transformation ratio in legacy applications where the relay cannot change the effective ratio by setting adjustment. As mentioned before, CT ratios have been standardized, and when other than a standard ratio is required, an auxiliary CT provides a convenient method of achieving the desired ratio. The auxiliary CT, however, makes its own contributions to the overall errors of transformation. In particular, the possibility that the auxiliary CT itself may saturate should be taken into consideration. Auxiliary CTs with multiple taps, providing a variable turns ratio, are also available. The burden connected into the secondary winding of the auxiliary CT is reflected in the secondary of the main CT, according to the normal rules of transformation: if the auxiliary CT ratio is $l{:}n$, and its burden is Z_1, it is reflected in the main CT secondary as Z_1/n^2.

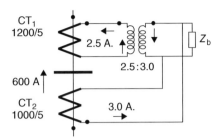

Figure 3.10 Auxiliary CT connections for Example 3.6.

Example 3.6 Consider the CT connection shown in Figure 3.10. CT_1 has a turns ratio of 1200:5, while CT_2 has a turns ratio of 1000:5. It is desired that when the primary current flows through the two lines as shown, the current in the burden be zero. Assume the primary current to be 600 A. The current in the secondary winding of CT_1 is 2.5 A and that in the secondary winding of CT_2 is 3 A. By inserting an auxiliary CT with a turns ratio of 3:2.5 or 1.2:1 in the secondary circuit of CT_1, the current in the auxiliary CT secondary becomes 3 A. With the polarity markings as shown, the burden current is zero.

The burden on CT_2 is Z_b, while that on CT_1 is $Z_b \times (1.2)^2 = 1.44Z_b$. The burden on the auxiliary CT is of course Z_b. CT connections such as these are used in various protection schemes to be discussed later, and utilize the fact that, assuming no auxiliary CT saturation, when the primary current flows uninterrupted through the two primary windings, the burden current remains zero, while if some of the primary current is diverted into a fault between the two CTs, the burden current is proportional to the fault current.

3.4.2 Wye and delta connections

In three-phase circuits, it is often necessary in legacy applications to connect the CT secondaries in wye or delta connections to obtain certain phase shifts and magnitude changes between the CT secondary currents and those required by the relays connected to the CTs. Consider the CT connections shown in Figure 3.11 [9]. The wye connection shown in Figure 3.11a produces currents proportional to phase currents in the phase burdens Z_f and a current proportional to $3I_0$ in the neutral burden Z_n. No phase shifts are introduced by this connection. The delta connection shown in Figure 3.11b produces currents proportional to $(I'_a - I'_b)$, $(I'_b - I'_c)$, and $(I'_c - I'_a)$ in the three burdens Z_f. If the primary currents are balanced, $(I'_a - I'_b) = \sqrt{3}\,|I'_a|\,\exp(j\pi/6)$, and a phase shift of 30° is introduced between the primary currents and the currents supplied to the burdens Z_f. By reversing the direction of the delta windings, a phase shift of $-30°$ can be obtained. The factor $\sqrt{3}$ also introduces a magnitude change that must be taken into consideration. We will discuss the uses of these connections as we study various relaying applications.

In modern applications using digital relays, the settings within the relays themselves may usually be used to apply phase shifts and ratio adjustments as needed.

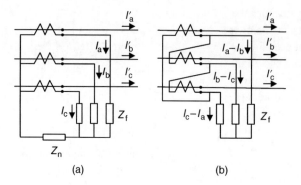

Figure 3.11 (a) and (b) Wye- and delta-connected CTs [9].

(a)   (b)

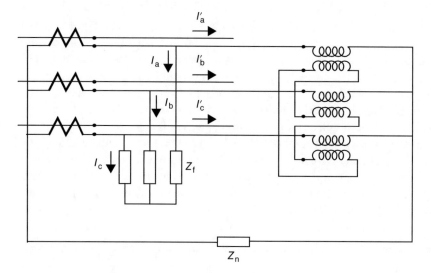

Figure 3.12 Zero-sequence current shunt.

3.4.3 Zero-sequence current shunts

Recall the wye connection of CT secondaries shown in Figure 3.11a. Each of the phase burdens Z_f carries phase currents, which include the positive-, negative-, and zero-sequence components. Sometimes (particularly in legacy electromechanical relay applications), it is desired that the zero-sequence current be bypassed from these burdens. This is achieved by connecting auxiliary CTs that provide an alternative path for the zero-sequence current. This is illustrated in Figure 3.12. The neutral of the main CT secondaries is not connected to the burden neutral. Instead, a set of auxiliary CTs have their primaries connected in wye and their secondaries in delta. The neutral of the auxiliary CTs is connected to the neutral of the main secondaries through the neutral burden Z_n. The secondary windings of the auxiliary CTs provide a circulating path for the zero-sequence current, and it no longer flows in the phase impedance burdens Z_f.

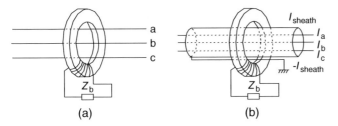

Figure 3.13   Flux-summing CT: (a) without and (b) with current in the cable sheath.

Note that modern digital relays are usually able to remove zero-sequence currents (if the application requires removal) by mathematical manipulation of phase currents. Therefore, zero-sequence shunts are not often required when digital relays are applied. In fact, the wye connection is preferred, so that the relay recording capabilities can capture each complete phase current for post-mortem disturbance analysis.

3.4.4   Flux-summing CT

It is possible to obtain the zero-sequence current using a single CT, rather than by connecting the secondaries of three CTs as in Figure 3.11a [9]. If three-phase conductors are passed through the window of a toroidal CT, as shown in Figure 3.13a, the secondary current is proportional to $(I_a + I_b + I_c) = 3I_0$. Since this arrangement effectively sums the flux produced by the three-phase currents, the CT secondary contains the true zero-sequence current. In a connection of three CTs as in Figure 3.11a, any mismatches between the three CTs will introduce an error in zero-sequence current measurement. This is entirely avoided in the present application.

However, it must be recognized that such a CT application is possible only in low-voltage circuits, where the three-phase conductors may be passed through the CT core in close proximity to each other. If the three-phase conductors are enclosed in a metallic sheath, and the sheath may carry some (or all) of the zero-sequence current, it must be compensated for by threading the sheath grounding lead through the CT core, as shown in Figure 3.13b. The ampere-turns produced by the sheath current are now canceled by the ampere-turns produced by the return conductor, and the net flux linking the core is produced by the sum of the three-phase currents. This sum being $3I_0$, the burden is once again supplied by the zero-sequence current.

3.5   Air-core CTs and electronic current transformers

Linear couplers [9, 10] are CTs without an iron core. The magnetizing reactance of these transformers is linear, and is very small compared to that of a steel-cored CT. Most of the primary current is used up in establishing the mutual flux in the linear coupler, and the secondary windings are very limited in the amount of current they can deliver. Indeed, the linear coupler operates as a current-to-voltage converter: the voltage in the secondary circuit is a faithful reproduction of the primary current. As long as the secondary current is very small, the transformation ratio is practically constant. A special type of air-core CT is the Rogowski coil that includes two wire loops wound in electrically opposite directions [11]. The Rogowski coil is designed to minimize the effect of external magnetic fields and reduce dependence on the physical position of the primary conductor within the coil. The main use of air-core CTs is in applications where saturation of the CT presents a major problem—as in the case of bus-protection applications to be discussed later. Air-core CTs are not much in use, since they must be installed in addition to CTs, which are needed for most relaying and metering functions.

A number of electronic CTs have been developed, which offer many advantages when compared to traditional CTs. Most of the practical electronic CTs are based upon the relationship between the magnetizing field produced by a current-carrying conductor and the plane of polarization of polarized light (Faraday effect). The conductor may pass through an optically active block or multiple turns of a fiber-optic cable placed around the conductor [12]. The angle through which the plane of polarization of the light rotates is detected at the receiving end (Figure 3.14). This angular shift is electronically converted to a voltage, which is proportional to the instantaneous value of the magnetizing force around the current-carrying conductor, and hence to the instantaneous value of the current. A low-energy analog output may then provide a replica of the current in the primary conductor [13, 14]. Alternatively, the voltage may be sampled at a suitable rate to provide a sampled data representation of the primary current. It should be clear that such an electronic CT is most suited to relays and meters that can utilize low-energy signals or sampled data of the signals. As will be seen later, this type of signal source is particularly suited for electronic relays and computer relays.

Electronic CTs are linear, and have a very wide dynamic range, that is, they are able to measure currents accurately at light loads as well as those corresponding to very heavy faults. Furthermore, since they do not include oil as an insulating medium, they do not constitute a fire hazard. They are also smaller in size and require less space in a substation. However, they do require a power supply to operate the various electronic circuits required

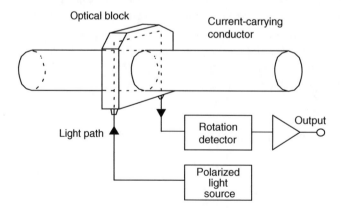

Figure 3.14 Principle of the magneto-optic current transformer (MOCT).

to sense and amplify the signals. Getting reliable power to these devices in the substation yard is a significant problem. Nevertheless, electronic CTs are starting to make a significant impact on relaying practices.

3.6 Voltage transformers

VTs—also known as potential transformers—are normal transformers with the primary winding connected directly to the high-voltage apparatus, and with one or more secondary windings rated at the standard voltage of 69.3 V for phase-to-neutral voltages or 120 V for phase-to-phase voltages. Their performance, equivalent circuit, and phasor diagrams are similar to those of a power transformer. The error of transformation of such a transformer is negligible for all practical purposes in its entire operating range—from zero to about 110% of its normal rating. We may consider such transformers to be error-free from the point of view of relaying. VTs are rather expensive, especially at extra high voltages: 345 kV or above. Consequently, they are usually found on low-, medium-, and high-voltage systems. At extra-high voltages, capacitive VTs, to be described in the next section, are the more usual sources for relaying and metering.

In passing, we may mention a possible problem with VTs when used on ungrounded (or non-effectively grounded) power systems [15]. As shown in Figure 3.15, when a ground fault occurs on such a system, the VTs connected to the unfaulted phases are subjected to a voltage equal to the phase-to-phase voltage of the power system. VTs with high flux capability ("fully fluxed") may be specified for application on non-effectively grounded systems to withstand phase-to-phase voltages without saturation. If they are not specified with high flux capability, the unfaulted phase voltage usually drives one of the transformers well into saturation, and, because of the excessive magnetizing current drawn by this transformer, may blow the protective fuse.

Figure 3.15 Blown fuse in a voltage transformer on ungrounded systems.

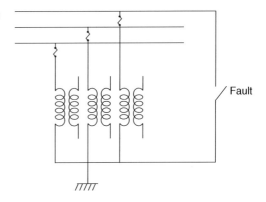

3.7 Coupling capacitor voltage transformers

One of the most common voltage sources for relaying—particularly at higher voltages—is the CCVT. A capacitor stack is usually connected to high-voltage transmission lines for the purpose of feeding the carrier signal as a pilot channel for transmission line relaying[1] (see Chapter 6). The same string of capacitors is used as a potential divider between the high-voltage apparatus and ground, and a tap provides a reduced voltage of about 1–4 kV, depending upon the particular choice made by the designer. The tap point is connected to a transformer through an inductance, as shown in Figure 3.16a. The turns ratio of the transformer is such that the secondary voltage is the standard voltage (69.3 or 120 V) required for relaying. The burden impedance is Z_b, and Z_f is a specially designed damping circuit for suppressing ferroresonance that may occur under certain conditions. We will discuss ferroresonance phenomena a little later.

Under normal steady-state operating conditions, the load current drawn by the parallel combination of Z_f and Z_b is relatively small. Nevertheless, as the transformer supplies the load current, there may be a phase shift between the primary voltage and the voltage appearing at the load. Consider the Thévenin equivalent circuit of the capacitive divider. The Thévenin voltage is given by $E_{th} = E_{pri}C_1/(C_1 + C_2)$, and the Thévenin source impedance is a capacitance of $(C_1 + C_2)$ (Figure 3.16b). If the primary and secondary currents in the transformer are I_1 and I_2, respectively, then

$$E_2 = E_{th} - I_1 \left[j\omega L + \frac{1}{j\omega(C_1 + C_2)} \right]. \tag{3.17}$$

Clearly, the secondary voltage will have a phase angle error, unless the inductance L is in resonance with $(C_1 + C_2)$ at the power system frequency ω. To avoid a phase angle error, an inductance of an appropriate size is introduced to satisfy the resonance condition:

$$L = \frac{1}{\omega^2} \frac{1}{(C_1 + C_2)}. \tag{3.18}$$

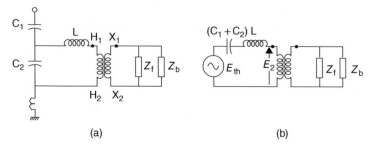

(a)   (b)

Figure 3.16 (a) and (b) CCVT connections and equivalent circuit.

---

1 The lower terminal of the capacitor stack is grounded through a small inductance, known as a drain coil. This offers a practically zero impedance to the power frequency voltage, and is effectively a solid ground connection. However, as discussed in Chapter 6, a carrier current of frequencies ranging from about 30 to 300 kHz is injected into the transmission lines through these capacitors. The drain coil offers a very high impedance to the carrier current that is effectively injected into the transmission lines without any loss of the signal to the ground connection.

Often, by careful design of the transformer, its leakage inductance can be made equal to L, and provides this compensation.

Since the Thévenin impedance of a CCVT is capacitive, the nonlinear magnetizing branch of the connected transformer may give rise to ferroresonant oscillations, especially under light loads. Unless these oscillations are eliminated, voltages of multiple frequencies—including sub-harmonic frequencies such as $\omega/3$—superimposed on the power frequency are likely to appear at the secondary terminals of the transformer. A special suppression circuit, represented by Z_f, is usually provided to damp these oscillations. In general, this circuit is a damped R, L, C circuit, a nonlinear resistor, a spark gap, or a combination of these elements. The actual configuration of the ferroresonance suppression circuit used depends upon the CCVT designer's preference.

Similar to CTs, VT windings are also marked to indicate their polarity. Terminals of like polarity may be identified by dots, or by terminal labels H_1 and H_2 (or P_1 and P_2), X_1 and X_2 (or S_1 and S_2). The primary voltage H_1 to H_2 is in phase with the secondary voltage X_1 to X_2. Both of these conventions are illustrated in Figure 3.16a. Similar to relay CT inputs, relay voltage inputs are also marked with defined polarities to facilitate correct polarity connections of the voltage inputs.

Windings of three-phase VTs may be connected in wye or in delta, as needed in each application. As in the case of CTs, the delta connection introduces a magnitude factor of $\sqrt{3}$, and a phase angle shift of $\pm30°$ depending upon the manner in which the delta is connected. In this respect, VTs are no different from normal power transformers. An open delta connection may be used to provide a three-phase voltage source with only two single-phase transformers. Delta and open delta connections cannot provide information about voltages with respect to ground or neutral. They are normally only applied on non-effectively grounded systems. The wye, delta, and open delta connections are illustrated in Figure 3.17.

A special VT secondary connection called open corner delta (not to be confused with open delta) connection is sometimes used to provide zero-sequence voltage to legacy electromechanical relays. This connection sums the three secondary voltages as shown in Figure 3.18.

Digital relays can sum three phase voltages mathematically to derive the zero-sequence voltage; so the open corner delta connection is not required in modern applications.

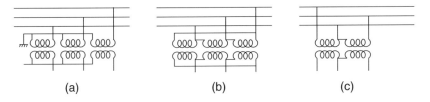

(a)   (b)   (c)

Figure 3.17 CCVT connections: (a) wye; (b) delta; and (c) open delta.

Figure 3.18 Open corner delta connection.

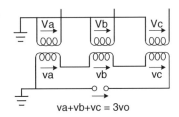

$$va + vb + vc = 3vo$$

Example 3.7 Consider the three VTs connected in delta on the primary and secondary sides as shown in Figure 3.19a. Assume that each of the three transformers has a leakage impedance of $(1 + j5)\,\Omega$. Let the burden impedances (also connected in a delta) be $50\,\Omega$. If the primary voltage is 69 kV and the turns ratio of each of the transformers is 69 000/120 = 575, the burden currents (which are the same as the secondary delta winding currents) can be found from the equivalent circuit shown in Figure 3.19a:

$$I_a' = \frac{E_{ab}}{Z_b} = \frac{E_{ab0} - I_a' Z_x}{Z_b},$$

where E_{ab} and E_{ab0} are the burden and source voltages, respectively. Solving for E_{ab} gives

$$E_{ab} = E_{ab0}\frac{Z_b}{Z_x + Z_b} = E_{ab0}\frac{50}{50 + 1 + j5} = E_{ab0} \times 0.976\angle 5.6°.$$

Thus, the ratio correction factor for this VT at this burden is $0.976\angle 5.6°$. The voltages across the three burdens are of course balanced and symmetric.

Now consider the two VTs connected in open delta, shown in Figure 3.19b. Using the line-to-line voltage relation at the burden and

$$E_{0ab} + E_{bc} + E_{ca} = 0,$$

and

$$E_{ab} = E_{ab0} - I_a' Z_x; \quad I_a' = \frac{1}{Z_b}(E_{ab} - E_{ca}) = \frac{1}{Z_b}(2E_{ab} + E_{bc}),$$

$$E_{bc} = E_{bc0} - I_b' Z_x; \quad I_b' = \frac{1}{Z_b}(E_{bc} - E_{ca}) = \frac{1}{Z_b}(2E_{bc} + E_{ab}).$$

Solving these equations for E_{ab} and E_{bc}, and remembering that

$$E_{bc0} = E_{ab0}e^{j2\pi/3},$$

$$E_{bc} = E_{bc0}\frac{\varepsilon^{-j2\pi/3}Z_x/Z_b - (1 + 2Z_x/Z_b)}{(Z_x/Z_b)^2 - (1 + 2Z_x/Z_b)^2},$$

$$E_{ab} = E_{ab0}\frac{\varepsilon^{+j2\pi/3}Z_x/Z_b - (1 + 2Z_x/Z_b)}{(Z_x/Z_b)^2 - (1 + 2Z_x/Z_b)^2}.$$

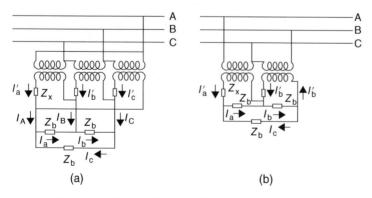

(a)   (b)

Figure 3.19 Connections for Example 3.7: (a) delta connection and (b) open delta connection.

Substituting the values of the burden and the leakage impedances gives

$$\frac{E_{ab}}{E_{ab0}} = 1.0254\angle - 9.85°,$$

$$\frac{E_{bc}}{E_{bc0}} = 0.8838\angle - 5.93°.$$

Also,

$$\frac{E_{ca}}{E_{ca0}} = -\left(\frac{E_{ab}}{E_{ab0}}\right) \times \left(\frac{E_{ab0}}{E_{ca0}}\right) - \left(\frac{E_{bc}}{E_{bc0}}\right) \times \left(\frac{E_{bc0}}{E_{ca0}}\right)$$

$$= -0.8838\angle - 5.93° \times \in^{+j2\pi/3} - 1.0254\angle - 9.85° \times \in^{-j2\pi/3}$$

$$= 1.017\angle - 1.1°.$$

Clearly, the three ratio correction factors are different from 1, and are also unequal. Thus, an open delta connection introduces unequal phase shifts in the three voltages, and may cause incorrect measurement of positive- and negative-sequence voltages through such a connection. On the other hand, the unbalances are generally quite small for all practical burdens. In any case, except for zero-sequence voltages which cannot be measured with delta or open delta primary connections, the errors are not significant for relaying applications.

3.8 Transient performance of CCVTs

The steady-state errors of a well-tuned CCVT should be negligible as far as relaying applications are concerned.[2] However, because of the tuned circuit used for compensation of the phase shift between the primary and secondary voltages, the CCVT produces secondary voltages during transient conditions, which may be significantly different from the primary system voltage. In particular, relays make use of CCVT output under faulted conditions when the power frequency voltages are low, and even a small transient component may cause problems for relaying applications. The main concern in relaying applications, therefore, is with the output voltage of a CCVT when a fault causes a step change in the voltage on the primary (power system) side of a CCVT. We will now derive expressions for the subsidence transient of a CCVT. As these phenomena are primarily produced by the linear parameters, we will not consider the nonlinearity introduced by the magnetizing branch. In fact, for the sake of simplicity, we will also omit the ferroresonance suppression circuit for this discussion.

Consider the equivalent circuit referred to the secondary winding as shown in Figure 3.20. We will assume the fault to be very close to the primary terminals of the CCVT, producing zero voltage at the primary terminals during the fault. As this causes the biggest possible

---

2 In practice, a CCVT is subject to a turns ratio error caused by changes in the capacitor sections. The capacitor values will change due to temperature by a significant amount, and if the relative changes in C_1 are different from those in C_2, a ratio error will result. Also, since the complete capacitor stack is made up of several small capacitors in series, as some of these small capacitors fail and get short-circuited over a period of time, the CCVT turns ratio changes from its design value.

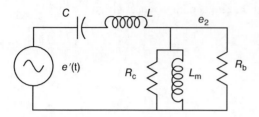

Figure 3.20 CCVT equivalent circuit for transient analysis.

change in the primary voltage of the CCVT, it produces the greatest possible subsidence transient. The source voltage is a power frequency sinusoid until the instant of the fault, and then it goes to zero:

$$e(t) = E_{\max} \cos(\omega t + \theta) \quad \text{for} \leq 0; \text{and } e(t) = 0 \quad \text{for } t > 0. \tag{3.19}$$

Since L and C are tuned to the frequency ω, the voltage across the load impedance R_b is also $e(t) = E_{\max} \cos(\omega t + \theta)$ until $t = 0$. We may use the principle of superposition by determining the response of the circuit to a voltage source $e'(t)$:

$$e'(t) = -E_{\max} \cos(\omega t + \theta) \quad \text{for } t \geq 0; \quad \text{and } e'(t) = 0 \quad \text{for } t < 0. \tag{3.20}$$

The Laplace transform of $e'(t)$ is given by Goldman [6]:

$$e'(s) = -E_{\max} \cos\theta \frac{s - \omega \tan\theta}{s^2 + \omega^2}. \tag{3.21}$$

After finding the response of the circuit to $e'(t)$, we may superimpose it on $e(t)$ to find the response of the circuit to $e(t)$. Solving the circuit of Figure 3.20 for the transient component of e_2, that is, $e_2'(s)$, gives

$$e_2'(s) = e'(s)e_2'(s) = e_2'(s)\frac{L_m}{\tau L} \frac{s^2}{s^3 + s^2(L + L_m)/\tau L + s\omega + \omega^2/\tau}, \tag{3.22}$$

where

$$\tau = \frac{L_m(R_c + R_b)}{R_c R_b}.$$

The time function $e_2'(t)$ may be found by taking the inverse Laplace transform of Equation (3.22). However, the result is somewhat messy, and not very illuminating. A simpler result is obtained for the case of infinite magnetizing inductance L_m. For this case,

$$e_2'(s) = e'(s)\frac{1}{\tau'} \frac{s}{s^2 + s/\tau + \omega^2}, \tag{3.23}$$

where

$$\tau' = \frac{L(R_c + R_b)}{R_c R_b}.$$

Substituting for $e'(s)$ and taking the inverse Laplace transform [6] gives

$$e_2'(t) = -E_{\max}\left[\cos(\omega t + \theta) - \cos\theta\sqrt{1 + (\cot\varphi + \operatorname{cosec}\varphi\tan\theta)^2} \times \varepsilon^{-\omega t\cos\varphi}\right.$$
$$\left. \times \sin(\omega t\sin\varphi + \psi)\right], \tag{3.24}$$

where

$$\psi = \arctan\left[-\sin\varphi/(\cos\varphi + \tan\theta)\right],$$

and

$$2\omega\tau' = \sec\varphi.$$

The actual voltage at the load is found by adding $e_2'(t)$ to its prefault value $e'(t)$ given by Equation (3.19).

Example 3.8 Consider the case of a fault occurring at zero voltage, that is, at θ equal to $\pi/2$. Assume the core loss resistance to be 1000 Ω and the load resistance to be 2000 Ω. Let the tuning inductance be 1.33 H. For this case,

$$\tau' = 0.002, \quad \cos\varphi = \frac{1}{2}\omega\tau' = 0.6613, \quad \sin\varphi cp = 0.7485,$$

$$\varphi = 48.46° = 0.8458 \text{ rad}$$

Also, since $\tan\theta = \infty$, $\psi = 0$. Substituting these values in the expression for $e_2'(t)$ gives

$$e_2'(t) = E_{max}\left[\cos\left(\omega t + \frac{\pi}{2}\right) - 1.336 \times \varepsilon^{-250t} \times \sin(282.2t)\right],$$

and, superimposing the prefault voltage, the secondary voltage is given by

$$e_2(t) = E_{max}\cos\left(\omega t + \frac{\pi}{2}\right) \quad \text{for } t \leq 0,$$

and

$$e_2(t) = 1.336E_{max} \times \varepsilon^{-250t} \times \sin(282.2t) \quad \text{for } t > 0.$$

The plots of $e(t)$ and $e_2(t)$ are given in Figure 3.21.

It should be remembered that Figure 3.21 has been obtained under some very simplified assumptions. First, we assumed that the ferroresonance suppression circuit is absent. Second, we assumed that the magnetizing inductance of the transformer is infinite, and that

Figure 3.21 CCVT transient response for Example 3.8.

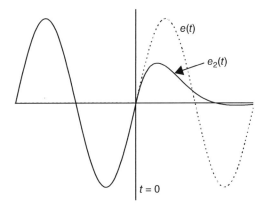

only the core loss component of the excitation current exists. Finally, we assumed that the load impedance is purely resistive. When none of these assumptions is made, it is too cumbersome to solve the problem in an analytic form using Laplace transform, although it can be done. A much simpler method is to use a network simulation program, such as Electromagnetic Transients Program (EMTP) [16], to simulate the behavior of the CCVT for different operating conditions. Other techniques of simulation are described in the literature, as are the results of actual field tests [17]. Results of more realistic simulations of the cases of a fault occurring at voltage maximum and at voltage zero are given in Figure 3.22a,b.

As mentioned previously, the performance of the CCVT during the transition from normal system voltage to a low value during a fault is of primary concern in relaying applications. This is referred to as a "subsidence transient" and it produces a transient voltage in the secondary that may be a damped oscillatory or unidirectional wave depending on the design of the CCVT, the connected burden, and the incidence point on the voltage wave. The apparent impedance to a relay may include errors in both magnitude and phase angle. There is a publication [18] that describes this effect in detail and lists the effect that the subsidence transient has on a variety of relays. In general, however, if time delay can be used, this will remove or reduce this error. Instantaneous relays such as zone 1 distance or pilot schemes may operate incorrectly. This effect is most pronounced during reclosing. At the instant of the initial fault, the usual cause of the fault is the breakdown of insulation, which is most likely to occur at a voltage maximum. As shown in Figure 3.22a, this results in a minimum subsidence voltage. During reclosing, however, if the fault still persists, the subsidence voltage could be any value depending on the point of the voltage wave when the reclose occurs. This condition is exacerbated by the fact that one of the mitigating conditions is the memory voltage provided with instantaneous relays. At the time of reclosing, this memory voltage has been significantly reduced or is gone completely. In about a cycle immediately after the occurrence of a fault, the output of a CCVT is significantly different from the actual primary voltage. In particular, the frequency of the output voltage is not equal to m, the fundamental power frequency. In designing a relay, the designer must take this phenomenon into account, and make the relay insensitive to these transient components. For faults that produce fundamental frequency voltages of greater than 5% of their nominal value, the transient components do not cause any significant problem.

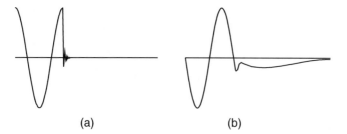

(a)   (b)

Figure 3.22 (a) and (b) CCVT response for a complete model circuit: (a) fault at voltage maximum and (b) fault at voltage zero.

3.9 Electronic voltage transformers

Electronic VTs are also available, though they have not been developed to the same extent as electronic CTs. They typically measure electric field strength using the Hall effect in which the direction of polarization light is changed by an electric field see references [12–14].

3.10 Summary

In this chapter, we have discussed the techniques of transforming power system voltages and currents to levels that are suitable for relaying and metering. We have described the appropriate industry standards for these transducers. We have discussed the importance of transient response of CTs and CCVTs to the relaying process, and we have analyzed these transient responses in some simple cases. We have also examined recent developments in the field of electronic transducers for power system applications.

Problems

3.1 Prove Equation (3.7). Recall that, for small E, $1/(1 - E)$ may be approximated by $1 + \in$. Under what conditions is R smaller than 1 in magnitude?

3.2 Write a computer program to calculate R for a CT with linear magnetizing impedance. Using the data given in Example 3.1, calculate and plot R (a complex quantity) as the burden impedance angle is varied from $+90°$ to $-90°$, while its magnitude is held constant at 2 Ω. Next, vary the magnitude of the burden impedance while the angle is held at $0°$. Make the plot in polar coordinates.

3.3 For a fixed burden impedance of 2 Ω resistive, calculate and plot the magnitude of R as the secondary current is varied between 0 and 50 A. Assume that the magnetizing characteristic is as given in Figure 3.3, and the 300:5 tap of the CT is being used. Assume further that the impedance angle of the magnetizing impedance is $60°$. On the same graph, plot the magnitude of R if the magnetizing impedance of the CT is assumed to be constant at a value corresponding to a secondary voltage of 60 V rms.

3.4 The secondary windings of two CTs with turns ratios of 300:5 are connected in parallel across a common burden of 1.0 Ω resistive. The lead impedance of one of the connections is 0.2 Ω, while that of the other is 0.0 Ω. What is the current in the burden when the primary windings of the two CTs carry a current of 3000 A? What is the current in each of the CT secondaries? Assume that the magnetizing characteristic is as given in Figure 3.3, and that the magnetizing impedance angle is $60°$.

3.5 Outputs of two CTs with designations of 10C200 and 10C400 are connected in parallel. What is the class designation of this combination considered to be a single CT?

3.6 Two ideal CTs with turns ratios of 300:5 and 600:5 are connected as shown in Figure 3.7a,b. If the primary current is 3000 A and the burden is 1 Ω, what are the voltages at the secondary terminals of the CTs in the two cases?

3.7 Three ideal CTs with turns ratios of 600:5 are connected in a wye and a delta configuration as shown in Figure 3.23a,b, respectively. For the primary currents shown in each case, what are the currents I_1, I_2, I_3, and I_0? (Figure 3.23).

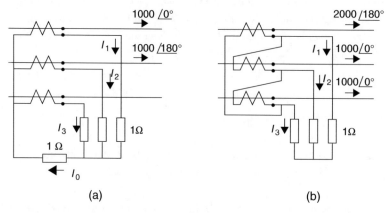

(a)   (b)

Figure 3.23   (a) and (b) Configuration for Problem 3.7.

3.8 Prove Equations (3.15) and (3.16). Derive similar expressions for the case of a purely inductive burden.

3.9 Using a network simulation program, set up a model of a realistic CT. Calculate the secondary current of the CT for various inception angles of a fault in the primary circuit. Also, simulate the effect of remanence in the CT core. Tabulate the time to saturate for a CT, and compare your results with those given by Goldman [6]. The current version of the program EMTP contains models of CTs and CCVTs.

3.10 The leakage impedance of a VT is $(1 + j6)$ Ω when referred to its secondary winding. If the secondary burden impedance is Z at an impedance angle of φ, calculate and plot its ratio correction factor in polar coordinates as the magnitude of Z is held constant at 200 Ω, and φ is varied between 0° and 360° (this is best done with a computer program). Generate similar plots for a range of magnitudes of Z; say varying between 50 and 100 Ω. Plots like these are of importance in metering applications, and have little significance in relaying.

3.11 Prove Equation (3.24). The proof is much facilitated by the use of a good table of Laplace transform pairs and a table of trigonometric identities.

3.12 A CCVT is connected to its rated primary source voltage of 138/J3 kV. The value of the capacitor from the high-voltage terminal to the tap point is 0.005 µF. The tap point voltage is 4 kV. What is the value of the capacitor connected between the tap point and the ground terminal? What is the value of the tuning inductance? If the power system frequency changes between 58 and 62 Hz (assuming the normal frequency to be 60 Hz), calculate and plot the ratio correction factor of this CCVT as a function of the power system frequency. Assume the burden connected to the secondary of the transformer to be 300 Ω resistive, and assume the transformer to be ideal.

References

1 Stevenson, W.D. Jr. (1982). *Elements of Power System Analysis*, 4e. New York: McGraw-Hill.
2 IEEE (2008). *IEEE C37.110 - IEEE Guide for the Application of Current Transformers Used for Protective Relaying Purposes, and Corrigendum 1 Corrections to Equation 18 and Equation 19.* Piscataway, NJ: IEEE Standards Association, IEEE.
3 Wright, A. (1968). *Current Transformers, Their Transient and Steady State Performance.* Chapman and Hall.
4 IEEE (2016). *IEEE/ANSI C57.13, Standard requirements for instrument transformers.* Piscataway, NJ: IEEE Standards Association, IEEE.
5 IEC Standard 61869-2 (2012). *Instrument Transformers – Part 2: Additional Requirements for Current Transformers.* International Electrotechnical Commission.
6 Goldman, S. (1966). *Laplace Transform Theory and Electrical Transients*, 416–424. New York: Dover Publications.
7 IEEE Power System Relaying Committee (2003). *CT Saturation Theory and Calculator.* http://www.pes-psrc.org/kb/published/reports/CT_SAT%2010-01-03.zip
8 IEEE Power System Relaying Committee (1977). Transient response of current transformers. *IEEE Trans. Power Appar. Syst.* **96** (6): 1809–1814.
9 Mason, C.R. (1956). *The Art and Science of Protective Relaying.* New York: Wiley.
10 Blackburn, J.L. and Domin, T.J. (2007). *Protective Relaying Principles and Applications*, 3e, 378–379. Boca Raton FL: CRC Press.
11 IEEE (2008). *IEEE C37.235 "IEEE Guide for the Application of Rogowski Coils Used for Protective Relaying Purposes".* Piscataway, NJ: IEEE Standards Association, IEEE.
12 IEEE (2018). *IEEE Std C37.241–2017 "IEEE Guide for Application of Optical Instrument Transformers for Protective Relaying".* Piscataway, NJ: IEEE Standards Association, IEEE.
13 IEEE (2005). *IEEE C37.92 Standard for Analog Inputs to Protective Relays From Electronic Voltage and Current Transducers.* Piscataway, NJ: IEEE Standards Association, IEEE.
14 Udren, E.A. et al. (2006). *Overview of IEEE C37.92–2005 standard for analog inputs to protective relays from electronic voltage and current transducers.* In: *IEEE PES Power Systems Conference and Exposition*, 2006 527–531.

15 Peterson, H.A. (1966). *Transients in Power Systems*, 300–302. New York: Dover Publications.

16 Dommel, H.W. (1969). Digital computer solution of electromagnetic transients in single and multiphase networks. *IEEE Trans. PAS.* **88** (4): 388–399.

17 Sweetana, A. (1970). Transient response characteristics of capacitive potential devices. *IEEE Trans. PAS.* **89**: 1989–1997.

18 IEEE Power System Relaying Committee (1981). Transient response of coupling capacitor voltage transformers. PSRC report. *IEEE Trans. PAS.* **100** (12): 4811–4814.

4

Nonpilot Overcurrent Protection of Transmission and Distribution Lines

4.1 Introduction

The study of transmission and distribution (T&D) line protection offers an opportunity to examine many fundamental relaying considerations that apply, in one degree or another, to the protection of all other types of power system equipment. Each electrical element, of course, will have problems unique to itself, but the concepts associated with nonpilot T&D system protection are fundamental to all other electrical equipment, and provide an excellent starting point to examine the implementation of power system protection.

T&D lines are primarily exposed to short circuits between phases or from phase to ground [1, 2]. This is also the main source of damage to all other electrical equipment. The range of the possible fault current, the effect of load, the question of directionality, and the impact of system configuration are all parts of the line protection problem. The solution to this problem, therefore, is a microcosm of all other relaying problems and solutions.

Since T&D lines are also the links to adjacent lines or connected equipment, the protection provided for the line must be compatible with the protection of all these other elements. This requires coordination of settings, operating times, and characteristics. The analytical techniques associated with short-circuit calculations, primarily the method of symmetrical components, are similarly applicable to all electrical equipment, but the variety of line configurations introduces additional complexity and sophistication.

The question of directionality was mentioned in Chapter 1. It is intimately associated with the design of the power system. A radial system, that is, one with a single generating source, can have fault current flowing in only one direction: from the source to the fault. Historically, distribution lines were always radial [2], but as discussed further in Section 4.5, distributed energy resources (DER) are increasingly being applied in modern times to allow bi-directional power and fault current flow. Additionally, some distribution lines are now being applied in loop or network configurations to increase reliability of supplies. Transmission lines are usually connected in networks with sources at each terminal. In a loop or network, since fault current can flow in either direction, the relay system must be able to distinguish between the two directions.

The length of the line, as one would expect, has a direct effect on the setting of a relay. Relays are applied, primarily, to protect a given line segment and, in addition, if possible, to back up the relays protecting adjacent line segments. It will be recalled from the discussion of zones of protection in Chapter 1 that a circuit breaker (CB) usually defines the boundary

Power System Relaying, Fifth Edition. Stanley H. Horowitz, Arun G. Phadke and Charles F. Henville.
© 2023 John Wiley & Sons Ltd. Published 2023 by John Wiley & Sons Ltd.

of a protected area and, with the circuit breaker contacts closed, there is no appreciable impedance between the end of one line segment and the beginning of the next. A relay, therefore, cannot be set on fault current magnitude alone in order to differentiate between a fault at the end of one zone or the beginning of the next. The problem is further complicated if the line is short as defined in Reference [1], that is, as shown in Figure 4.1, its impedance is much less than the source impedance. In such a case, there is very little difference in current magnitude for a fault at one end of the line compared to a fault at the other. It is then difficult to set a relay so that it only protects its own line and does not overreach into the next. Also, as shown in Figure 4.1, a long line, that is, one with an impedance greater than four times the source impedance, has another problem. It can be set easily to reach well into its own line segment without overreaching, but the further it "sees," the closer the fault current magnitude approaches load current, and the ability to make this important distinction is more difficult.

The voltage class of a transmission line must also be considered when applying a relay system. The higher voltage levels would normally be expected to have more complex, hence expensive, relay systems. This is so because higher voltages have more expensive equipment associated with them and one would expect that this voltage class is more important to the security of the power system than lower voltages. The higher relay costs, therefore, are usually more easily justified. This is not always the case; however, nor is the specific voltage level, by itself, an indication of its importance. Some systems have very important transmission systems at the low end of the high voltage (HV) or extra high voltage (EHV) range and very little higher voltage transmission. These lines should, therefore, justify higher relay costs. Of course, if a higher voltage system is installed, or an installed EHV system matures, the justification for more sophisticated relays becomes easier. In other situations, with a mature operating EHV system, the same relay costs for the HV system might not be justified. In order of ascending cost and complexity, the protective devices available for transmission line protection are:

1) Fuses.
2) Sectionalizers and reclosers.
3) Instantaneous overcurrent.
4) Inverse, time delay, and overcurrent.
5) Directional overcurrent.
6) Distance.
7) Pilot.

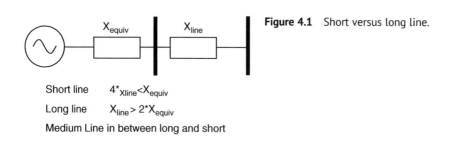

Short line $4^{*}X_{line} < X_{equiv}$

Long line $X_{line} > 2^{*}X_{equiv}$

Medium Line in between long and short

Figure 4.1 Short versus long line.

In this chapter and Chapter 5, we will examine nonpilot line protection systems, also known internationally as graded or relatively selective or nonunit systems.

In these systems, the relays base their decisions solely on the measurement of electrical quantities at the relay end of the protected line section. In this chapter, we will study time delay and instantaneous overcurrent relays. In Chapter 5, our interest will be on distance relays using nonpilot systems. In Chapter 6, we will examine pilot systems in which quantities are measured at all terminals of a protected line section and communicated to every terminal.

4.2 Fuses, sectionalizers, and reclosers

Fuses, sectionalizers, and reclosers are more frequently applied on distribution systems, since they are that system's primary protective devices [2]. Distribution is normally defined as the system level energizing the final step-down transformer, that is, the distribution transformer, serving the industrial, commercial, or residential customer. The following discussion in this section provides an overview of the key issues associated with distribution system protection. For more information, the reader may review Reference [2].

The distribution system is divided into mains and laterals. The mains are three-phase systems providing the backbone of the distribution service, and the laterals are single-phase taps connected to the mains. Industrial and commercial customers that require three-phase service are fed from the mains. Residential and smaller industrial customers are usually serviced by the laterals. It is the function of the distribution planning engineer to equalize the single-phase loads, so the load at the substation is essentially balanced. Figure 4.2 shows a single-line representation of a typical distribution circuit. Commonly, the horizontal feeder would be a three-phase main and each tap would be a single-phase load, each load coming from a different phase.

Except for increasingly common networked systems or with DER that will be discussed at the end of this section, a distribution system was almost exclusively radial, that is, it had a source at only one end. Its operating voltages are between 2.4 and 34.5 kV.

In North America, most distribution transformers are pole-mounted, although underground residential distribution (URD) is gaining in popularity and use. URD transformers are usually pad-mounted, that is, installed on a concrete pad on the ground, usually in a remote corner of a residential area. In such an installation, there is some concern over the safety and potential physical damage that could occur from a violent type of failure. This has led to the application of current-limiting (CL) fuses that drastically reduce the "let-through" energy for a high-current fault compared to other types of fuse. A CL fuse consists of one or more silver wires or ribbon elements suspended in an envelope filled with sand. When operating against a high current, the fusible element melts almost instantaneously over all its length. The resulting arc loses its heat energy rapidly to the surrounding sand. The rapid loss of heat energy limits the current to a small value known as the "let-through" current.

The most commonly used protective device in a distribution circuit is the fuse. Fuse characteristics vary considerably from one manufacturer to another, and the specifics must be obtained from manufacturers' appropriate literature. The time–current characteristic (TCC)

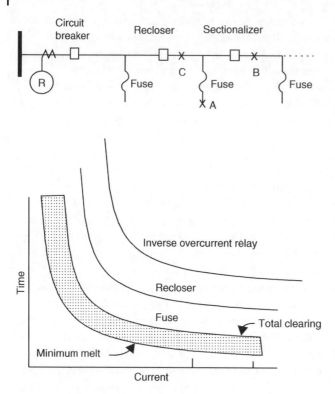

Figure 4.2 Overcurrent device coordination.

curves of fuses are presented in the form of minimum melt and total clearing times (TCTs), as shown in Figure 4.2. Minimum melt is the time between initiation of a current large enough to cause the current-responsive element to melt and the instant when arcing occurs. TCT is the total time elapsing from the beginning of an overcurrent to the final circuit interruption, that is, TCT = minimum melt + arcing time. It is important not to mix fuse types such as American National Standards Institute (ANSI) "K" or "T." The operational characteristics are different enough that loss of coordination can occur. This is the same precaution that should be followed in applying other protective devices such as overcurrent relays. Although two fuses or relays may be generically similar, there are differences in manufacture and subtle differences in the operating characteristics that will cause coordination difficulties.

In addition to the different melting curves, fuses have different load-carrying capabilities that must be recognized. Manufacturers' application tables show three load current values: continuous, hot-load pickup, and cold-load pickup. Continuous load is the maximum current that is expected for 3 h or more and for which the fuse will not be damaged. Hot load is the amount that can be carried continuously, interrupted, and immediately reenergized without blowing. Cold load follows a 30-min outage and is the high current that is the result of a loss of diversity when service is restored. Since the fuse will also cool down during this period, the cold-load pickup and the hot-load pickup may approach similar values.

Owing to the large volume of equipment involved in the distribution system, construction standards and equipment specifications are rarely custom-tailored for a specific situation or location. Instead, universal standards and specifications are developed for ease of replacement and installation. The special hardware that is associated with distribution systems such as potheads (the transition connectors between overhead and underground cable), poles, crossarms, and insulators is used in such large quantities that cost and replacement stock, instead of specific application, become the controlling factors. The interrupting devices, in addition to the fuse itself, are sectionalizers and reclosers. It will be seen that in Figure 4.2, there is no TCC for a sectionalizer. This is because a sectionalizer cannot interrupt a fault. It "counts" the number of times it "sees" the fault current and opens after a preset number while the circuit is de-energized. The sectionalizer provides a way of improving selectivity of fault clearing without introducing another TCC that would cause slowing down of upstream protective devices. A recloser has limited fault-interrupting capability and recloses automatically in a programed sequence.

As the fault current and interrupting requirements increase with system voltage, the protection shifts from a fuse and/or recloser to a relay for the detection of a fault and to a circuit breaker for its interruption. At the location in the system where this shift occurs, the relays must coordinate with the reclosers, sectionalizers, and fuses. Figure 4.2 shows how this coordination is accomplished. The fuse has the controlling TCC, that is, $I^2t = k$. The recloser and relay operating characteristic must have the same shape with sufficient time delay between the curves, at the same current magnitude, to allow each downstream device to clear the fault. The longer times allow the adjacent upstream device to provide backup in the event the fault is not cleared in that device's operating time. Since the operating time around the minimum current is not precise, it is common practice to stagger the pickup current, as shown in Figure 4.2.

Referring to Figure 4.2, a permanent fault at A should be cleared by the branch fuse, leaving service to the main line and to the other branches undisturbed. A fault at B should be cleared by the sectionalizer, but, since the sectionalizer cannot interrupt a fault, the actual clearing is performed by the recloser. The sectionalizer "sees" the fault current, however, and registers one count. The recloser also sees the fault and trips, de-energizing the line. If the sectionalizer setting is "1," it will now open, allowing the recloser to reclose and restore service to the rest of the system. If the sectionalizer setting is more than "1," for example, "2," the sectionalizer will not open after the first trip. Instead, the recloser recloses a second time. If the fault is still on, the sectionalizer will see a second count of fault current. The recloser will trip again, allowing the sectionalizer now to open, removing the fault, and the recloser will successfully reclose, restoring service up to the sectionalizer. For a fault at C, the recloser trips and recloses as it is programed to do. The sectionalizer does not see the fault and does not count. Note that automatic reclosing is only applied on air-insulated overhead lines. If the insulation is any material other than air, automatic reclosing will not be successful, and the insulation will need to be repaired after a failure causing a short circuit.

In the previous paragraph, we noted that a permanent fault at A should be cleared by the branch fuse. A permanent fault is one that will remain on a line whether or not it is energized. Such faults might be caused by a pole crossarm failure or by solid- or liquid-insulation failure or contact by tall objects such as cranes or falling trees. However, most faults on air-insulated overhead lines are temporary. Such faults are caused by a variety of issues such as:

- Lightning-caused transient overvoltages resulting in insulator flashovers.
- Wind or vehicle accidents causing conductors to slap together.
- Flying objects such as small tree branches momentarily short-circuiting phases.
- Small animals such as squirrels or birds that fall away after causing a short circuit.

If a line suffering from a temporary fault is de-energized, the fault will extinguish, the air will regain its insulating properties, and the line may be successfully re-energized shortly afterward.

Some distribution companies choose to implement a practice called "fuse saving" whereby a reclosable protective device on the source side of a fuse may open temporarily to try to clear a temporary fault. For instance, referring to Figure 4.2, for a fault at A, the recloser on the main line could open (via a fast tripping characteristic) before the branch fuse is damaged. The recloser would close again shortly afterward and if the fault was temporary, supply would be restored to the previously faulted branch, and the undamaged fuse would continue to protect the branch. A recloser may trip via its fast-tripping characteristic once or twice before switching to a slow-tripping characteristic after subsequent reclosure(s). If the fault was permanent, the slow-tripping characteristic of the recloser would allow the branch fuse to trip and clear the permanent fault. Figure 4.3 shows example time–current characteristics of a recloser and two example downstream fuses. Note that there is an upper limit of current for which fuses can be "saved" without damage. At this upper current limit, the fault-clearing time of the fast-tripping device should be less than 75% of the minimum melting time of the fuse.

This practice is called "fuse saving" and has some advantages and disadvantages that may impact reliability and performance indices. Advantages include:

- Greater continuity of service to the maximum number of customers for an improvement in the system average interruption duration index (SAIDI) [3].
- Cost savings due to reduced need to replace fuses.

An important disadvantage of fuse saving is more frequent short interruptions of service to a larger number of customers for degradation of the momentary average interruption frequency index (MAIFI) [3].

The alternative to fuse saving is called "fuse blowing" or "fuse sacrificing," in which the fuse nearest the fault is always the first to operate and isolate the fault. Utility decisions with respect to the application of fuse-saving or fuse-blowing schemes and automatic reclosing depend on a variety of factors. Fuse saving may be applied where maximum service continuity is more important than avoidance of momentary interruptions. Automatic reclosing may be normally enabled for maximum service continuity. Fuse blowing may be applied where it is important to minimize momentary interruptions. If automatic reclosing is normally enabled, it will be disabled when hotline work is being performed on the feeder. Alternatively, automatic reclosing may be normally disabled, and only enabled during storm conditions. Automatic reclosing is never applied on cables or any equipment that is not air-insulated.

Also shown in Figure 4.3 is the maximum current at which the 30 A fuse will coordinate below the 65 A fuse. If the fault current is higher than shown (3000 A), the 65 A fuse might be damaged even if the 35 A fuse clears the fault before the minimum melt time of the 65 A

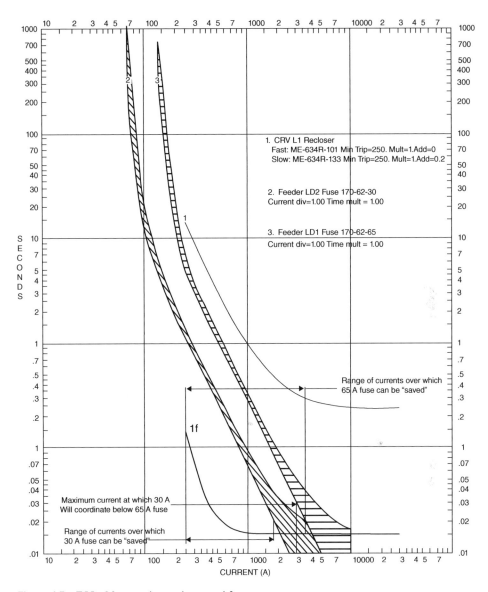

Figure 4.3 TCC of fuse-saving recloser and fuses.

fuse is exceeded. The 3000 A fault current is the current at which the total clearing time of the 30 A fuse is 75% of the minimum melting time of the 65 A fuse.

Note from Figure 4.3 that TCCs are normally plotted with axes using log–log scales. This is because there is such a large range of currents and times. The logarithmic scales normalize the quantities and make the various characteristics and differences between them more easily visually discernable at high and low currents and times.

In recent years, the use of customer-owned generators (co-generators), independent power producers (IPPs), or nonutility generators (NUGs), and feeder-to-feeder switching

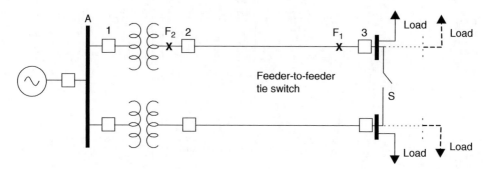

Figure 4.4 Feeder-to-feeder switching.

and networking has resulted in the distribution system protection, taking on more of the characteristics of transmission system protection. NUGs introduce energy sources within the distribution system that result in fault currents that differ in magnitude and direction from the fault currents produced by the utility's sources and over which they have control. As a result, the relays at the distribution station must now recognize these potential variations in short-circuit current in the same way as the relays in a networked system.

Feeder-to-feeder switching provides a backup source in the event a substation transformer is out of service or a segment of the distribution system must be de-energized. A simple application of this concept is shown in Figure 4.4.

Normally the tie switch S is open, and each station transformer feeds its own load as described in the earlier text. For a permanent fault at F_1, the sectionalizers or reclosers on the transformer side will open automatically; the line must then be de-energized by opening the downstream breaker (3) and closing switch S, shifting the remaining load to the other transformer. For a transformer zone fault at F_2, the station breakers (1) and (2) are opened and the entire load can be fed from the other station. The shift from one substation to another affects the magnitude and direction of fault current, and this must be taken into account when applying and setting the line protection devices. If the tie switch is normally closed, coordination problems on the feeders arise. A directional relay at breaker (3) will be required to resolve such problems. Similarly, the use of co-generators on the distribution system introduces another source of energy, independent and remote from the utility's substation, which also affects the magnitude and direction of fault current. Normally this switching is done manually, by sending personnel to the various locations.

Obviously, if the switching could be done automatically, there would be significant savings in time and money. This concept of distribution automation (DA) is being increasingly applied to improve service reliability. Difficulties in its implementation are being resolved by the application of digital technology and improved communications systems [2, 4]. Clearing a fault in a radial system is relatively simple, since directional relays are not required. If the fault can be seen by relays in either direction, then directional relays are required, which means that a potential polarizing source must be provided at many switching devices. This is covered in greater detail in Section 4.5. Note that DA systems may use current and voltage sensors that are less bulky and more economical than substation instrument transformers. In many cases, these sensors are no larger than distribution line post insulators [4].

4.3 Inverse, time-delay overcurrent relays

4.3.1 Application

The principal application of non-directional overcurrent relays is on a radial system where they provide both phase and ground protection.

Example 4.1 Given the one-line diagram shown in Figure 4.5a, see the associated alternating current (AC) and direct current (DC) connections for legacy two-phase and three-phase relays and one ground overcurrent relay.

A basic complement of time-delay overcurrent relays would be two phases and one ground relay. This arrangement protects the equipment for all combinations of phase and ground faults, uses the minimum number of relays, but provides only the minimum of redundancy. Adding the third phase relay provides complete backup protection. Several important design considerations are shown in this example. The conventions and practices used to depict elements of relays and their associated circuits were introduced in Sections 2.7 and 2.8 of Chapter 2. In one-line, elementary, schematic, and wiring diagrams, use is made of identifying device function numbers that are shown in Figure 4.5a–c. These numbers are based on a system adopted as standard for automatic switchgear by the Institute of

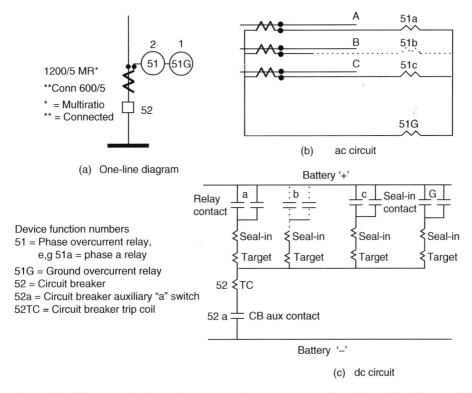

Figure 4.5 Effect on redundancy of two versus three overcurrent relays. (a) one-line diagram; (b) AC circuit; and (c) DC circuit.

Electrical and Electronics Engineers (IEEE) and incorporated in ANSI/IEEE C37.2. The system is used in connection diagrams, instruction books, and specifications. It provides a readily identifiable reference to the function of a device in a circuit to which additional descriptive numbers or letters can be added to differentiate between similar functions for different devices. A list of most of the commonly used device function numbers is given in Appendix A.

Figure 4.5a shows a single-line representation of the three-phase power system showing the number of current transformers (CTs), their polarity marks, the fact that it is multi-ratio (MR), the total number of turns (1200/5), the actual connected ratio (600/5), and the number and type of relays connected to the CTs. On a fully completed one-line diagram, there would be comparable notes describing the circuit breaker (52) type and mechanism as well as similar notes for other power equipment. Figure 4.5b shows the three-phase secondary AC circuit of the CTs and relays, and Figure 4.5c shows the relay DC tripping circuit details. The advantage of the third relay is shown by the dotted line. All faults are covered by at least two relays and so any one relay can be removed for maintenance or calibration, leaving the other relay in service. The extra phase relay also provides a degree of redundancy, so that complete protection is still available if any of the relays develop a defect. A modern implementation of the legacy protection system shown in Figure 4.5 is shown in Figure 4.6a–c. This implementation uses a single multifunction digital relay instead of three or four independent electromechanical relays described in Example 4.1. Several important differences should be noted in this digital implementation.

The device is named 51F, because it provides all the protection for the feeder. Individual functions within 51F provide protection for each of the three phases, and a more sensitive ground fault function. In future references to protection elements within a multifunction relay, we will refer to the protection "function" rather than the protection "relay." The fault detection functions are not independent of each other. There is a single relay, usually powered by a DC source, providing complete protection. Installation costs and maintenance costs are significantly reduced. Flexibility is increased, and built-in analog and digital recording features greatly enhance disturbance analysis. However, because only a single device provides all the protection, if the power supply or the relay fails, there is no independent means of detecting the fault. This means that a separate, independent backup protection

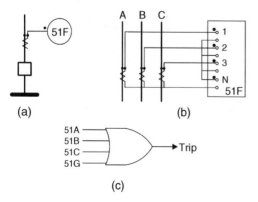

(a)

(b)

(c)

Figure 4.6 Digital implementation of non-redundant overcurrent protection. (a) one-line diagram; (b) AC circuit; and (c) logic diagram.

system will normally be required for acceptable protection dependability. Usually the backup system will be provided in one of two ways:

i) An independently powered multifunction separate (redundant) protection system may be applied to provide the same degree of protection.
ii) Alternatively, a separate backup protection further away from the protected system will be depended upon. This is a more economical alternative, but will be less sensitive, and slower than the primary protection. It will also usually trip a larger zone than the primary.

The degradation in dependability provided by the multifunction device is mitigated by the increased monitoring capability of a digital relay so that operators will receive an alarm if a relay or power supply fails. Thus, the unavailability of the digital primary protection will be much less than that of any one of the three or four independent electromechanical relays in Example 4.1.

Overcurrent relays are also used in industrial systems and on subtransmission lines that cannot justify more expensive protection such as distance or pilot relays. The time-delay overcurrent relay is a typical example of the level detector described in Section 2.2 of Chapter 2. The construction of the electromechanical version of an overcurrent relay is shown in Figure 2.9 of Chapter 2. It should be noted that, historically, the characteristic curve of an induction disk relay has not been precisely reproducible by solid-state circuits, nor its shape defined by a simple formula. In solid-state analog relays, a complex combination of filters is required to produce the characteristic shape. In the case of digital relays, the performance has been standardized [5, 6] so that the shapes of their characteristics can be described by formulae with various standard constants. The standards provide two equations—the time to reset when the current is less than the pickup current, and the time to trip when the current is greater than the pickup current. Note that the delayed reset time is an option that is not necessarily applied. The equations have the form:

For $0 > M < 1$ as an option,

$$t(I) = \frac{t_r}{M^p - 1},\qquad (4.1)$$

where

I is the current measured,
$t(I)$ is the reset time as a function of time,
t_r is the total reset time when $M = 0$,
$M = $ multiples of pickup,
p is a constant.

For $M > 1$,

$$t(I) = \left(\frac{A}{M^p - 1} + B\right),\qquad (4.2)$$

where

A, B, and p are constants.

Table 4.1 Constants for IEEE Standard time–current characteristics [6].

Curve shape	A	B	p	t_r (s)
Moderately inverse	0.0515	0.1140	0.0200	4.85
Very inverse	19.61	0.4910	2.000	21.6
Extremely inverse	28.2	0.1217	2.000	29.1

Enabling of the reset time delay is an optional setting in digital time–overcurrent functions that allow the reset time of the virtual disk to emulate the reset time of an electromechanical relay. The delayed reset assists coordination with legacy electromechanical relays during automatic reclosures, with short dead times. In most applications, the reset time is chosen to be instantaneous whenever the current drops below the pickup setting.

The constants A, B, and p determine the shape of the time–current tripping characteristic according to Table 4.1 [6]. The total reset time t_r determines the optional reset time when $M = 0$.

The standard performance of digital overcurrent relays also requires them to emulate an electromechanical relay when $M > 1$. In this case, the speeds of rotation of the mechanical disk and the virtual disk are both a function of the current. If the current changes during the fault, the speed of rotation also changes. The digital implementation manages this emulation by integrating small changes of disk position until it reaches a trip threshold given by Equation (4.3):

$$\int_0^{T_0} \frac{1}{t(I)}\, dt = 1, \tag{4.3}$$

where

T_0 is the total operating time. $t(I)$ is the operating time as a function of current.

4.3.2 Setting rules

There are two settings that must be applied to all time-delay overcurrent functions: the pickup and the time delay. A third parameter that must also be considered is the shape of the TCC.

4.3.2.1 Pickup setting

The first step in applying inverse, time-delay overcurrent functions is to choose the pickup setting of the function, so that it will operate for all short circuits in the line section for which it is to provide protection. This is its fundamental function and it must be set so it will always operate for faults in that zone of protection. This will require margins above normal operating currents and below minimum fault currents. The considerations necessary to determine these margins are discussed in Example 4.4.

If possible, this setting should also provide backup for an adjacent line section or adjoining equipment such as a line-terminated transformer. It should be emphasized, however, that the backup function is a secondary consideration. The primary function of protecting its own line section should not be compromised to provide this backup feature. The pickup of a function (as shown in Figure 2.12 of Chapter 2) is the minimum value of the operating current, voltage, or other input quantity reached by progressive increases of the operating parameter that will cause the function to reach its completely operated state when started from the reset condition. This is usually a contact closure, operation of a silicon-controlled rectifier (SCR), or similar action. For example, an overcurrent function set for 4.0 A pickup will operate only if the current through the function is 4 A or greater. This pickup value is determined by the application of the function and must be available on the specific relay type being used. The setting is first calculated, considering the maximum load and minimum fault currents in terms of the primary current, and then, using the CT ratio, the secondary current is calculated. It is then specified in terms of the pickup of the function being used; so the relay can be set and calibrated.

4.3.2.2   Time-delay setting

Most overcurrent relays have a range of adjustment to make them adaptable to as wide a range of load and fault values as possible. This time-delay feature is an independent parameter that is obtained in a variety of ways depending on the design of the relay. Figure 4.7 shows an example of a set of curves at various time dial settings from 1 to 10.

This range of adjustment is limited for electromechanical relays, because of the physical limitations within the relay and to simplify the relay construction. A family of electromechanical relays is therefore designed to cover the range of fault values. In the case of digital relays, much wider setting ranges are available and a single model of relay will cover most applications. It is important to note that the actual relay settings must be based on the specific manufacture's relay model and curves.

The time-delay feature of the function is an independent parameter that is obtained in a variety of ways, depending on the design of the relay. It is easiest to visualize in an electromechanical relay. As shown in Figure 2.9 of Chapter 2, a time dial is provided that positions the moving contact relative to a fixed contact. The dial is marked from a setting of 1/2 to 10, fastest to slowest operating times, respectively. The operating speed of the disk is determined by the magnitude of the operating current, and the operating time is determined by the distance the moving contact has to travel before it reaches the fixed contact. This is an inverse time–current relationship, that is, the greater the operating current, the less time it takes to travel from the reset position to the operating position. In an electromechanical relay, there are other time-delay mechanisms such as clock movements, bellows, or diaphragms. These may be inverse or fixed timing devices. In solid-state analog relays, timing is achieved by electrical circuits using R–L–C timing circuits.

In digital relays, the time delay is a setting that modifies Equations (4.1) and (4.2) as shown in the following text:

For $0 > M < 1$ as an option,

$$t(I) = \text{TD}\left(\frac{t_r}{M^p - 1}\right), \tag{4.4}$$

where

I is the current measured,
TD is the time dial setting in per unit of the maximum time dial setting available on
   the relay,
$t(I)$ is the reset time as a function of the current (M),
t_r is the total reset time at the maximum time dial and when $M = 0$,
M = multiples of pickup current,
p is a constant.

For $M > 1$,

$$t(I) = \text{TD}\left(\frac{A}{M^p - 1} + B\right),$$

(4.5)

where

A and B are constants applicable at specific time dial setting,
p is a constant as given in Table 4.1, or as otherwise specified by the relay manufacturer,
TD is the time dial setting in per unit of the time dial for which the A and B constants are
   applicable. Note that the TCCs shown in Figure 4.7 are for the A and B constants applicable at a time dial of 5.

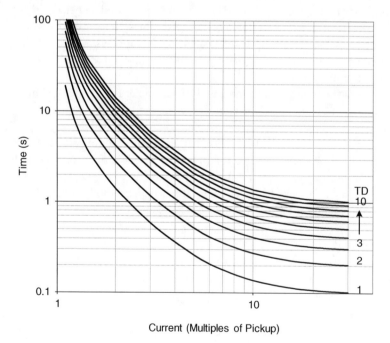

Figure 4.7   Time-delay overcurrent relay-operating characteristic at various time dial (TD) settings.

The purpose of the time-delay setting is to enable relays to coordinate with each other. A family of curves must be provided; so two or more relays, seeing the same fault as defined by the multiples of pickup, can operate at different times. Examples 4.2 and 4.3 show how the pickup, time-delay setting, and fault current relate to each other.

The effect of adding independent time settings is to convert the single characteristic shown in Figure 2.12 of Chapter 2 into a family of curves as shown in Figure 4.7. As explained previously, the curves are plotted in terms of multiples of the pickup (pu) value, and not in terms of the actual pickup value of current (or voltage, etc.). This allows the same curves to be used, regardless of the actual input value corresponding to a specific pickup setting. It is important to note that actual relay settings must be based on the specific manufacturer's curves for the relay in question. Appendix D shows several manufacturers' curves for some legacy electromechanical relays commonly used in North America.

Example 4.2 Figure 4.7 is a very inverse TCC using standard constants [6] with A and B based on a time dial setting of 5. Therefore, the formula used to calculate the operating time is given by the following:

$$T_{OP} = \frac{TD}{5}\left(\frac{19.61}{M^2 - 1} + 0.491\right), \tag{4.6}$$

where

T_{op} is the operating time,
TD is the time dial setting,
M is the multiple of pickup current.

Determine the operating time for a relay with a 4.0 A pickup, time dial setting of 1.0, and 12.0 A operating current.

The input current of 12 A corresponds to a value of $M = 12/4 = 3$ pu. Using this value and a TD setting of 1.0, the operating time is calculated from Equation (4.6) as 0.59 s.

As another example, for a relay with 5.0 A pickup, time dial setting of 2.0, and 15.0 A operating current, the operating current (M) is 3 pu, and the operating time corresponding to a time dial setting of 2 is 1.18 s. The operation of the relay is not consistently repeatable when the operating current is only slightly above its pickup setting. In electromechanical relays, the net operating torque is so low at this point that any additional friction or slight errors in calibration may prevent operation. In fact, most manufacturers' curves do not extend below 1.5 pu. Solid-state analog or digital relays are more precise, and can be applied with closer tolerances. Nevertheless, it is usual to calibrate the time–overcurrent function at some multiple of pickup (such as 3 or 4) to get consistent results. Similarly, the actual operating time may not correspond exactly to the time dial setting at any given multiple of pickup as given on the manufacturer's curve. The conservative approach is to calibrate the relay with the desired current (converted to pu by dividing the operating current in secondary amperes by the pickup current) and to set the required time by small adjustments around the nominal dial setting.

Example 4.3 Referring again to Figure 4.5, determine the TD setting required to achieve an operating time of 1.0 s for a function set at 10.0 A pickup and an operating current of 50.0 A.
  Equation (4.6) can be rearranged as

$$\text{TD} = \frac{5T_{\text{OP}}}{\left(\dfrac{19.61}{M^2 - 1} + 0.491\right)}. \tag{4.7}$$

The input current of 50.0 A corresponds to a value of $M = 50/10 = 5$ pu. Using this and the operating time of $T_{\text{OP}} = 1.0$ s in Equation (4.7), the calculated value of TD is 3.8.

As another example, consider a function with a pickup setting of 5.0 A and the same operating time of 1.0 s and fault current of 50 A. The operating current is $50/5 = 10$ pu and the operating time is 1.0 s, so a time dial setting of 7.3 is calculated from Equation (4.7).

The two time dial settings calculated in this example are not integral values like those shown in Figure 4.7 or in Appendix D, in fact, they have been rounded to one decimal place. Time dial settings for electromechanical relays are usually calculated from curves like those of Appendix D. Settings calculated from formulae are more precise. In all cases, when settings are first applied, or changed, operating times are determined by function test calibration.

4.3.2.3 Selection of shape of TCC

The standard shapes defined in Reference [6] are as noted in Table 4.1, moderately inverse, very inverse, and extremely inverse. Other shapes also exist such as inverse, short time inverse, long time inverse, and inverse definite minimum time, and these vary according to different standards and manufacturers. A constant I^2t shape (with or without a definite minimum operating time) is also available in some models of relays.

Figure 4.8 shows the TCC of four shapes of curves: the moderately inverse, very inverse, and extremely inverse using the formulae of Equation (4.2) and the constants of Table 4.1. The $I^2t =$ constant shape up to 20 pu is drawn using the extremely inverse constant for A, and a value of $B = 0$. Above 20 pu, a definite minimum time has been added to the $I^2t =$ constant TCC.

Note that since a fuse TCC is determined by heating of a fusible element, it closely approximates an I^2t shape except for some flattening when the clearing time approaches one cycle. Therefore, the extremely inverse function TCC most closely approximates the fuse TCC, and this curve is often chosen for relays immediately upstream of a fuse. Other relays further upstream may be chosen with less inverse characteristic to provide better coordination. In applications with weak sources, or on short lines when the fault current does not change very much over the length of the protected line, less inverse curves such as moderately inverse are more appropriate.

In the case of very weak sources such as ground overcurrent protection for resistively grounded neutral systems, a definite time–overcurrent characteristic may be applied. This characteristic has a fixed time for any current above its pickup, as shown in Figure 4.9.

In the case of electromechanical relays, individual models of a relay can have only a single shape of TCC. Different models have different shapes, such as the examples in Appendix D. In such relays, the characteristics of the curves are provided graphically. In the case of

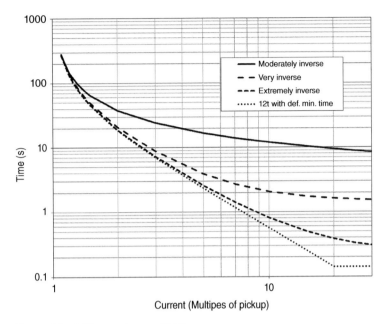

Figure 4.8 Different shapes of TCC.

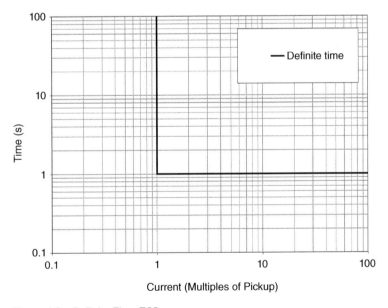

Figure 4.9 Definite Time TCC.

digital relays, the shape of the TCC may be chosen by a setting from a variety of shapes provided by the manufacturer with formulae for calculation of operating times.

Having established the mechanics of specifying the pickup tap and time dial setting of the function, we must now examine how these parameters relate to the actual load and fault currents on the system.

For phase relays, it is necessary to establish the maximum load current, which requires engineering judgment based upon planning studies and reasonable system contingencies. Then, the minimum phase fault current must be calculated. This will usually be a phase-to-phase fault at some reduced generation and system configuration. The relay should then be set somewhere between twice maximum load current and a third of the minimum fault current value. The ratio of minimum fault current to relay pickup current may be considered as the "dependability factor." The ratio of relay pickup current to maximum load current may be considered as the "security factor." These are, of course, rough guidelines. Dependability and security factors are not standardized and depend significantly on user experience. The actual setting must be an evaluation of setting closer to the maximum load, increasing dependability and decreasing security, or achieving less than three times pickup for the minimum fault, decreasing dependability and increasing security.

A ground relay must "see" all phase-to-ground faults within its zone of protection, and under conditions that give a minimum fault current. Note that in calculating ground current, it is the zero-sequence current that is of interest. (ground current $= 3I_0$). There is no concern for load current, but normal phase and load unbalance and CT errors must be considered and the relays set above these values. Again, setting between twice the "normal" ground current and a third of the minimum fault value is desirable. In the absence of any other information, for transmission systems where the load is well balanced, the normal ground current may be taken to be 10% of the maximum load current.

For North American commercial and residential distribution systems, there are usually a high percentage of single-phase loads connected phase to neutral. In addition, there may be many single-phase laterals supplied from a three-phase main branch. Further, in many cases, feeders may be switched from one source to another though manual operation with staggered operation of single-phase switches. All of these factors may result in temporary or steady-state unbalance currents flowing through the neutral and/or ground. These load unbalances require significantly higher ground currents than 10% of the maximum load current. The criterion to set the pickup higher than the "normal" ground current as well as lower than the minimum fault current considering fault impedance becomes more difficult than in the case of transmission line protection which normally only serves balanced three-phase loads. Reference [2] provides information about practices that may be used by various utilities. The reference identifies four practices that may be used to determine a minimum pickup setting considering the possibility of high fault resistance:

i) Choose a certain percentage of the minimum bolted (zero fault resistance) current.
ii) Use a company-established value of fault resistance at the end of the protected line.
iii) Use a fault-resistance value that provides a minimum fault current equal to the continuous current-carrying capability of the line conductor. This alternative notes that lower fault currents may not be detected.
iv) Use a percentage of three-phase fault currents.

Whatever the criterion that is used, there is a concern that there could be single line-to-ground faults with such high impedance that they may not be detected by conventional overcurrent protection. In fact, the definition of a high-impedance fault (HIF) is "HIFs that occur do not produce enough fault current to be detectable by conventional overcurrent relays or fuses" [7]. Although advanced protection systems to detect high-impedance faults on distribution lines do exist, their use is not yet formally established [7].

These rules for setting relays, and others, have been established over many years of experience, to cover a variety of unknowns. For instance, short-circuit studies performed to determine fault currents for relaying purposes do not normally include the resistance component of the impedances. Since relaying studies generally precede construction, the line lengths are usually anticipated values, and are not the actual installed values. Maximum and minimum load flows are, at best, good engineering estimates. Equipment impedances are usually contract values, not test impedances. Finally, the relay accuracy itself is subject to specific relay designs and manufacturers' tolerances. As a result, these rules include conservative margins built into the settings. Solid-state and digital relays can be set and calibrated to closer tolerances, and some leeway is allowed in their setting.

An additional benefit of multifunction relays is that an alarm can be added to let operators know if normal steady-state phase or ground currents approach pickup settings of the overcurrent relays. For example, an alarm could be initiated if the phase current stays above 50% of the phase overcurrent pickup setting for several minutes. Similarly, an alarm could be initiated if the state ground current ($3I_0$) stays above 75% of the pickup of the ground overcurrent pickup setting for several minutes.

Example 4.4 Referring to Figure 4.10, determine the CT ratio, pickup, and time dial settings for the relay at breaker (1), assuming that no coordination with any other relay is required. Assume that the maximum load is 95 A, minimum fault is 600 A, and the maximum fault is 1000 A. Select a CT ratio to give 5.0 A secondary current for maximum load, that is, 95/5 = 19:1.

Since this is not a standard CT ratio (refer to Table 3.1 of Chapter 3), we select the nearest CT ratio of 20:1 or 100:5. The relay pickup setting should be bracketed by twice the maximum load and one-third of the minimum fault. Using the actual CT ratio, twice maximum load is 190 A divided by 20, or a relay current of 9.5 A.

Assuming the relay has a pickup range of 0.05–100 A secondary, in increments of 0.1 A, we would select a setting of 10.0 A, giving a primary current relay pickup of 200 A. Dividing by 95 A, load current results in a margin of 2.1 pu to prevent false operation (security). The minimum fault is 600 A divided by the relay pickup of 200 A, which gives 3 pu to ensure correct operation (dependability). For this configuration, no coordination is required; so one can set the time dial close to the lowest setting (fastest time) of 1.

The principle of relay coordination can be explained by reference to Figure 4.11, which shows a series of radial lines and the time–distance characteristics of the associated inverse-time relays. These are relay-operating curves selected for each of the relays, plotted as a function of fault location. Since the

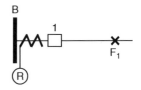

Max. load = 95 amperes
Min. fault = 600 amperes
Max. fault = 1000 amperes

Figure 4.10 Time-delay overcurrent relay setting.

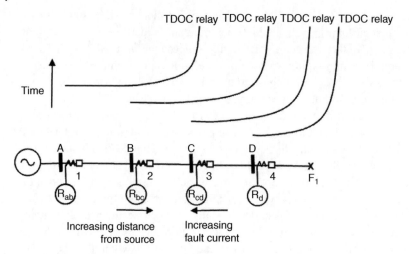

Figure 4.11 Relay coordination principles.

magnitude of the fault current decreases as the fault moves away from the source, these curves appear "reversed" when compared to those in Figure 4.7.

For fault F_1 at the end farthest from the generating source, relay R_d, at tripping breaker (4), operates first; relay R_{cd} at breaker (3) has a higher time dial setting which includes a coordinating time delay S to let breaker (4) trip if it can; similarly, relay R_{bc}, at breaker (2), coordinates with the relay at breaker (3) by having a still longer time delay (including the same coordinating time S); and finally, relay R_{ab} at breaker (1) has the longest time delay and will not trip unless none of the other breakers trips, provided it can see the fault, that is, provided the fault current is greater than its pickup setting. Should a fault occur between breakers (3) and (4), relay R_d will receive no current and therefore will not operate; relay R_{cd} will trip, since its operating time is faster than that of relay R_{bc}. For the settings shown in Figure 4.11, relay R_{ab} will not see this fault. Relay R_{bc} must still provide backup relaying for this fault, as discussed before.

Referring again to Figure 4.11 and to the time sequence shown in Figure 4.12, a fault at the end of the line should be cleared by R_d and CB_4. R_{cd} sees the same fault as R_d and starts to close its contacts. If the circuit breaker clears the fault, assuming electromechanical relays, R_{cd} will reset after some overtravel. The exact overtravel time can be obtained for specific relays. Contact overtravel is included in the margin of safety. Overtravel does not apply to solid-state or digital relays that do not have moving parts, although some coordination security margin would be provided. If the fault is not cleared by R_d or CB_4, R_{cd} continues to close its contacts and initiates a trip of CB_3 at the end of its operating time. To be sure that R_{cd} does not close its contacts before the fault is cleared by CB_4, it must be set longer than R_d operating time (U), plus CB_4 clearing time (V), and a coordinating security factor (X) (including overtravel, W). It is usual to add 0.3–0.5 s coordinating time to the operating time of R_d, which is calculated at its maximum fault current. The same fault current is used to determine the operating time of relay R_{cd}.

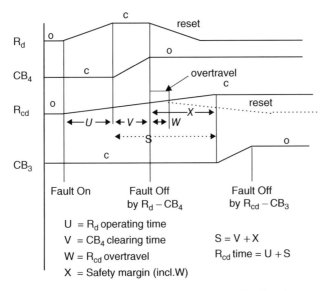

Figure 4.12 Time-delay overcurrent relay coordinating time.

Example 4.5 Assume we have a radial system with two adjacent line segments as shown in Figure 4.13. The segment farthest from the source is set as shown in Example 4.4. The relay protecting the next line segment closest to the source must protect its own line and, if possible, back up the relays protecting the next line. The pickup should therefore be for the same primary current as the downstream relay and the time setting must coordinate with it. This time setting should be made with maximum current conditions, that is, assuming a three-phase fault, with maximum generation behind the relay. In a radial system, all relays for which coordination is required must be examined for operation at this same primary current (Example 4.6 shows how to coordinate relays in a looped system). When coordination is achieved at maximum current, the shape of the inverse curves, provided they are all of the same family of inverseness, will ensure coordination at all lesser current values.

From Example 4.4, the pickup setting of R_b is 200 A primary or 10.0 A secondary considering a CT ratio of 20:1, and the time dial setting is 1. Theoretically, to ensure that R_{ab} backs up R_b, it should be set for the same pickup current that is, it sees the same faults, but is set at a slower (higher) time dial.[1] The operating time of R_b is determined from Equation (4.6) at

---

1 In practice, for proper selectivity with the adjacent relays, as we move from the fault location to the source, each phase relay should have a slightly higher pickup (in terms of primary current) than its preceding relay. This is to avoid the situation where, with unexpected heavy load current or a fault with resistance, the relay current could be at this pickup point and small errors in calibration or CT output would allow the upstream relay to drift closed while the downstream relay remained inoperative. For instance, referring to Figure 4.7 where both R_{ab} and R_b are supposedly set at a pickup of 200 A secondary current, either the relays or their CTs could be slightly in error so that R_{ab} had a 190 A setting and R_b had a 210 A setting. If then a high-resistive fault occurred so that the fault current was 200 A, R_{ab} would drift closed while R_b stayed open.

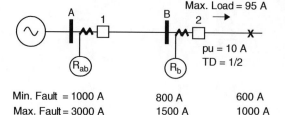

Figure 4.13 Time-delay overcurrent relay setting and coordination.

		Max. Load = 95 A

pu = 10 A
TD = 1/2

Min. Fault = 1000 A   800 A   600 A
Max. Fault = 3000 A   1500 A   1000 A

the maximum fault current at bus B (1500 A) divided by its pickup setting (20 × 10 A) or 7.5 pu and the 1 time dial. This is found to be 0.17 s. Add 0.3 s coordinating time and R$_{ab}$ operating time should be 0.47 s. If we make reference now to Equation (4.7), at the same maximum fault current of 1500 A at bus B and pu of 20 × 10 A, the multiple of the tap setting is 7.5 pu and the operating time is 0.47 s. The required time dial setting is 2.8 (the actual time dial should be determined by test).

In a network, the coordination of time-delay overcurrent relays is complicated by the problem of infeed and outfeed. Although this condition is most frequently associated with distance relays, as is discussed in Chapter 5, it is also an important consideration when setting time-delay overcurrent relays. The condition occurs when, for a given fault, due to the system configuration, there is a different current in the downstream relays than there is in the relay being set.

Example 4.6   Referring to Figure 4.14, there is more current in the relays at CB 5 with all lines in service than there is with lines 3–4 out of service (8.33 vs 5.88 A). This helps the coordination between relays at CB 5 and relays at CB 1. With lines 3–4 out of service, the infeed is no longer present and the current decreases in the relays at CB 5 (8.33–5.88 A) and increases in the relays at CB 1 (4.165–5.88 A). This is the critical case for coordination.

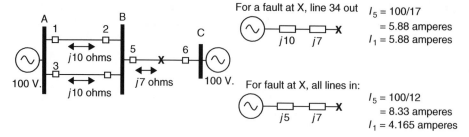

Figure 4.14   Effect of infeed.

4.4 Instantaneous overcurrent relays

4.4.1 Application

The design and construction of the electromechanical and solid-state instantaneous overcurrent relays are described in Section 2.4 (Figure 2.7) and Section 2.4 (Figure 2.14) of Chapter 2. In order to properly apply the instantaneous overcurrent relay, however, there must be a substantial reduction of short-circuit current, as the fault is moved away from the relay toward the far end of the line. The relay must be set not to overreach the bus at the remote end of the line and there still must be enough of a difference in the fault current between the near- and far-end faults to allow a setting for the near-end fault. This will prevent the relay from operating for faults beyond the end of the line and, at the same time, will provide high-speed protection for an appreciable portion of the circuit.

Figure 4.11 also shows why simple inverse-time–overcurrent relays cannot be used without additional help. The closer the fault is to the source, the greater the fault current magnitude, yet the longer the tripping time. The addition of instantaneous overcurrent relays makes this system protection viable. If an instantaneous relay can be set to see almost up to, but not including, the next bus, all the fault-clearing times can be lowered, as shown in Figure 4.15.

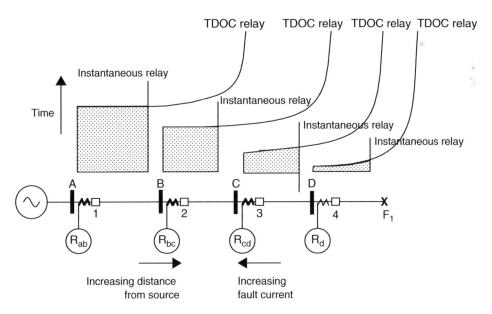

Figure 4.15 Application of instantaneous and time-delay overcurrent relays.

4.4.2 Setting Rules

The inaccuracies of the short-circuit studies discussed in setting an inverse, time-delay over-current relay apply equally to the instantaneous overcurrent relays. Since the instantaneous relay must not see beyond its own line section, the values for which it must operate are very much higher than even emergency loads. Therefore, load is not usually a consideration for the instantaneous relay setting. Also, for an electromechanical instantaneous relay, the distinction between "go" and "no-go" is much better defined than for a time-delay overcurrent relay, particularly an induction disk relay. As a result, there is no need to set an instantaneous overcurrent relay with margins such as 200% of load and one-third of fault current. However, in addition to the inaccuracies of the relay itself, there is a factor called "transient overreach" that must be considered. Transient overreach is the tendency of a relay to instantaneously pick up for faults farther away than the setting would indicate. When discussing the parameter T in Chapter 3 (Section 3.3), we noted that this factor is related to the time constant of the DC decay of the fault current; the slower the decay, the more overreach is possible. High-voltage transmission systems are more susceptible to transient overreach than lower-voltage distribution systems because the latter have a lower X/R ratio in their line impedances. The tendency is also more pronounced in electromagnetic attraction relays than in induction-type relays. Transient overreach is usually only a concern for instantaneous or zone-1 relays. Their reach settings are more critical than backup relays, and backup relays have a time delay that allows the offset to decay. It is therefore common to set an instantaneous relay about 125–135% above the maximum value for which the relay should not operate, and 90% of the minimum value for which the relay should operate. Solid-state or digital relays can be set closer, for example, 115% above the maximum no-go value.

Example 4.7   Using Figure 4.13 and the same fault currents of Example 4.5, set the instantaneous relays at buses A and B.

Setting R_{bc}
To avoid overreaching the terminal at the end of the line, R_b should be set at 135% of the maximum fault current at that location, that is, $R_b = 135\% \times 1000\,A = 1350\,A$. Dividing by the 20:1 CT ratio to get relay current, the pickup of R_b is $1350/20 = 67.5\,A$.

Check to see how the relay will perform for a minimum fault at bus B. The minimum fault at bus B is 800 A, divided by the relay pickup of 1350 A, giving a multiple of pickup of 0.59: **the relay will not operate.**

Check to see how the relay performs at the maximum current of 1500 A. 1500/1350 gives 1.11: **barely above pickup.**

This provides very little line protection and an instantaneous relay would not be recommended in this situation.

Setting R_{ab}
Avoid overreaching bus B by setting $R_{ab} = 135\%$ times the maximum fault current at that bus, that is, $1.35 \times 1500\,A = 2025\,A$.

Check the secondary current by dividing the primary current by the CT ratio: $2025/20 = 101.25\,A$. Current over 100 A may cause saturation of the magnetic components and is too high for electromechanical relays. Depending upon the specific design, this current may also be too high for both solid-state and digital relays.

Referring again to Table 3.1 of Chapter 3, select the next highest standard CT ratio of 40:1. The relay current then becomes $2025/40 = 50.63$ A. Depending upon the actual breaker and CT configuration, this change in CT ratio may require that the setting of the time–overcurrent relays, for example, Example 4.5, may have to be recalculated.

Check the performance of the relay at minimum and maximum currents at bus A. Pickup at 1000 A is $1000/2025 = 0.49$. **The relay will not pick up.** Pickup at 3000 A, however, is $3000/2025 = 1.48$ and the relay will pick up.

The decision whether or not to use an instantaneous relay depends upon the clearing time of the time-delayed overcurrent relay of Example 4.5. The fact that the clearing time of the instantaneous relay will only occur at the maximum bus fault also indicates that there may be very little advantage to be gained with the instantaneous relay. However, if the source connected to bus A is an actual generator, not a system equivalent, then faster clearing at the maximum fault current is justified and an instantaneous relay would be used.

4.5 Directional overcurrent relays

4.5.1 Application

Directional overcurrent relaying is necessary for multiple source circuits, when it is essential to limit relay tripping for faults in only one direction. It would be impossible to obtain correct relay selectivity through the use of a nondirectional overcurrent relay in such cases. If the same magnitude of the fault current could flow in either direction at the relay location, coordination with the relays in front of, and behind, the nondirectional relay cannot be achieved except in very unusual system configurations. Therefore, overcurrent relaying is made directional to provide relay coordination between all of the relays that can see a given fault. Directional relays require two inputs, the operating current and a reference, or polarizing, quantity (either voltage or current) that does not change with fault location. The operating torque of a two-input relay, which may be used to provide the directional feature, is given by Equation (2.15) in Chapter 2.

To illustrate the need for directionality, refer to Figure 4.16. As a radial system (switch X open), circuit breakers (4) and (5) receive no fault current for a fault at F_1. In fact, for this system configuration, breaker (4) is not required. In the loop system (switch X closed), we cannot set relays at (4) above those at (5) to be selective for a fault at F_2, and still maintain coordination between (4) and (5) for a fault at F_1. Directional relays are required. Occasionally, a point in the loop can be found when there is sufficient difference between a fault in the forward direction and one in the backward direction so that the settings alone can discriminate between them. For this to be a safe procedure, usually a ratio of 4:1 between forward and reverse faults would be required.

Example 4.8 Referring to Figure 4.16, with switch X closed, assume that the current through (4) and (5) for a fault at F_1 is 100 A and for a fault at F_2 is 400 A. Setting the relay at (4) for pickup at 25 A gives 4 pu for the fault at F_1 and 16 pu for the fault at F_2. This relay must, therefore, be directional to see faults only in the direction from breaker (4) to breaker

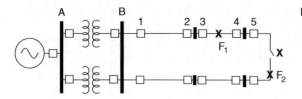

Figure 4.16 Radial versus loop circuit.

(3). Setting the relay at (5) at 125 A, however, allows it to have 3.2 pu for the fault in its protected zone at F_2, but less than 1.0 pu for the fault at F_1. It therefore does not have to be directional. However, such a condition may change with system growth and pass unnoticed until a false trip occurs. It is therefore good practice to use directional relays at both locations.

There are two approaches to providing directionality to an overcurrent relay:

1) Directional control: As shown in Figure 4.17a, the design of the relay is such that the overcurrent element will not operate until the directional element operates, indicating that the fault is in the tripping direction. In electromechanical relays, this is done using the directional element contact to allow starting of the overcurrent element coil; so no torque can be developed until the directional element contact is closed. In solid-state and digital relays, this is accomplished by the logic circuitry or the algorithm.
2) Directional supervision: As shown in Figure 4.17b, this relay has output supervision of the trip function. Both the directional and overcurrent elements must operate before a trip output is obtained.

The directional control relay is more secure. Referring to Figure 4.17c, with the directional supervision design, if a fault occurs in the non-trip direction, as it does for breaker (4), the overcurrent element can pick up from the contribution in that direction. Only the directional element prevents a trip of breaker (4). If we assume that breaker (2) opens

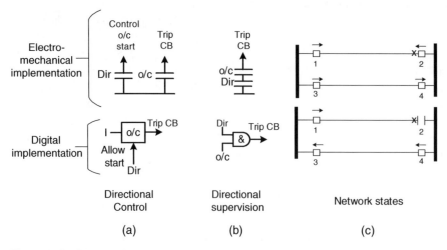

Figure 4.17 Directional control versus directional supervision overcurrent relays.

before breaker (1), which is possible, then the reversal of fault current through circuit breaker (4) will cause a race between the overcurrent element opening and the directional element closing.

If the directional element wins the race, that is, it operates before the overcurrent relay resets, there will be a false trip.

With the directional control design, this situation cannot occur, since the overcurrent relay is controlled by the directional element and will not start when the fault is in the non-trip direction. At breaker (3), although the directional element will operate with all breakers closed, the overcurrent element will not time out, since it must coordinate with breaker (2). When breaker (3) opens, and the current reverses, the overcurrent element will not be able to time out any further, since it will be blocked by reset of the directional element.

4.6 Polarizing

As described in Section 2.2 in Chapter 2, the ability to differentiate between a fault in one direction or another is obtained by comparing the phase angle of the operating current phasor, which varies directly with the direction of the fault, and some other system parameter that is not dependent on the fault location. This constant parameter is referred to as the polarizing quantity. For phase relays, the polarizing quantity is, almost invariably, the system voltage at the relay location. Depending on the voltage and current connections, a directional relay will operate either for normal load current or fault current. In an electromechanical relay, the maximum torque on an induction disk occurs when the two torque-producing fluxes are 90° apart in time and space. The space criterion is easily obtained by the location of the flux-producing coils around the relay. The time criterion is achieved by creating an appropriate phasor difference between the two operating quantities. In a digital implementation, a simple comparison of phase angles of operating current and polarizing voltage will be sufficient to determine direction.

4.6.1 Power directional relays

For protective applications where we are concerned with conditions other than short circuits, power directional relays are required. These relays operate under conditions of balanced load and relatively high power factor. The voltage and current phasors are approximately in phase with each other and of constant magnitude, as shown in the balanced load case of Figure 4.18a. Three-phase power relays are usually polarized by the positive-sequence voltage and measure the product of the voltage and in-phase component of positive-sequence current. They will give an output when the product of the current and voltage exceeds the set threshold as shown in Figure 4.18b. Single-phase power relays are polarized by individual-phase voltages and currents. The threshold for operation for a three-phase power relay shown in Figure 4.18b is reached when $V_{pos} \times I_{pos} \times \cos(\theta) \geq S_0$.

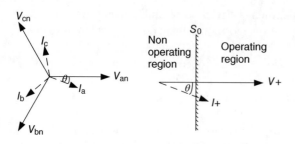

Figure 4.18 (a) Phasors for power relays. (b) Threshold for operation of three-phase power relay.

Balanced three-phase load with lagging power factor

(a)

Threshold for operation

(b)

4.6.2 Fault directional relays

The more usual case, however, involves protection during short circuits. We consider phase and ground directional relays separately.

4.6.2.1 Directional phase overcurrent protection

For directional relays applied to operate during a fault, it is important to remember two facts:

1) The system voltage will collapse at the point of the short circuit. Therefore, to obtain sufficient torque under fault conditions, the polarizing voltage of directional phase overcurrent relays must not be only the faulted phase.
2) The fault power factor is low, that is, the current lags the voltage by nearly 90°. There are a variety of polarizing voltages available for directional discrimination for multiphase faults. One common technique uses the phase-to-phase voltage of opposite phases shifted by 90° as shown in Figure 4.19a (quadrature polarization).

The directional overcurrent function shown in Figure 4.19b shows a minimum operating current "I_{min}." Most directional elements will require a minimum operating current to make a directional decision. However, this minimum operating current of a directional element is not necessarily the same as the pickup current of an overcurrent function. The directional overcurrent element in a multifunction relay may be used for a variety of functions, so it will have a dedicated, independently adjustable minimum pickup current to ensure correct operation in the presence of low signal-to-noise ratio. As shown in Figure 4.19b, the minimum operating current will have an adjustable angle setting (Φ) for the phase angle with respect to the polarizing voltage. In electromechanical relays, this angie is called the "maximum torque" angle. In solid-state relays, technically this angle is the "characteristic" angle, but is frequently also called the "maximum torque" angle.

As shown in Figure 4.19b, the A phase directional element will operate when $I_a(\text{Cos}(\theta\text{-}\Phi)) > I_{min}$.

In the case of close in three-phase faults, all voltages will be very small, or zero. Most modern directional phase overcurrent elements will include a memory function that enables

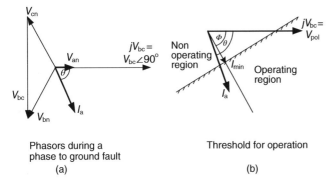

Figure 4.19 (a) Phasors during a phase fault. (b) Threshold for operation of directional phase overcurrent relay.

them to use pre-fault voltages for polarizing for a short time if all fault voltages are too small for reliable directional discrimination.

4.6.2.2 Potential polarizing for ground directional overcurrent protection

For ground faults, the operating current is often derived from the residual circuit of the phase CTs ($I_n = 3I_0$). Since the current can be derived from any phase that is faulted, it is necessary to obtain a related voltage to obtain a correct directional response. As with phase relays, whenever a directional relay is involved, two facts must be known: the magnitude and the direction of the fault. Both factors may be combined in a single-function product-type relay or in separate elements such that the directional element controls the overcurrent element.

For ground directional indication, a reference is necessary to determine whether the line current is into or out of the protected line. The zero-sequence voltage ($3E_0$) can be used, since it is always in the same direction regardless of the location of the fault. This is called potential polarizing. In the case of electromechanical relays, the zero-sequence voltage can be obtained across the open corner of a wye-grounded, broken-delta potential transformer, as shown in Figure 3.18 of Chapter 3. In the case of digital relays, the three phase-to-neutral voltages are simply summed together. The vector sum of the three line-to-neutral voltages $V_a + V_b + V_c = 3V_0$, which is zero for balanced conditions or for faults not involving ground. The magnitude of $3V_0$ during line-to-ground faults depends on the location of the fault, the impedance of the ground circuit, and the ratio of the zero-sequence impedance to the positive-sequence impedance. Potential transformers are usually connected wye-grounded–wye-grounded to provide metering and relaying potential and can pass high-voltage zero-sequence components to the low-voltage side. An important difference between directional phase and voltage-polarized ground overcurrent elements is the phase-angle relationship between the polarizing voltage and operating current.

Figure 4.20 shows the voltage profiles in the various sequence networks and the phase domain, as the measuring point moves away from the source toward a fault. At the source, the phase voltage and positive-sequence voltage are at nominal values. At the fault (assuming zero fault resistance), they are zero. On the other hand, there are no source voltages in

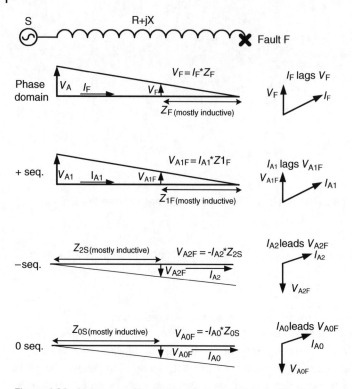

Figure 4.20 Voltage profile changes from source to fault point.

the negative- and zero-sequence networks. During unbalanced faults, these voltages will rise from zero at the source to a non-zero magnitude at the fault point. During balanced three-phase faults, of course there will be no negative- or zero-sequence components.

It can be seen from Figure 4.20 that the phase and positive-sequence voltages at an intermediate measuring point (not at the fault location) will be somewhere between nominal and zero magnitude. The phase and positive-sequence currents will lag the measured voltages by the angle of the impedance between the measuring point and the fault. The negative- and zero-sequence networks have no voltage sources. Their voltages at the measuring point will be somewhere between zero and the value at the fault location. Further, the negative- and zero-sequence voltages will be approximately opposite polarity from the phase and positive-sequence voltages. Figure 4.19 shows that the zero-sequence current leads the zero-sequence voltage by an angle of $180°$—source impedance angle. The relationship of negative-sequence current and voltage is similar to the zero-sequence.

Figure 4.21a shows the operating characteristic of a zero-sequence-voltage-polarized directional ground overcurrent relay. Some ground relays may use $-3V_0$ for polarizing, and then the characteristic would be as shown in Figure 4.21b. In both cases, the angles shown in Figure 4.21 have the same meaning as in Figure 4.19 (except in this case they refer to zero-sequence quantities).

Directional ground overcurrent relays need not necessarily use zero-sequence quantities. Negative-sequence quantities are also present during ground faults. Additionally, negative-sequence quantities are also present during phase-to-phase faults not involving

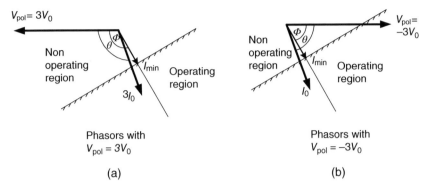

Figure 4.21 (a) and (b) Operating characteristics of a zero-sequence-voltage-polarized directional ground overcurrent relay.

ground. This gives such relays a wider scope of application than zero-sequence-voltage-polarized relays.

It has been found that zero-sequence mutual coupling between parallel lines may sometimes cause an unreliable directional declaration [8]. Negative-sequence directional overcurrent is not subject to misoperation due to zero-sequence mutual coupling and may offer better sensitivity in certain cases where a strong zero-sequence source could result in low levels of zero-sequence voltage. Negative-sequence-voltage-polarized directional overcurrent relays have similar characteristics to zero-sequence-voltage-polarized directional ground overcurrent relays.

4.6.2.3 Current polarizing for ground directional overcurrent protection

Another method of obtaining a directional reference is to use current in the neutral of a wye-grounded/delta power transformer. Figure 4.22 shows the operating characteristic of a current-polarized directional ground overcurrent relay.

Where there are several transformer banks at a station, it is common practice to parallel all of the neutral CTs to be sure that, if at least one of the banks is in service, the necessary polarization is provided. Three-winding banks can be used as long as there is a delta winding and one wye-grounded winding. If there are two wye-grounded windings, then both neutrals must have their CTs paralleled with their ratios inversely proportional to the bank voltage ratings. The grounded neutral of an autotransformer with delta tertiary may or may not be a suitable polarizing source. The autotransformer is different from the three-winding transformer, in that some of the fault current flows directly from one voltage system to the other while the rest of the fault current is transferred through the transformer. It is the direct flow of current in the common winding to the neutral that presents the problem. For a high-side fault, if IH_0 represents the high-side zero-sequence fault current and IL_0 represents the low-side zero-sequence fault current (both currents in amperes), then the neutral current $= 3(IH_0 - IL_0)$, $IL_0 = IH_0 \times n \times k$, where n is the voltage ratio and k is the current distribution factor that

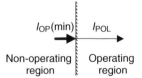

Figure 4.22 Operating characteristic of a current-polarized directional ground relay.

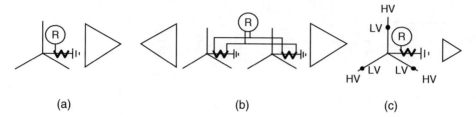

Figure 4.23 Current polarizing. (a) Single wye–delta transformer; (b) two separate wye–delta transformers; and (c) autotransformer.

gives the zero-sequence current contribution of the low-voltage system. This expression reduces to

$$I_{neutral} = 3IH_0(1 - kn).$$

If this current is positive, current flows up the neutral; if negative, it flows down the neutral. In other words, if kn is 1, the neutral current is zero; if greater than 1, current flows down the neutral; if less than 1, current flows up the neutral. Although the current can be up or down the neutral for a high-side fault, in practice, with the usual system and transformer parameters, the current is always up the neutral for low-voltage faults [9].

Figure 4.23 shows the connections for a variety of transformer installations. Note that in the case of Figure 4.23b, two separate wye–delta transformers are shown. CT connections for a three-winding wye–wye–delta would be similar, but only one transformer delta winding would be shown in the figure. Either potential or current polarization may be used, or both simultaneously. However, since the zero-sequence voltage always has its maximum magnitude at the fault and decreases from the fault point to the grounded system neutral, it is possible that there will not be sufficient zero-sequence voltage at the station to give adequate polarizing. Current polarizing is commonly used at stations where transformer banks are solidly grounded.

4.6.2.4 Other polarizing methods for ground directional overcurrent protection

There are a variety of other techniques that may be used to directionalize ground overcurrent protection. They include:

1) Dual polarizing using both zero-sequence current and voltage polarizing so that either the element can determine direction whether or not the ground current sources are available.
2) Negative- or zero-sequence-current-compensated voltage polarizing. In this case, the current through an impedance can reinforce the voltage polarizing signal to provide greater sensitivity for forward faults.
3) Virtual polarizing, in which faulted-phase-selection logic identifies the faulted phase. Then the logic ignores the faulted phase voltage and sums the voltages of the other two phases to provide a large voltage which is substantially 180° out of phase with the zero-sequence voltage. This virtual zero-sequence voltage is then used as a polarizing voltage as in Figure 4.21b.
4) Negative- or zero-sequence impedance measurement, in which the directional decision is made by determining the magnitude and angle of the negative- or zero-sequence impedance. For forward direction faults, the measured impedance will be negative, or

if positive, less than a set threshold. For reverse direction faults, the measured impedance will be positive and more than a set threshold.

More discussion of these other polarization techniques and relevant factors concerning their application can be found in Reference [8].

4.7 Summary

In this chapter, we have examined the protection of a transmission line with time-delay and instantaneous overcurrent relays using only the information available at the relay location. These are the most commonly used relays for the protection of radial and low-voltage subtransmission lines, and industrial or commercial installations. Time-delay relays coordinate with each other by grading the relay-operating currents, in primary amperes, as the fault is moved from the load to the source end and by allowing enough time between the relay operating times so that the relays closest to the fault will operate first. Instantaneous relays must be set so as not to overreach their intended zone of protection. In a networked system, relays must be directional so they will operate only for faults in the desired direction and allow coordination by current and time in that one direction. Coordination must be made by comparing the current in each relay, taking due account of the distribution of current for a given fault. Directionality is obtained by comparing the operating current with a voltage or current that is constant, regardless of the location of the fault.

Problems

4.1 (a) Draw the one-line diagram of a 138/12 kV, 10 000 kVA, 15%, grounded-wye/delta transformer, 12 kV circuit breaker, 138 kV motor-operated switch, and 138 kV 1200/5 MR CTs connected 600/5. Add the following relays and meters: three instantaneous overcurrent relays, two phase time-delay overcurrent relays, three ammeters, and one time-delay overcurrent ground relay. Show the appropriate ANSI device function numbers. (b) Draw the associated three-phase AC CT secondary, relay, and meter connections. (c) Draw the DC tripping circuit showing the relay contacts and the circuit breaker trip coil and auxiliary switch.

4.2 (a) For the system shown in Figure 4.24, calculate the current for a three-phase fault at bus A and again at bus B. (b) Repeat (a) for the system shown in Figure 4.25.

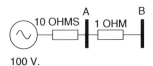

Figure 4.24 System diagram for Problem 4.2a.

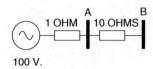

Figure 4.25 System diagram for Problem 4.2b.

4.3 (a) Referring to Figure 4.24, set an instantaneous relay at bus A. Calculate the pickup for faults on buses A and B and discuss the application of the instantaneous relay for this configuration. (b) Repeat (a) for Figure 4.25.

4.4 Repeat Problems 4.2 and 4.3 for a system where the two segments are both equal to 5.

4.5 (a) Referring to the relay characteristics of Appendix D(D.1)–(D.3), determine the operating time for each relay with a 5.0 A pickup, time dial setting of 3, and a fault current of 15.0 A. (b) Which relay, and what time dial setting, should be used if the relay is set at 5.0 A pickup, with a fault current of 15.0 A, and operating time of 1.5 s? (c) Select two different relays, their pickup, and time dial settings when the fault current is 20.0 A and the relay must operate in 2 s.

4.6 For the radial system shown in Figure 4.26, calculate the instantaneous and time-delay overcurrent relay settings at each bus. Assume that the transformer must not be de-energized and that the relays at bus B are "looking into" a transformer differential and do not need to coordinate with it. Assume that any pickup tap is available, but use the relay characteristic of Figure 4.5.

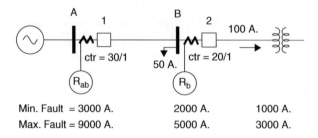

Min. Fault = 3000 A.	2000 A.	1000 A.
Max. Fault = 9000 A.	5000 A.	3000 A.

Figure 4.26 System diagram for Problem 4.6.

4.7 The current in both instantaneous relays of Problem 4.6 exceeds 100 A. Revise the CT ratios and recalculate both time-delay and instantaneous relay settings at buses A and B.

4.8 Given the subtransmission system shown in Figure 4.27, calculate the currents for faults at F_1 and F_2 and the relay pickup settings at breakers (3) and (4). Determine which relays must be directional.

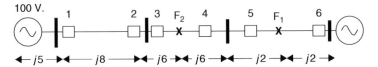

Figure 4.27 System diagram for Problem 4.8.

4.9 Given the system shown in Figure 4.28, calculate the currents in breakers (1), (3), and (5) for a fault at X for the following three conditions: (a) system normal (SIN); (b) line 1–2 open; and (c) line 3–4 open. Select and set the appropriate relays from Appendix D for the relays at breakers (1), (3), and (5). Neglect load, and assume that breaker (5) needs no coordination.

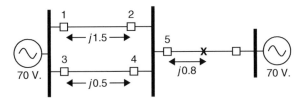

Figure 4.28 System diagram for Problem 4.9.

4.10 You are given that the system shown in Figure 4.29 has a 110/220 kV autotransformer. The positive- and zero-sequence impedances in ohms or percent are as shown in the figure, the zero-sequence impedances being in parentheses. Assume that the low-voltage system is solidly grounded. For a phase-a-to-ground fault at the midpoint of the transmission line, calculate the transformer current I_n in the neutral and the phase a currents I_a and $I_a{}^t$ on the high and low sides of the transformer. If the source on the low-voltage side is to be grounded through a reactance, determine the value of the grounding reactance for which the transformer neutral current becomes zero. As the grounding reactance changes around this value, the direction of the neutral current will reverse, and will affect the polarizing capability of the neutral current for ground faults on the high side. Can faults on the low-voltage side ever cause the neutral current to reverse?

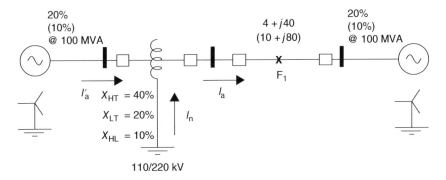

Figure 4.29 System diagram for Problem 4.10.

References

1 IEEE (2015). *Standard C37.113-2015 IEEE Guide for Protective Relay Applications to Transmission Lines*. Piscataway: IEEE Standards Association, IEEE.

2 IEEE Std C37.230 (2020). *IEEE Guide for Protective Relay Applications to Distribution Lines*. Piscataway: IEEE Standards Association, IEEE.

3 IEEE (2012). *Standard Std 1366-2012 IEEE Guide for Electric Power Distribution Reliability Indices −2012*. Piscataway: IEEE Standards Association, IEEE.

4 IEEE Committee Report (2013) "Effect of Distribution Automation on Protective Relaying" IEEE PES Resource Center PS-TR35

5 IEC 60255-151: (2009); Measuring relays and protection equipment – Part 151: Functional requirements for over/under current protection.

6 IEEE Std C37.112-2018 *IEEE Standard for Inverse-Time Characteristics Equations for Overcurrent Relays*. Piscataway: IEEE Standards Association, IEEE.

7 IEEE Committee Report (1996) "High Impedance Fault Detection Technology" IEEE Power System Relaying and Control Committee On Line at http://www.pes-psrc.org/kb/published/reports/High_Impedance_Fault_Detection_Technology.pdf

8 IEEE Committee Report (2014), "Considerations in Choosing Directional Polarizing Methods for Ground Overcurrent Elements in Line Protection Applications" IEEE PES Resource Center PS-TR33

9 Blackburn, J.L. (1952). Ground relay polarization. *AIEE Trans. PAS* **71**: 1088–1095.

5

Nonpilot Distance Protection of Transmission Lines

5.1 Introduction

Distance relays are normally used to protect transmission lines [1]. They respond to the impedance between the relay location and the fault location. As the impedance per mile of a transmission line is fairly constant, these relays respond to the distance to a fault on the transmission line—and hence their name. As will be seen shortly, under certain conditions, it may be desirable to make distance relays respond to some parameter other than the impedance, such as the admittance or the reactance, up to the fault location. Any of the relay types described in Chapter 2 can be made to function as a distance relay by making appropriate choices of their design parameters. The R–X diagram is an indispensable tool for describing and analyzing a distance relay characteristic, and we will examine it initially with reference to a single-phase transmission line. Similar principles are applicable in case of a three-phase transmission line, provided that appropriate voltages and currents are chosen to energize the distance relay. This matter of energizing voltages and currents in three-phase systems will be considered in detail later.

5.2 Stepped distance protection

Before describing the specific application of stepped distance protection, the definitions of underreach and overreach must be addressed. "Underreaching" protection is a form of protection in which the relays at a given terminal do not operate for faults at remote locations on the protected equipment [2]. This definition states that the relay is set so that it will not see a fault beyond a given distance (e.g., an instantaneous relay should not see the remote bus, as discussed in Section 4.4 of Chapter 4). The distance relay is set to underreach the remote terminal. The corollary to this definition, of course, is that the relay will see faults less than the setting. "Overreaching" protection is a form of protection in which the relays at one terminal operate for faults beyond the next terminal. They may be constrained from tripping until an incoming signal from a remote terminal has indicated whether the fault is beyond the protected line section [2]. Note the added restriction placed on overreaching protection to avoid loss of coordination.

Power System Relaying, Fifth Edition. Stanley H. Horowitz, Arun G. Phadke and Charles F. Henville.
© 2023 John Wiley & Sons Ltd. Published 2023 by John Wiley & Sons Ltd.

The zone of distance relays is open at the far end. In other words, the remote point of reach of a distance relay cannot be precisely determined, and some uncertainty about its exact reach must be accepted. This uncertainty of reach is typically about 5% of the setting. Referring to Figure 5.1a, the desired zone of protection is shown with a dotted line. The ideal situation would be to have all faults within the dotted area trip instantaneously. Owing to the uncertainty at the far end, however, to be sure that we do not overreach the end of the line section, we must accept an underreaching zone (zone 1). It is customary to set zone 1 between 85 and 90% of the line length and to be operated instantaneously. It should be clear that zone 1 alone does not protect the entire transmission line: the area between the end of zone 1 and bus B is not protected. Consequently, the distance relay is equipped with another zone, which deliberately overreaches beyond the remote terminal of the transmission line.

This is known as zone 2 of the distance relay, and it must be slowed down so that for faults in the next line section (F_2 in Figure 5.1a), zone 1 of the next line is allowed to operate before zone 2 of the distance relay at A. This **coordination delay** for zone 2 is usually of the order of 0.3 s, for the reasons explained in Chapter 4.

The reach of the second zone is generally set to at least 120–150% of the line length AB. It must be borne in mind that zone 2 of relay R_{ab} must not reach beyond zone 1 of relay R_{bc}, otherwise some faults may exist simultaneously in the second zones of R_{ab} and R_{bc}, and may lead to unnecessary tripping of both lines. This concept, of coordination by distance as well as by time, leads to a nesting of the zones of protection, and is illustrated in Figure 5.1b.

It should be noted that the second zone of a distance relay also backs up the distance relay of the neighboring line. However, this is true for only part of the neighboring line, depending upon how far the second zone reaches. In order to provide a backup function for the entire line, it is customary to provide yet another zone of protection for the relay at A. This is known as the third zone of protection, and usually extends to 120–180% of the next line section. The third zone must coordinate in time and distance with the second zone of the

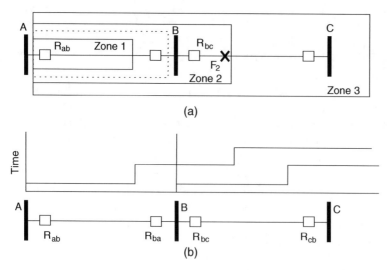

(a)

(b)

Figure 5.1 (a) and (b) Three-zone step distance relaying to protect 100% of a line and back up the neighboring line.

neighboring circuit, and usually the operating time of the third zone is of the order of 1 s. The three zones of protection of the two line sections AB and BC are shown in Figure 5.1b.

It should also be mentioned that it is not always possible to have acceptable settings for the two overreaching zones of distance relays. Many of these issues will be discussed in greater detail in the later sections. However, it is worth noting some of the limiting causes at this time. First, a complication is caused by dissimilar lengths of adjacent lines. If the length of a downstream line is less than 20% of the line being protected, its zone 2 will certainly overreach the first zone of the shorter line. Similarly, the zone 3 of the first line may overreach the zone 2 of the next line. The guidelines for setting the reach of zones mentioned in the earlier text must be considered to be approximate, and must be adjusted to meet a specific situation at hand. Zone 3 was originally applied as a remote backup to zones 1 and 2 of an adjacent line in the event that a relay or breaker failure prevented clearing the fault locally [3]. The reach setting, however, is a complex problem and is the subject of many ongoing studies and suggestions that will be discussed in detail in this chapter and Chapters 10 and 11. Briefly, however, the zone 3 characteristic must provide protection against faults, but should not operate for normal, albeit unusual, system conditions such as heavy loads or stability swings. Computer relaying makes provision for identifying heavy loads or stability swings through its load encroachment feature that is discussed further in Chapter 11. Another consideration is the effect of the fault current contributions from lines at the intermediate buses. This is the problem of infeed, and will be discussed in greater detail later.

Example 5.1   Consider the transmission system shown in Figure 5.2. The relay R_{ab} is to be set to protect the line AB and back up the two lines BC and BD. The impedances of the three lines are as shown in Figure 5.2. (Note that these impedances are in primary ohms—i.e., actual ohms of the transmission lines. Normally, the settings are expressed in secondary ohms, as will be explained in Section 5.3.) Zone 1 setting for R_{ab} is $0.85 \times (4 + j30)$ or $(3.4 + j25.5)\,\Omega$. Zone 2 is set at $1.2 \times (4 + j30)$ or $(4.8 + j36)\,\Omega$. Since the relay R_{ab} must back up relays R_{bc} and R_{bd}, it must reach beyond the longer of the two lines. Thus, zone 3 is set at $[(4 + j30) + 1.5 \times (7 + j60)]$ or at $(14.5 + j120)\,\Omega$. The time delays associated with the second and third zones should be set at about 0.3 and 1.0 s, respectively.

It should be noted that if one of the neighboring lines, such as line BD, is too short, then the zone 2 setting of the relay R_{ab} may reach beyond its far end. For the present case, this would happen if the impedance of line BD is smaller than $[(4.8 + j36) - (4.0 + j30)] = (0.8 + j6)\,\Omega$. In such a case, one must set zone 2 to be a bit shorter to make sure that it does not overreach zone 1 of R_{bd}, or, if this is not possible, zone 2 of the relay R_{ab} may be set longer

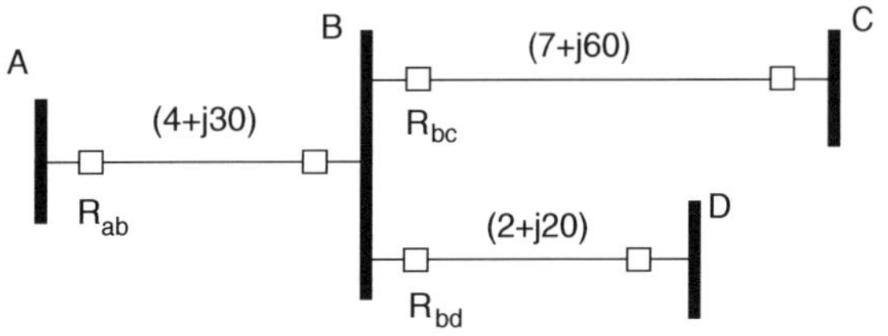

Figure 5.2   System for Example 5.1.

than zone 2 of relay R_{bd} or it may be dispensed with entirely, and only zone 3 may be employed as a backup function for the two neighboring lines.

The control circuit connections to implement the three-zone distance relaying scheme are shown in Figure 5.3. Figure 5.3a is a legacy implementation with electromechanical relays. In that implementation, the seal-in unit contact shown is typical for all three phases, and the seal-in coil may be combined with the target coils in some designs. The three distance-measuring elements Z_1, Z_2, and Z_3 close their contacts if the impedance seen by the relay is inside their respective zones. The zone 1 contact activates the breaker trip coil(s) immediately (i.e., with no intentional time delay), whereas the zones 2 and 3 contacts energize the two timing devices T_2 and T_3, respectively. Once energized, these timing devices close their contacts after their timer settings have elapsed. These timer contacts also energize the breaker trip coil(s). Should the fault be cleared before the timers run out, Z_2, Z_3, T_2, and T_3 will reset as appropriate in a relatively short time (about 1–4 ms).

Figure 5.3b shows a simplified implementation with a modern digital protection system. This figure is a hybrid logic diagram and control circuit connection diagram. It shows the logical implementation of the three-zone tripping and timing. It also shows the protection

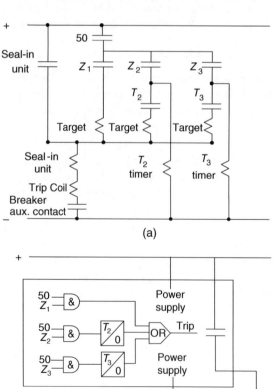

Figure 5.3 Control circuits for a three-zone step distance protection: (a) electromechanical implementation; (b) simplified digital implementation.

(a)

(b)

system power supply and physical trip output relay contact that trips the circuit breaker. The simplified diagram does not show the breaker auxiliary contact input to the protection system that may seal in the trip output. Instead of a sealed-in trip output, some implementations may have a logical seal-in for a minimum trip duration output and conditioned with a minimum current detector. Also not shown in the simplified logic diagram are local trip indicators, and internal event recording facilities and possible other protection and control functions.

Note also in both panels (a) and (b) of Figure 5.3 that a supervising instantaneous overcurrent function (device 50) is shown. This function prevents any tripping by a distance function unless the current is above a level determined by the setting of function 50. On a simplistic level, the distance function trips if the measured value of (V/I) is less than a set amount. The current supervision function ensures that there is a certain minimum current flow before a fault is declared. Without such supervision, there could be a concern, for instance, that the function might trip when the line is de-energized as the voltage drops to zero and any noise in the current signal could cause an undesirable trip, since ($V/I = 0/0$) is indeterminate.

We should remember that the zone settings for zones 2 and 3 are affected by the contributions to the fault current made by any lines connected to the intervening buses, that is, buses B and C in Figure 5.1. This matter will be dealt with in the discussion of infeed and outfeed in Section 5.6. The problem is caused by the different currents seen by the relays as a result of the system configuration. As shown in Example 5.6, the operating currents in the upstream relays change significantly due to infeed or outfeed. Note that Example 5.6 does not discuss the effect of infeed our outfeed at the remote terminal, but from a third terminal on the protected line. However, similar considerations apply on the two-terminal line as well due to the impact of infeed or outfeed at the remote terminal on the impedances seen by overreaching zone 2 or zone 3 elements.

5.3   *R–X* diagram

In general, all electromechanical relays respond to one or more of the conventional torque-producing input quantities: (i) voltage, (ii) current, (iii) product of voltage, current, and the angle θ between them, and (iv) a physical or design force such as a control spring [4]. Similar considerations hold for solid-state relays as well. For the product-type relay, such as the distance relay, analyzing the response of the relay for all conditions is difficult because the voltage varies for each fault or varies for the same fault but with different system conditions.

To resolve this difficulty, it is common to use an *R–X* diagram to both analyze and visualize the relay response. By utilizing only two quantities, R and X (or Z and θ), we avoid the confusion introduced using the three quantities E, I, and θ. There is an additional significant advantage, in that the *R–X* diagram allows us to represent both the relay and the system on the same diagram.

Consider an ideal (zero resistance) short circuit at location F in the single-phase system shown in Figure 5.4. The distance relay under consideration is located at line terminal A. The primary voltage and current at the relay location are related by

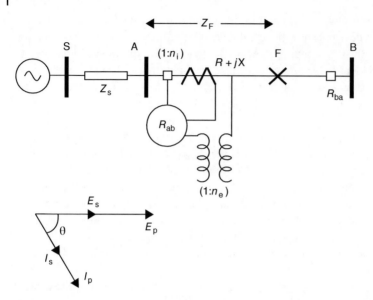

Figure 5.4 Voltage, current, and impedance as seen by the relay.

$$Z_{f,p} = \frac{E_p}{I_p}, \tag{5.1}$$

where the subscript "p" represents primary quantities. In terms of the secondary quantities of voltage and current transformers (CTs), the relay sees the primary impedance $Z_{f,p}$ as $Z_{f,s}$, where

$$Z_{f,s} = \frac{E_s}{I_s} = Z_{f,p}\frac{n_i}{n_e}, \tag{5.2}$$

where n_i and n_e are the CT and voltage transformer (VT) turns ratios. It is customary to suppress the subscript "s," with the understanding that the secondary quantities are always implied. Thus, we will mean $Z_{f,s}$ when we use Z_f.

Example 5.2 Consider a single-phase distance relaying system utilizing a CT with a turns ratio of 500:5 and a VT with a turns ratio of 20 000:69.3. Thus, the CT ratio n_i is 100, while the VT ratio n_e is 288.6. The impedance conversion factor n_i/n_e for this case is (100/288.6), or 0.3465. All primary impedances must be multiplied by this factor to obtain their secondary values. Thus, the line in Example 5.1 with an impedance of $(4 + j30)$ Ω primary would appear to be $(1.386 + j10.395)$ Ω secondary.

The zone 1 setting for this line would be 85% of this impedance or $(1.17 + j8.84)$ Ω secondary. Of course, the actual setting used would depend upon the nearest value that is available on a given relay.

Although we have defined Z_f under fault conditions, it must be borne in mind that the ratio of E and I at the relay location is an impedance under all circumstances, and when a fault occurs, this impedance assumes the value Z_f. In general, the ratio E/I is known as the apparent impedance "seen" by the relay. This impedance may be plotted as a point on the complex *R–X* plane. This is the plane of (apparent) secondary ohms. One could view the impedance as the voltage phasor, provided that the current is assumed to be the reference phasor, and of unit magnitude. This way of looking at the apparent impedance seen by a relay as the voltage phasor at the relay location is often very useful when relay responses to changing system conditions are to be determined. For example, consider the apparent impedance seen by the relay when there is normal power flow in the transmission line. If the load current is of constant magnitude, and the sending-end voltage at the relay location is constant, the corresponding voltage phasor, and hence the impedance, will describe a circle in the *R–X* plane. Lighter loads—meaning a smaller magnitude of the current—produce circles of larger diameters. Similarly, when the load power factor is constant, the corresponding locus of the impedance is a straight line through the origin. Figure 5.5 shows these contours for varying load current magnitudes and power factors. Note that when the real power flows into the line, the corresponding apparent impedances lie in the right half of the plane, while a reversed power flow maps into the left half-plane. Similarly, lagging power-factor load plots in the upper half-plane, while a leading power-factor load plots in the lower half-plane. Zero power transfer corresponds to points at infinity. A line open at the remote end will have leading reactive current, and hence the apparent impedance will map at a large distance along the negative *X* axis.

Now consider the fault at location F as shown in Figure 5.4. The corresponding apparent impedance is shown at F in Figure 5.5. As the location of the fault is moved along the transmission line, the point F moves along the straight line AB in Figure 5.5. Thus, the transmission line as seen by the relay maps into the line AB in the *R–X* plane. The line AB makes an angle θ with the *R* axis, where θ is the impedance angle of the transmission line. (For an overhead transmission line, θ lies between 70° and 88°, depending upon the system voltage, the larger angles being associated with higher transmission voltages.) When the fault is on the transmission line, the apparent impedance plots on the line AB, for all other faults or

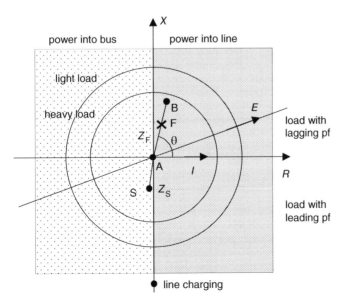

Figure 5.5 *R–X* diagram as a special case of the phasor diagram (pf, power factor).

loading conditions, the impedance plots away from the line AB. Often it is convenient to plot the source impedance Z_s also on the R–X diagram, as shown in Figure 5.5.

Example 5.3   Let the rated load for the single-phase transmission line shown in Figure 5.4 be 8 MVA. This corresponds to 400 A at the rated voltage of 20 000 V. The apparent impedance corresponding to this load is $(20\,000/400) = 50\,\Omega$ primary. In terms of secondary ohms, this impedance becomes $50 \times 0.3465 = 17.32\,\Omega$. Thus, a load of 8 MVA at 0.8 power factor (pf) lagging is $17.32 \times (0.8 + j0.6) = (13.86 + j10.4)\,\Omega$ secondary. This is shown as L_1 in Figure 5.6. A load of 8 MVA with a leading power factor of 0.8 is $(13.86 - j10.4)\,\Omega$ secondary, which maps as point L_2. Similarly, 8 MVA flowing from B to A maps into L_3 and L_4 for leading and lagging power factors, respectively.

The line impedance of $(1.39 + j10.4)\,\Omega$ secondary maps into point B, while the zone 1 setting of $(1.17 + j8.84)\,\Omega$ maps into the point Z_{1a}. A similar relay located at B would have its zone 1 map at Z_{1b}. If we assume the equivalent source impedances as seen at buses A and B to be $j10$ and $j8\,\Omega$ primary, respectively, they will be $j3.46$ and $j2.77\,\Omega$ secondary, respectively, as shown by points S_1 and S_2 in Figure 5.6. If the line charging current is 15 A, the apparent impedance seen by the relay at A when the breaker at terminal B is open is $-j(20\,000/15) = -j\,1333\,\Omega$ primary, or $-j461.9\,\Omega$ secondary. This is shown as the point C, on a telescoped y axis in Figure 5.6. The zones of protection of a relay are defined in terms of its impedance, and hence it is necessary that they cover areas in the immediate neighborhood of the line AB. As the load on the system increases, the possibility of it encroaching upon the protection zones becomes greater. Ultimately, at some values of the load, the relay is in danger of tripping. The R–X diagram offers a convenient method of analyzing whether this is the case. A fuller account of the loadability of a distance relay is considered in Section 5.11.

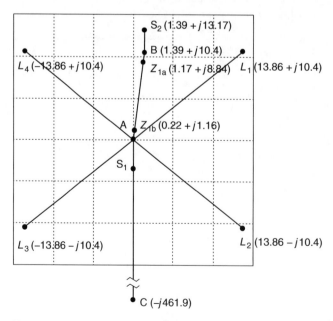

Figure 5.6 R–X diagram for Example 5.3.

5.4 Three-phase distance relays

On a three-phase power system, there are 10 distinct types of possible faults: a three-phase fault, three phase-to-phase faults, three single-phase-to-ground faults, and three double-phase-to-ground faults. The equations that govern the relationship between voltages and currents at the relay location are different for each of these faults. We should therefore expect that it will take several distance relay functions, each of them measuring a different pair of voltage and current inputs to measure the distance to the fault correctly. It is a fundamental principle of distance relaying that, regardless of the type of fault involved, the voltage and current measured by the appropriate relay are such that the relay will measure the positive-sequence impedance to the fault [5, 6]. Once this is achieved, the zone settings of all relays can be based upon the total positive-sequence impedance of the line, regardless of the type of the fault. We will now consider various types of fault, and determine the appropriate voltage and current inputs to be used for the distance relay functions responsible for each of these fault types.

5.4.1 Phase-to-phase faults

Consider a fault between phases b and c of a three-phase transmission line. We may consider this to be the fault at location F in Figure 5.4, provided we view that figure as a one-line diagram of the three-phase system. The symmetrical component representation for this fault is shown in Figure 5.7. The positive- and negative-sequence voltages at the fault bus are equal, and are given by

$$E_{1f} = E_{2f} = E_1 - Z_{1f}I_1 = E_2 - Z_{1f}I_2, \tag{5.3}$$

where E_1, E_2, I_1, and I_2 are the symmetrical components of voltages and currents at the relay location, and the positive- and negative-sequence impedances of the transmission line are equal. It follows from Equation (5.3) that

$$\frac{E_1 - E_2}{I_1 - I_2} = Z_{1f}. \tag{5.4}$$

Figure 5.7 Symmetrical component circuit for b–c fault.

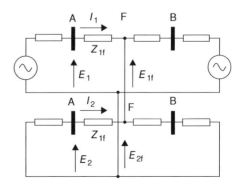

Also, since the phase quantities at the relay location are given by

$$E_b = E_0 + \alpha^2 E_1 + \alpha E_2 \text{ and } E_c = E_0 + \alpha E_1 + \alpha^2 E_2, \tag{5.5}$$

then

$$(E_b - E_c) = (\alpha^2 - \alpha)(E_1 - E_2) \text{ and } (I_b - I_c) = (\alpha^2 - \alpha)(I_1 - I_2). \tag{5.6}$$

Substituting Equation (5.6) in Equation (5.4) gives

$$\frac{E_b - E_c}{I_b - I_c} = \frac{E_1 - E_2}{I_1 - I_2} = Z_{1f}A. \tag{5.7}$$

Thus, a distance function which measures the line-to-line voltage between phases b and c, and which measures the difference between the currents in the two phases, will measure the positive-sequence impedance to the fault, when a fault between phases b and c occurs. Similar analysis will show that, for the other two types of phase-to-phase fault, when the corresponding voltage and current differences are measured by the functions, the positive-sequence impedance to the fault will be measured.

The symmetrical component diagram for a phase-b-to-c-to-ground fault is shown in Figure 5.8. It should be clear from this figure that for this fault also, the performance equations for the positive- and negative-sequence parts of the equivalent circuit are exactly the same as those for the b-to-c fault. Finally, for a three-phase fault at F, the symmetrical component diagram is as shown in Figure 5.9. For this case,

$$E_1 = E_a = Z_{1f}I_1 = Z_{1f}I_a,$$
$$E_2 = E_0 = 0, \tag{5.8}$$
$$I_2 = I_0 = 0.$$

Also, for this case, $E_a = E_1$, $E_b = \alpha^2 E_1$, and $E_c = \alpha E_1$, and similar relations hold for the phase currents. Consequently, for a three-phase fault,

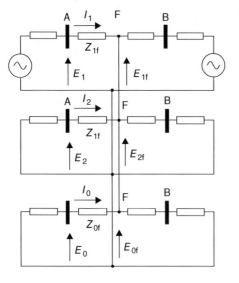

Figure 5.8 Symmetrical component circuit for b–c–g fault.

Figure 5.9 Symmetrical component circuit for a three-phase fault.

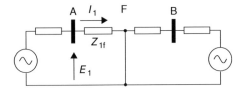

$$\frac{E_a - E_b}{I_a - I_b} = \frac{E_b - E_c}{I_b - I_c} = \frac{E_c - E_a}{I_c - I_a} = Z_{1f}. \tag{5.9}$$

The differences of phase voltages and currents used in Equation (5.9) are known as "delta" voltages and currents, and we see that functions measuring the delta voltages and currents respond to the positive-sequence impedance to a multiphase fault. A complement of three-phase distance functions covers the seven multiphase faults between them. For double-phase or double-phase-to-ground faults, one of the three functions measures the positive-sequence impedance to the fault, while for a three-phase fault, all three functions measure the correct impedance. The connections for legacy types of phase relays are shown schematically in Figure 5.10a. Modern digital protection systems are connected to all three phase VTs and CTs which are connected in wye configuration as shown in Figure 5.10b. The delta quantities for the different functions are computed mathematically.

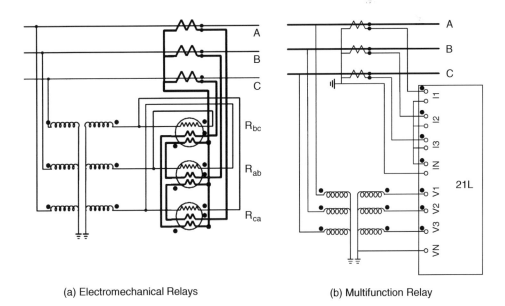

(a) Electromechanical Relays   (b) Multifunction Relay

Figure 5.10 (a) and (b) Current transformer and voltage transformer connections for distance relays for phase faults.

5.4.2 Ground faults

For a fault between phase a and ground, the symmetrical component connection diagram is as shown in Figure 5.11. The voltages and currents at the relay location for this case are given by

$$
\begin{aligned}
E_{1f} &= E_1 - Z_{1f}I_1, \\
E_{2f} &= E_2 - Z_{1f}I_2, \\
E_{0f} &= E_0 - Z_{0f}I_0.
\end{aligned}
\tag{5.10}
$$

The phase a voltage and current can be expressed in terms of the symmetrical components, and the voltage of phase a at the fault point can be set equal to zero:

$$
\begin{aligned}
E_{af} &= E_{0f} + E_{1f} + E_{2f} \\
&= (E_0 + E_1 + E_2) - Z_{1f}(I_1 + I_2) - Z_{0f}I_0 = 0
\end{aligned}
\tag{5.11}
$$

Or

$$
E_a - Z_{1f}I_a - (Z_{0f} - Z_{1f})I_0 = 0,
\tag{5.12}
$$

where I_a has been substituted for the sum $(I_0 + I_1 + I_2)$ in Equation (5.12). Finally, a new current I'_a is defined as follows:

$$
I'_a = I_a + \frac{Z_{0f} - Z_{1f}}{Z_{1f}}I_0 = I_a + \frac{Z_0 - Z_1}{Z_1}I_0 = I_a + mI_0,
\tag{5.13}
$$

where, in Equation (5.13), Z_0 and Z_1 are the zero- and positive-sequence impedances of the entire line. The factor m is known as a compensation factor, which compensates the phase current for the mutual coupling between the faulted phase and the other two unfaulted

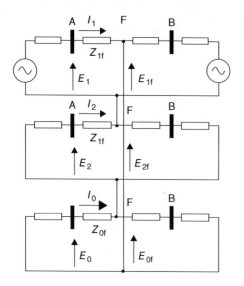

Figure 5.11 Symmetrical component circuit for an a–g fault.

Figure 5.12 Current transformer and voltage transformer connections for distance relays for ground faults.

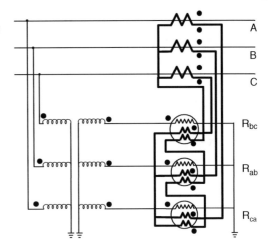

phases. The compensation factor m is also sometimes known as KN or $K0$. It follows from Equation (5.12) that for a phase-a-to-ground fault,

$$\frac{E_a}{I'_a} = Z_{1f}. \tag{5.14}$$

Thus, if the distance relay is energized with the phase a voltage and the compensated phase a current, it also measures the positive-sequence impedance to the fault. The factor m for most overhead transmission lines is a real number, and varies between 1.5 and 2.5. A good average value for m is 2.0, which corresponds to Z_0 of a transmission line being equal to $3Z_1$. As in the case of phase relays, it takes three ground distance relays to cover the three single-phase-to-ground faults. It should be noted that for a three-phase fault, the compensated phase current becomes I_a, since there is no zero-sequence current for this fault. For this case, Equation (5.14) is identical to Equation (5.18), and hence the three ground distance relays also measure the correct distance to the fault in the case of a three-phase fault. A schematic connection diagram for the three electromechanical ground distance relays is shown in Figure 5.12. A full complement of three zones of electromechanical phase and ground distance relays will require six distance-measuring elements connected as shown in Figures 5.10 and 5.12. A single modern multifunction connected as shown in Figure 5.10b would perform the same function, and more.

Example 5.4 Consider the simple system represented by the one-line diagram in Figure 5.13. The system nominal voltage is 13.8 kV, and the positive- and zero-sequence impedances of the two elements are as shown in the figure. The zero-sequence impedances are given in parentheses. We will verify the distance calculation Equations (5.9) and (5.14) for three-phase, phase-to-phase, and phase-to-ground faults.

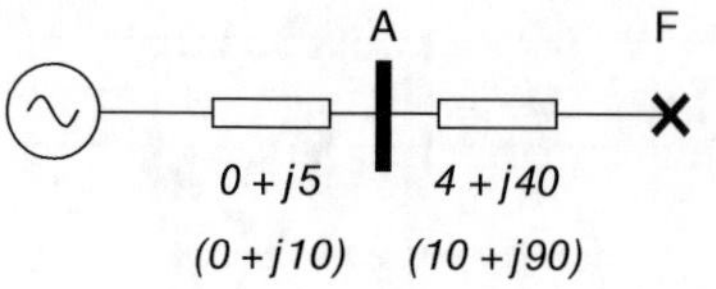

Figure 5.13 System for fault impedance calculation.

5.4.2.1 Three-phase fault

For this case, only the positive-sequence current exists, and is also the phase a current. It is given by

$$I_a = I_1 = \frac{7967.4}{4 + j45} = 176.36\angle - 84.92°,$$

$7967.4 = (13\,800/\sqrt{3})$ is the phase-to-neutral voltage. The phase a voltage at the relay location is given by

$$E_a = E_1 = 7967.4 - j5 \times 176.36\angle - 84.92° = 7089.49\angle - 0.63°.$$

Thus, the fault impedance seen by the relay in this case is

$$Z_f = \frac{E_a - E_b}{I_a - I_b} = \frac{E_a}{I_a} = \frac{7089.49\angle - 0.63°}{176.36\angle - 84.92°} = 4 + j40\ \Omega.$$

5.4.2.2 Phase-to-phase fault

For a b–c fault,

$$I_1 = -I_2 = \frac{7967.4}{2 \times (4 + j45)} = 88.18\angle - 84.92°.$$

Also, $I_b = -I_c = I_1(\alpha_2 - \alpha) = -j\sqrt{3}I_1 = 152.73\angle - 174.92°$. In addition, $(I_b - I_c) = 305.46 \angle - 174.92°$. The positive- and negative-sequence voltages at the relay location are given by

$$E_1 = 7967.4 - j5 \times 88.18\angle - 84.92° = 7528.33\angle - 0.3°,$$
$$E_2 = j5 \times 88.18\angle - 84.92° = 44090\angle 5.08°,$$

and the voltages of phases b and c at the relay location are

$$E_b = \alpha^2 E_1 + \alpha E_2 = 7528.33\angle - 120.3° + 440.90\angle 125.08°$$
$$= -4051.3 - j6139.3,$$

$$E_c = \alpha E_1 + \alpha^2 E_2 = 7528.33\angle 119.7° + 440.90\angle - 114.9°$$
$$= -3916.09 + j6139.3.$$

Thus, $E_b - E_c = 12\,279.37\angle - 90.63°$, and

$$\frac{E_b - E_c}{I_b - I_c} = \frac{12\,279.37\angle - 90.36°}{305.46\angle - 174.92°} = 4 + j40\ \Omega.$$

5.4.2.3 Phase-a-to-ground fault

For this fault, the three symmetrical components of the current are equal:

$$I_1 = I_2 = I_0 = \frac{7967.4}{(0 + j10) + 2 \times (0 + j5) + (10 + j90) + 2 \times (4 + j40)}$$
$$= 41.75\angle - 84.59°.$$

The symmetric components of the voltages at the relay location are

$$E_1 = 7967.4 - j5 \times 41.75\angle - 84.59° = 7759.58 - j19.68,$$
$$E_2 = -j5 \times 41.75\angle - 84.59° = -207.82 - j19.68,$$
$$E_0 = -(0 + j10) \times 41.75\angle - 84.59° = -415.64 - j39.36.$$

And the phase a voltage and current at the relay location are

$$E_a = E_1 + E_2 + E_0 = 7136.55\angle - 0.63°,$$
$$I_a = I_1 + I_2 + I_0 = 125.25\angle - 84.59°.$$

The zero-sequence current compensation factor m is given by

$$m = \frac{Z_0 - Z_1}{Z_1} = \frac{1 - j90 - 4 - j40}{4 + j40} = 1.253\angle - 1.13°,$$

and the compensated phase a current is $I'_a = I_a + mI_0 = 177.54\angle - 84.92°$; and, finally,

$$\frac{E_a}{I'_a} = \frac{7136.55\angle - 0.63°}{177.54\angle - 84.92°} = 4 + j40 \ \Omega.$$

5.4.3 Relays in unfaulted phases

Although there are three phase distance and three ground distance functions in use for protection against all 10 types of fault on a three-phase system, only one of these functions measures the correct distance to the fault for a specific fault type, and it is interesting to see what the remaining functions measure. In general, they measure impedances that are greater than the impedance to the fault. However, under certain conditions—such as for close-in faults—the distance measured by the other functions may be such that an erroneous operation of some of the other functions may result. Of course, if the protection system is designed to trip all three phases for every fault, the fact that one or more functions may operate for one fault is of no practical significance, although erroneous targets produced by some of the functions may lead to unnecessary confusion in postmortem analysis of the fault, and is to be avoided if at all possible. Only when single-phase tripping for phase-to-ground fault is to be used does it become a matter of concern if a phase distance function responds (erroneously) to a ground fault and causes a three-phase trip. We will therefore consider the operation of the phase distance functions for a fault on phase a. Similar analysis can be carried out for all the distance functions in the unfaulted phases for each of the fault types [7].

Consider the phasor diagram for a phase-a-to-ground fault on a radial system as shown in Figure 5.14. The pre-fault voltages and currents are represented by the unprimed quantities, while the faulted quantities are shown with primes. This being an unloaded radial system,

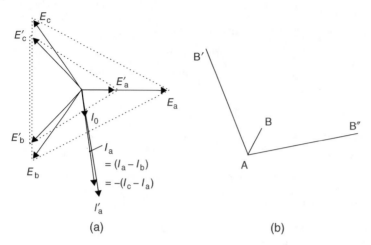

Figure 5.14 Phasor diagram for phase-a-to-ground fault: (a) voltages and currents and (b) apparent impedances.

no pre-fault current exists. The phase a voltage at the relay location drops to a small value during the fault, while the current in phase a lags the phase a voltage by the impedance angle of the combination $(Z_1 + Z_2 + Z_0)$. The voltages of the unfaulted phases will change in magnitude and angle as shown in the phasor diagram. The compensated phase current as defined by Equation (5.13) for this case (since $I_a = 3I_0$) is given by $I'_a = (1 + m/3)I_a$. The phase distance functions measure delta currents, and since $I_b = I_c = 0$, the three delta currents for this fault are $(I_a - I_b) = I_a$, $(I_c - I_a) = -I_a$, and $(I_b - I_c) = 0$. These delta currents and voltages are also shown in Figure 5.14.

Remember that the impedance seen by any distance function is equal to its voltage when the corresponding current is taken as a reference phasor of unit magnitude. Since the delta current for the b–c function is zero for this fault, it sees an infinite impedance and will not misoperate. As the delta currents for the a–b and c–a relays are I_a and $-I_a$, respectively, we can visualize their response to this fault by redrawing the delta voltages E_{ab} and E_{ca} with I_a and $-I_a$ as the reference unit phasors and adjusting (increase) the magnitude of the two voltage phasors by the factor $(1 + m/3)$, in relation to the phase a voltage. This last adjustment is necessary because the phase a voltage is seen as the fault impedance with $I'_a = (1 + m/3)I_a$ as the unit of measurement, whereas E_{ab} and E_{ca} are the fault impedances seen by those two relays with $\pm I_a [= \pm I'_a/(1 + m/3)]$ as the units of measurement. Thus, the impedances seen by the a–g, a–b, and c–a relays for the a–g fault are seen to be AB, AB$'$, and AB$''$, respectively, as shown in Figure 5.14. It should be clear that for a ground fault near the relay location, the a–b and c–a relays may misoperate, if the protection zone covers AB$'$ and AB$''$ for small values of AB.

The analysis presented in the earlier text must be modified to include the effect of pre-fault load. The load current will change the current phasors and hence the impedances seen by the relays. Such effects are of less importance in a qualitative discussion, and the reader is referred to Reference [7] for a more detailed treatment of the subject.

5.4.4 Fault resistance

In developing distance relay equations, we assumed that the fault under consideration was an ideal (i.e., zero resistance) short circuit. In reality, for multiphase faults, the fault arc will be between two high-voltage conductors, whereas for ground faults, the fault path may consist of an electrical arc between the high-voltage conductor and a grounded object—such as the shield wire, or the tower itself or a tree.

In any case, the fault path may have a resistance in it, which may consist of an arc resistance or an arc resistance in series with the tower footing or other resistance in the case of a ground fault. In the case of lines with a lightning shield wire, the tower footing resistance is practically constant during the fault and ranges between 5 and 50 Ω. However, in cases where there is no shield wire, and the towers are on rocky ground, the footing resistance may be much higher even up to hundreds of ohms [8]. Also, in the case of a fault to a tree or a tall object such as a crane, the fault resistance may also be much higher. The voltage across an arc is roughly proportional to the arc length. This means that the arc resistance is roughly proportional to the length and inversely proportional to the current flowing in it. Note, however, that the length changes with time due to wind, convection, and magnetic forces as the fault current continues to flow. During the early period of the arc, say in the first few milliseconds, the arc length could be assumed to be the distance between conductors. However, as the arc channel gets elongated in time, the arc resistance increases. For relaying considerations, it is sometimes assumed that the arc resistance is given by an empirical formula [8]:

$$R_{\text{arc}} = 1804 \times \frac{L}{I}, \tag{5.15}$$

where 1804 is a constant,
L is the arc length in meters, and
I is the fault current in the arc in amperes.

Since the length changes very shortly after the fault starts are not possible to estimate, the arc resistance cannot usually be estimated confidently. For multiphase faults, the resistance is often small enough to be neglected, or calculated using Equation (5.15) with an arc length assumed to be determined by the distance between conductors multiplied by a dependability factor [8]. For single-line–to-ground faults to the tower, if lightning shield wire is applied, the footing resistance maybe assumed to be the dominating factor (given above as up to 50 Ω). For faults to foreign objects such as trees, the resistance will be unknown. Ground overcurrent protection will usually provide more sensitive protection than distance relays. Setting of ground overcurrent protection was discussed in Chapter 4.

The fault resistance introduces an error in the fault distance estimate, and hence may create an unreliable operation of a distance relay. Consider the single-phase transmission system shown in Figure 5.15a, and assume that the fault resistance is equal to R_f. If the contribution to the fault from the remote end is I_r, the fault current $I_f = I + I_r$, and the voltage at the relay location is given by

$$E = Z_f I + R_f(I + I_r). \tag{5.16}$$

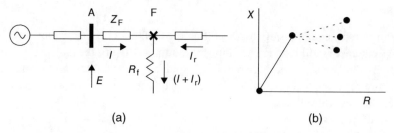

Figure 5.15 (a) and (b) Fault path resistance and its effect on the R–X diagram.

The apparent impedance Z_a seen by the relay is

$$Z_a = \left(\frac{E}{I}\right) = Z_f + R_f\left(\frac{I_r}{I} + 1\right). \tag{5.17}$$

Since I_r may not be in phase with I, the fault resistance may contribute an error to the resistance as well as the reactance of the faulted line segment. This is illustrated in the R–X diagram in Figure 5.15b. In order to accommodate the resistance in the fault path, it is necessary to shape the trip zone of a distance relay in such a manner that the region surrounding the apparent impedance is included inside the zone. It will be seen in Section 5.11 that different types of distance relays have differing ability of accommodating the fault resistance. It should be remembered that a larger area for the protection zone in the R–X plane accommodates greater fault path resistance, while it also affects the loadability of the relay.

Example 5.5 Assume a single-phase circuit as shown in Figure 5.15a, and let the line impedance to the fault point be $(4 + j40)\ \Omega$, while the fault resistance is $10\ \Omega$. Let the current to the fault in the line in question be $400 \angle -85°$ A, while the current contribution to the fault from the remote end is $600 \angle -90°$ A. Then, the apparent impedance seen by the relay, as given by Equation (5.17), is

$$Z_a = (4 + j40) + 10\left(1 + \frac{600\angle-90°}{400\angle-85°}\right) = 28.94 + j38.69\ \Omega.$$

If the receiving-end current contribution is in phase with the sending-end current, the error in Z_a will be in the real part only. That is not the case here, and hence the reactance also is in error.

5.5 Distance relay types

Distance protection functions may be classified according to the shape of their zones of operation. Traditionally, all zone shapes have been circular, because an electromechanical relay, with the torque shown in Equation (2.15) of Chapter 2, produces a circular boundary for the zones of operation. Some of the terminology used in describing the zones (e.g., "the line of maximum torque") dates back to the electromechanical origins of distance relays. However, far more complex zone shapes can be achieved with modern solid-state and

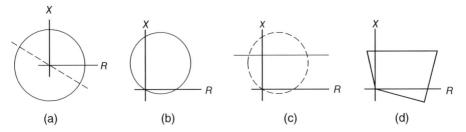

Figure 5.16 (a)–(d) Types of distance protection characteristics.

computer relays, although some of the older terminology continues to be used in describing the latter relays.

Four general distance protection types are recognized according to the shapes of their operating zones (characteristics): (a) impedance characteristic (with directional limiter), (b) admittance or mho characteristic, (c) reactance characteristic (with mho limiter), and (d) quadrilateral characteristic. These four relay characteristic shapes are illustrated in Figure 5.16. The impedance relay has a circular shape centered at the origin of the R–X diagram and is usually applied with a directional limiter to look forward or backward. The admittance (or mho) characteristic has a circular shape that passes through the origin. In some applications, the mho characteristic may be offset. The reactance characteristic has a zone boundary defined by a line parallel to the R axis. The zone would extend to infinity in three directions, so it is applied with a mho function as well to limit its reach. The quadrilateral characteristic, as the name implies, is defined by four straight lines. This last characteristic is only available in solid-state or computer relays. More complex shapes can be obtained using one or more of the abovementioned characteristics, in a logical combination to provide a composite tripping zone boundary.

5.6 Relay operation with zero voltage

It will be recalled that the generalized torque equation for an electromechanical relay, as given by Equation (2.15) in Chapter 2, is adapted to a directional relay or a mho relay by appropriate choices of various constants appearing in that equation. Thus, the balance point equation for a directional relay is equivalent to [$VI \sin (0 + \varphi) = 0$], whereas for a mho relay, the corresponding equation is [$V = IZ_r \sin (0 + \varphi)$]. At $V = 0$, both of these equations lead to uncertainties of operation. The directional relay equation is satisfied by any value of I and θ for $V = 0$, while for the mho relay equation, if V is zero, the angle between V and IZ_r cannot be determined, and the operation of the relay becomes uncertain. The region around the origin in both of these characteristics is poorly defined as to the performance of the relay (see Figure 2.13b,c of Chapter 2). Consequently, a zero-voltage fault is likely to be misjudged as to its direction by both of these relays.

It is possible to design relays that will overcome the problem of zero-voltage faults. A common technique is to provide a memorized pre-fault voltage to supplement the polarizing voltage. The memorized voltage will sustain the pre-fault voltage in the polarizing circuit for a short while after the occurrence of the zero-voltage fault. The phase angle of the voltage impressed by the memory circuit on the voltage signal is very close to that of the voltage before the occurrence of the fault. In the digital implementation, there is no difficulty in sustaining the memory polarization for a long time. However, there is a problem that if the system frequency changes quickly, whether or not a fault is still present, the memorized voltage phasor rotating at pre-fault frequency will not provide accurate measurement with fault current and voltage phasor at a different frequency. Unexpected operations of memory-polarized relays have been observed [9, 10]. In most implementations of memory polarization, it will only be used or sustained for any significant duration if the real-time polarizing voltage at the relay is very low. Even when memory polarization is in effect, it will only be effective for a duration of less than a fraction of a second.

Memory action requires that the pre-fault voltage seen by the relay is normal, or close to normal. If there is no pre-fault voltage in the primary circuit (as would be the case if the transmission line is being energized after having been de-energized for some time, and line-side potential is being used for relaying), no memory action is available. This is a common situation when live-tank circuit breakers are being used (Section 1.5 of Chapter 1). It should be remembered that a true zero-voltage fault is somewhat rare and can only be caused if grounding chains or switches are left connected by maintenance crews on unenergized power apparatus. To provide adequate protection against a zero-voltage fault, an instantaneous overcurrent relay can be used to reach just beyond the relay location. The impedance for a fault just beyond the reach of the instantaneous relay will provide enough of a voltage drop to provide reliable operation of the impedance unit. There is, of course, a setting problem with an instantaneous relay. Since it is nondirectional, there must be inherent discrimination between forward and backward faults.

Since the instantaneous relay is not directional, it may not be possible to set it to discriminate between faults in the protected zone and faults behind the relay. A common solution is to use an instantaneous overcurrent function that is normally inoperative, but is made operative as soon as the breaker closes. The function remains in the circuit for 10–15 cycles, allowing the breaker to trip in primary or breaker-failure time, after which the function is removed from service. It is, of course, assumed that there is little likelihood of a backward fault occurring during the short time that the instantaneous relay is in the circuit. This type of protection is called switch-on-to-fault (abbreviated as SOTF) protection and is usually applied when the relay is supplied with line-side VTs.

5.7 Polyphase relays

The distance functions discussed so far have all been essentially single-phase functions, that is, every fault will result in the operation of one or more functions depending upon the voltage and current inputs being measured by each function. It would seem that a polyphase function would be more appropriate than a collection of single-phase functions for protecting a three-phase system. Electromechanical polyphase relays have been described in the literature [11], and have been in use on three-phase systems for many years. Computer-based polyphase relays have also been developed [12]. Electromechanical polyphase relays utilize all three-phase voltages and currents to develop a torque that changes direction at the balance point (zone boundary) of the relay. A suitable combination of voltages and currents must be chosen, so that this torque reversal occurs for all types of fault that may occur on a three-phase system. Reference [11] cited in the earlier text describes the operation of a polyphase electromechanical relay in greater detail. One advantage of a polyphase relay is that for most zero-voltage faults, at least some of the voltages are not zero and a proper directionality is maintained by the relay. Of course, for a three-phase, zero-voltage fault, it is still necessary to include a memory action or a nondirectional overcurrent function as described in Section 5.7. Usually, two polyphase relays are needed to protect against phase and ground faults. In the case of computer-based polyphase relays [12], it is possible to use a single performance equation to cover all fault types.

5.7.1 Zone versus phase packaging

We have examined the individual elements of a three-phase distance relay protection scheme without regard to the way in which each of the elements is combined in one or more relay cases. Although the packaging has no effect on the protection of the line per se, as an historical issue it did have an impact on the cost of the installation and the complexity associated with testing and maintaining the relays.

In the historical context, a complete phase protection package consisted of an element for each pair of phases, that is, phases a– b, b–c, and c–a, and separate elements for zones 1, 2, and 3, plus the required timers and interfaces to the communication equipment. The timers and communication equipment were mounted separately, each in its own relay case, very often on separate panels or cabinets. The phase and zone connections could be combined in a variety of ways. All of the zones for a given pair of phases could be mounted in one relay case, referred to as "zone-packaged" or all of the phase pairs for a given zone could be mounted in one case and referred to as "phase-packaged." The advantages and disadvantages of the two schemes are related to the then common practice of testing and calibrating relays without taking the associated line out of service. That meant that during testing adequate protection must be left in service. With phase packaging, all of the zones of the pair of phases being tested, that is, the instantaneous zone 1 and the backup zones 2 and 3, were removed. The remaining phases, and their zones, remain in service. With zone packaging, each of the zones is removed, one at a time, for example, zone 1 of all of the phase pairs is out of service, then zone 2, and then zone 3. Unless the number of zones is not equal to 3, there is no difference in the number of relay cases between the two packaging schemes, but the panel arrangement may differ and, in association with the control switches and testing facilities, the wiring will differ. Ground protection was always provided by independent relays (either ground distance or directional overcurrent). Ground relays were mounted separately, so that they remained in service when any of the phase relays are tested, and all of the phase relays remained in service when the ground protection was tested. In addition, the testing facilities usually had provision for restoring some of the internal elements of a relay case to service while other elements are removed. The choice between the two packaging schemes is not clear, but usually depended upon the previous practices and personal preferences of the utility's engineering and testing personnel.

The discussion in the earlier text relates to historical packaging of electromechanical relays and before reliability regulations required redundant protection systems to be applied to bulk electric system transmission lines. It is not an issue with solid-state or digital relays applied in redundant systems. Solid-state relays have various protection functions on the same card or are connected to the same backplane within a common case. The testing facilities usually allow each element to be tested or calibrated separately without removing the other functions from service. Modern digital relays are installed as complete protection systems. The complete system operates as a single multifunction system that must be taken out of service for testing or maintenance. Of course, internal monitoring and self-testing features of digital relays significantly reduce the needed frequency of testing. Packaging is not an issue in modern applications.

5.8 Relays for multiterminal lines

Occasionally, transmission lines may be tapped to provide intermediate connections to loads or to reinforce the underlying lower-voltage network through a transformer. Such a configuration is known as a multiterminal line, and is often built as a temporary, inexpensive measure for strengthening the power system. Although the resulting power system configuration is inexpensive, it does pose some special problems for the protection engineer. When the multiterminal lines have sources of generation behind the tap points, or if there are grounded neutral wye–delta power transformers at more than two terminals, the protection system design requires careful study [13]. Consider the three-terminal transmission line shown in Figure 5.17. For a fault at F, there is a contribution to the fault current from each of the three terminals. Consider the relaying current and voltage at relay R_1. (For the sake of simplicity, we will assume this to be a single-phase system, although it must be understood that we are interested in three-phase systems, and we must consider the actual distance evaluations for each type of fault. This aspect of multiterminal line protection is no

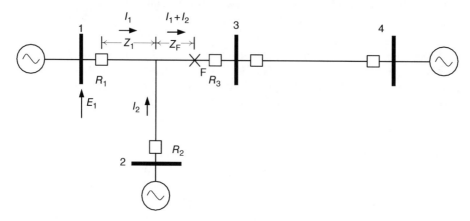

Figure 5.17 Effect of infeed on zone settings of distance relays.

different from the usual considerations of faults on a three-phase system.) The voltage at bus 1 is related to the current at bus 1 by the equation:

$$E_1 = Z_1 I_1 + Z_f(I_1 + I_2), \tag{5.18}$$

and the apparent impedance seen by relay R_1 is

$$Z_{app} = \frac{E_1}{I_1} = Z_1 + Z_f\left(1 + \frac{I_2}{I_1}\right). \tag{5.19}$$

The current I_2, the contribution to the fault from the tap, is known as the infeed current when it is approximately in phase with I_1 and as the outfeed when its phase is opposite to that of I_1. Of course, completely arbitrary phase relationships are also possible, but in most cases the phase relationship is such that the current I_2 is an infeed current. Equation (5.19) shows that the apparent impedance seen by relay R_1 is different from the true impedance to the fault: $(Z_1 + Z_f)$. When the tap current is an infeed, the apparent impedance is greater than the correct value. Thus, if we set the zone 1 setting of relay R_1 at about 85% of the line length 1–2, many of the faults inside the zone of protection will appear to be outside the zone, and the relay will not operate. We must accept this condition, since, when the tap is out of service, the correct performance of the relay is restored. It would be insecure to set zone 1 of the relay to a high value, in order that the apparent impedances for all faults inside the 85% point lie inside the zone setting. With such a setting, if the tap source should be out of service for some reason, faults beyond the 85% point will cause zone 1 operation.

On the other hand, zones 2 and 3 of relay R_1 must reach beyond buses 2 and 3, respectively, under all possible configurations of the tap. Thus, for these (overreaching) zone settings, we must set the zones with all the infeeds in service. Then, if some of the infeeds should be out of service, the impedance seen by the relay will be smaller, and will be definitely inside the corresponding zones.

As noted in the aforementioned discussion, though less common than infeed, an outfeed condition is also possible in the case of multiterminal lines. All that is required is a sufficiently strong external connection between the two terminals. In such a case, outfeed is also a possibility as shown in Figure 5.18. The path for outfeed is shown in Figure 5.18 as a cloud.

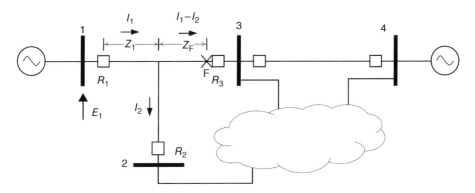

Figure 5.18 Effect of outfeed on zone settings of distance relays.

In the case of outfeed, the distance functions in the relay at R1 will tend to overreach. When the tap current is an outfeed, the apparent impedance for a fault beyond the tap is less than the correct value. Thus, if we set the zone 1 setting of relay R1 at about 85% of the line length 1–2, some the faults beyond Station 3 will appear to be inside the zone, and the relay may undesirably trip to isolate an unfaulted line. In this case, the zone 1 reach must be reduced to less than 85% of the line length 1–2.

We may summarize simply the principle of protecting a multiterminal line as follows: In the case of infeed, underreaching zones are set with infeeds removed from consideration, and overreaching zones are set with the infeeds restored. In the case of outfeed, underreaching zones are set with infeeds included in consideration. Overreaching zones must be set with outfeeds removed from consideration. The interested reader may refer to Reference [13] for more information. These ideas are illustrated in the following infeed example.

Example 5.6 Consider the system shown in Figure 5.19. We may assume that the relative magnitudes of I_1, I_2, and I_3 remain unchanged for any fault on the system between the buses A through G. This is clearly an approximation, and in an actual study, we must use appropriate short-circuit calculations for each of the faults. We are required to set the three zones of the relay R_b. It is assumed (as determined by the short-circuit study) that $I_2/I_1 = 0.5$.

5.8.1 Zone 1

This must be set equal to 85% of the smaller of the two impedances between buses B and D, and B and G. Also, we will consider the infeed to be absent for setting zone 1. Thus, the zone 1 setting is $0.85 \times (4 + j40 + 1 + j10) = 4.25 + j42.5 \ \Omega$.

5.8.2 Zone 2

This is set equal to 120% of the longer of the two impedances between buses B and D, and B and G. The infeed will be considered to be present, and will apply to the impedance of the segment C–D. Thus, the zone 2 setting is $1.2\,[4 + 40 + 1.5 \times (2 + j20)] = 8.4 + j84 \ \Omega$.

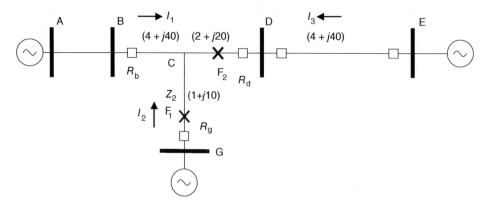

Figure 5.19 System with infeed for Example 5.6.

5.8.3   Zone 3

Assuming that line D–E is the only one needing backup by the relay R_b, the zone 3 setting is obtained by considering the infeed to be in service. The apparent impedance of the line B–D with the infeed is $(4 + j40) + 1.5 \times (2 + j20) = 7 + j70$ Ω. To this, 150% of the impedance of line D–E must be added, duly corrected for the infeed. Thus, the zone 3 setting of R_b is $7 + j70 + 1.5 \times 1.5 \times (4 + j40) = 16 + j160$ Ω.

5.9   Protection of parallel lines

Transmission lines that are on the same tower, or paralleled along the same right of way, present unique problems to the associated line relays. The difficulty stems from the fact that the lines are mutually coupled in their zero-sequence circuits. The small amount of negative- and positive-sequence mutual couplings can usually be neglected. The zero-sequence coupling causes an error in the apparent impedance as calculated by Equation (5.14).

Consider the fault at F on one of the two mutually coupled lines as shown in Figure 5.20. For a phase-a-to-ground fault, the symmetrical component voltages at the fault location R_1 are given by (Equations (5.10)–(5.14)):

$$E_{1f} = E_1 - Z_{1f}I_1,$$

$$E_{2f} = E_2 - Z_{1f}I_2, \tag{5.20}$$

$$E_{0f} = E_0 - Z_{0f}I_{01} - Z_{0mf}I_{02},$$

where I_{01} and I_{02} are the zero-sequence currents in lines 1 and 2, respectively, and Z_{0mf} is the zero-sequence mutual impedance in the faulted portion of the transmission line. As before, the voltage of phase a at the fault point can be set equal to zero:

$$E_{af} = E_{0f} + E_{1f} + E_{2f} = (E_0 + E_1 + E_2) - Z_{1f}(I_1 + I_2) - Z_{0f}I_{01} - Z_{0mf}I_{02} = 0 \tag{5.21}$$

Or,

$$E_a - Z_{1f}I_a - (Z_{0f} - Z_{1f})I_{01} - Z_{0mf}I_{02} = 0. \tag{5.22}$$

Figure 5.20   Fault on a mutually coupled transmission line.

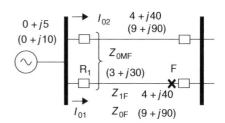

I'_a of Equation (5.13) must now be replaced by the following:

$$
\begin{aligned}
I'_a &= I_a + \frac{Z_{0f} - Z_{1f}}{Z_{1f}} I_{01} + \frac{Z_{0mf}}{Z_{1f}} I_{02} \\
&= I_a + \frac{Z_0 - Z_1}{Z_1} I_{01} + \frac{Z_{0m}}{Z_1} I_{02} = I_a + m I_{01} + m' I_{02}.
\end{aligned}
\tag{5.23}
$$

Giving a definition of m' being the compensation factor for the impact of the zero-sequence mutual coupling with the unfaulted parallel line ($m' = Z_{0m}/Z_1$).

And finally, in terms of this modified compensated phase current, the impedance to the fault point is given by

$$
\frac{E_a}{I'_a} = Z_{1f}.
\tag{5.24}
$$

It should be noted that the current from a parallel circuit must be made available to the relay, if it is to operate correctly for a ground fault. This can be accomplished if the mutually coupled lines are connected to the same bus in the substation. If the two lines terminate at different buses, this would not be possible, and in that case one must accept the error in the operation of the ground distance function and adjust the set reach of the various zones to accommodate the error.

Example 5.7 Let the system shown in Figure 5.20 represent two mutually coupled transmission lines, with impedance data as shown in the figure. The zero-sequence impedances are given in parentheses, and the mutual impedance in the zero-sequence circuits of the two transmission lines is $(3 + j30)$ Ω. The rest of the system data are similar to those of Example 5.4. For a phase-a-to-ground fault at F, the transmission line impedance is divided by 2 because of the parallel circuit of equal impedance. Thus, the positive- and negative-sequence impedance due to the transmission lines in the fault circuit is $(2 + j20)$ Ω, while the zero-sequence impedance is $0.5 \times (9 + j90 + 3 + j30) = (6 + j60)$ Ω. The symmetric components of the fault current are given by

$$
\begin{aligned}
I_1 &= I_2 = I_0 = \frac{7967.4}{2 \times (0 + j5) + 0 + j10 + 2 \times (2 + j20) + 6 + j60} \\
&= 66.169 \angle -85.23°.
\end{aligned}
$$

The currents seen by the relay are half these values because of the even split between the two lines, and $I_a = 3 \times 1/2 \times I_1 = 99.25 \angle -85.23°$. The zero-sequence currents in the two lines are

$$
I_{01} = I_{02} = 33.085 \angle -85.23°.
$$

The zero-sequence compensation factors m and m' are given by (Equation (5.23))

$$
m = \frac{9 + j90 - 4 - j40}{4 + j40} = 1.25 \quad \text{and} \quad m' = \frac{3 + j30}{4 + j40} = 0.75.
$$

The compensated phase a current, as given by Equation 5.23, is

$$
I'_a = I_a + m I_0 + m' I_{02} = 165.45 \angle -85.23°.
$$

The symmetrical components of the voltages at the relay location are given by

$$E_1 = 7967.4 - j5 \times 66.169\angle -85.23° = 7637.7 - j27.51,$$
$$E_2 = -j5 \times 66.169\angle -85.23° = -329.7 - j27.51,$$
$$E_0 = -j10 \times 66.169\angle -85.23° = -689.4 - j55.02,$$

and the phase a voltage at the relay location is

$$E_a = E_1 + E_2 + E_0 = 6649.51\angle -0.95°.$$

Finally, the impedance seen by the relay R_a is

$$\frac{E_a}{I'_a} = \frac{6649.5\angle -0.95°}{165.42\angle -85.23°} = 4 + j40 \ \Omega.$$

Besides the effect of mutual coupling on the performance of the ground distance relay, some other relay operations may also be affected in the case of parallel transmission lines. Consider the two examples given in the following text.

5.9.1 Incorrect directional ground relay operation

Consider the system shown in Figure 5.21. A ground fault on line 2 induces a zero-sequence current I_{01} in line 1. This current, in turn, circulates through the grounded neutrals at the two ends of the transmission line. The current at the B terminal of this line is out of the line and into the bus. However, if the transformer neutral current is used for polarization of the directional ground relay, the operating and the polarizing currents will be in the same direction. This condition is identical to that corresponding to a fault on the line. Similarly, the condition at the end A of the line, with current in the transformer neutral and the line current once again in phase with each other, is indistinguishable from that due to an internal fault. The ground directional relays at both ends may be fooled into seeing this condition as an internal fault. The same false operating tendency would exist if potential polarizing were used. In other words, the phase of the polarizing quantity is not independent of the direction of current flow, as it is when a short circuit occurs. This problem is known as zero-sequence voltage reversal and is further discussed in Reference [9]. Negative-sequence polarized directional ground overcurrent relays are not susceptible to the problem of zero-sequence

Figure 5.21 Current polarization error caused by induced current in a coupled line.

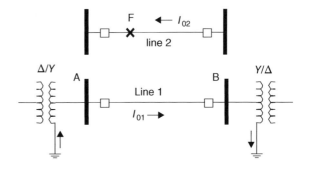

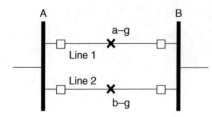

Figure 5.22 Simultaneous faults on parallel transmission lines.

voltage reversal and are often used for directional ground fault protection in applications where zero-sequence mutual coupling may be a concern.

5.9.2 Incorrect phase distance relay operation

This problem is encountered when phase and ground distance relaying are used for protecting parallel transmission lines that are connected to common buses at both ends [14]. Consider the case of a simultaneous fault between phase a and ground on line 1, and between phase b and ground on line 2, as shown in Figure 5.22. This is known as a cross-country fault. It may be caused by the fault arc from the first ground fault expanding with time, and involving the other transmission line in the fault or by a lightning strike to a double-circuit tower causing a backflash to both circuits simultaneously. Such a fault produces fault current contributions in both phases a and b of both circuits, and may be detected as a phase a–b–g fault on both lines.

A multiphase fault will cause a three-phase trip of both circuits. This problem is particularly severe when single-phase tripping and reclosing are used. In this case, the correct and desirable operation would of course be a single-phase trip on each of the circuits, maintaining a three-phase tie between the two ends of the lines, although the impedances would be unbalanced. Reference [14] examines the calculations involved and proposes a solution involving a digital relay that uses currents and voltages from all six phases. Single-phase tripping and reclosing were discussed in Chapter 1. The possibility of failure of distance relays for such faults has been well recognized in Europe, where both double-circuit towers and single-phase tripping are common.

5.10   Effect of transmission line compensation devices

We have assumed, so far, that the transmission system is relatively uncomplicated, and that overcurrent, distance, or directional relays can be applied in a straightforward manner to provide reliable protection. There are primary transmission elements, however, that upset this arrangement. In particular, series capacitors that are installed to increase load or stability margins, or series reactors that are used to limit short-circuit currents, can significantly affect the transmission line protection. Our concern here is not the protection of the devices themselves: that will be the subject of later chapters. Our concern here is how these devices affect the transmission line protection itself.

5.10.1 Series capacitors

A series capacitor can upset the basic premise upon which the principles of distance and directional relaying are founded. Thus, we normally assume that fault currents reverse their direction only for faults on two sides of a relay, and that the ratio of voltage to current at a relay location is a measure of the distance to a fault. A series capacitor introduces a discontinuity in the reactive component of the apparent impedance as the fault is moved from the relay to, and beyond, the capacitor (Figure 5.23). Depending upon the size and location of the capacitor, distance relay settings may or may not be possible. Consider a fault at F_1 in Figure 5.23. The fault current now leads the voltage, and is indistinguishable from the conditions resulting from a fault in the reverse direction. As the fault moves toward F_2, which is at the zone 1 boundary, the apparent impedance follows the trajectory shown on the *R–X* diagram. In this situation, the distance relay at R_{ac} will fail to operate

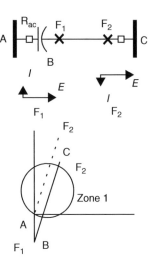

Figure 5.23 *R–X* diagram with a series capacitor.

for F_1. It should be remembered that most series capacitors are fitted with protective devices across their terminals, such as protective gaps, surge suppressors, or circuit breakers. The effect of these protective devices is to short-circuit, or bypass, the capacitors in the presence of faults. Thus, the distance relays may be made slow enough so that the capacitors are first removed from service by the protective devices, and the proper operation of the distance relays is restored. An alternative scheme for protecting transmission lines with series capacitors is to use line current differential or phase comparison relaying, which will be covered in Chapter 6. Reference [8] provides more detailed information on protection problems caused by series compensation.

5.10.2 Series reactors

Series reactors introduce impedance into the line, but since the angle of the reactor is almost 90°, as is the transmission line, there is very little discontinuity in the *R–X* diagram. This is illustrated in Figure 5.23. If the reactor can be switched in or out of service, the line impedance will change, and must be taken care of by changing the relay zone settings. The series reactors will also affect the settings of overcurrent relays, since the short-circuit currents are affected by the series reactor. However, as the series reactors are usually required for achieving safe short-circuit current levels, they are seldom taken out of service without taking the line out of service also. Consequently, the reactor can be considered to be present, and the line relays set accordingly. However, in the unlikely event that the reactor is removed from service, the line impedance will be reduced. Assuming that there are no changes in the relay setting, the relay will now see beyond its original zone of protection, that is, it will now overreach its desired protective zone. For instance, if the reactor shown in Figure 5.24 has an impedance of $Z_{AB} - 10\ \Omega$ and the line section BC has an impedance of $j40$ (ignoring the resistance), zone 1 at both ends would be set at $0.85 \times (40 + 10) = 42.5$. If the reactor is

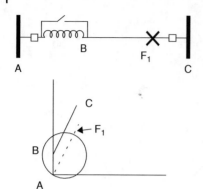

Figure 5.24 R–X diagram with a series reactor.

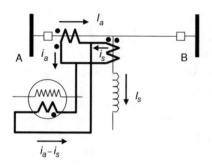

Figure 5.25 Compensation of shunt current in a relay.

removed from service, the line impedance would be 40 Ω. With the relays set for 42.5 Ω, they would see faults beyond the original zone of protection. This may be acceptable if the zone 1 overreach does not extend into the next line section.

5.10.3 Shunt devices

Shunt capacitors and reactors are installed for entirely different reasons, and even if tied to the line itself usually do not have a significant impact on the transmission line relays. There is a steady-state load current associated with the shunt devices, which is seen by the line relays, but the margins used in differentiating between load and short-circuit currents are usually sufficient to avoid any problems. If a problem does exist, it is not too difficult to connect the CTs of the shunt devices, so that the load current is removed from the line relay measurement (Figure 5.25).

5.11 Loadability of relays

Recall the discussion in Section 5.3 of the apparent impedance seen by a relay as the load on the line changes. As the load on a transmission line increases, the apparent impedance locus approaches the origin of the R–X diagram. For some value of line loading, the apparent impedance will cross into a zone of protection of a relay, and the relay will trip. The value of load MVA at which the relay is on the verge of operation is known as the loadability limit of the relay. Consider the characteristics of a directional impedance relay, and a mho relay, with a zone setting of Z_r secondary ohms as shown in Figure 5.26.

If the load power factor angle is assumed to be φ and the angle of maximum torque is assumed to be θ, the value of the apparent impedance at which the loading limit will be reached is Z_r for the directional impedance relay and $Z_r \cos(0 + \varphi)$ for the mho relay (note that a lagging power factor angle is considered to be negative). If the primary voltage of the line is E kilovolts (phase to neutral), and VT and CT ratios are n_v and n_i, respectively, the loadability limit for the directional distance relay is (in MVA)

$$S_{1,\text{imp}} = 3\frac{E^2}{Z_p} = 3\frac{E^2 n_i}{Z_r n_v}, \tag{5.25}$$

where Z_p is the primary impedance, and for the mho relay is (in MVA)

Figure 5.26 Loadability of distance relays with different characteristic shapes.

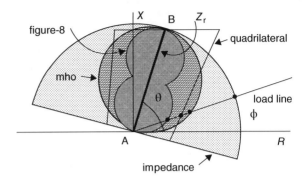

$$S_{1,mho} = 3 \frac{E^2 n_i}{Z_r \cos{(\theta + E\varphi)} n_v}. \tag{5.26}$$

It is clear from Figure 5.26 that the loadability of a mho relay is significantly greater than that of a directional impedance relay. The loadability of a relay can be further increased using a figure-of-eight characteristic (offset mho), or a quadrilateral characteristic, as shown in Figure 5.26. The latter characteristic is achievable with solid-state or computer-based relays.

With the increasing use of computer relays, the application of relays with mho-type configuration has changed dramatically. The possible number of zones has increased to four or five zones with the increased circular characteristic that, in turn, increases the relays susceptibility to load encroachment. This, in turn, has produced solutions addressed by the various relay manufacturers.

Example 5.8 We will consider the loadability of the zone 1 setting of the relay from Example 5.2. (This will illustrate the principle of checking the loadability, although one must realize that the critical loadability, which provides the smaller limit, is that associated with the third zone.) The CT and VT ratios for the relay were determined to be $n_i = 100$ and $n_v = 288.6$. The zone 1 setting is $1.17 + j8.84 = 8.917 \angle 82.46°$. From Equation (5.25), the loadability of an impedance relay is given by (the phase-to-neutral voltage is 20 kV)

$$S_{1,imp} = 3 \times \frac{20^2 \times 100}{8.917 \times 288.6} = 46.63 \, \text{MVA}.$$

In the case of a mho relay, we must calculate the loadability at a specific power factor. Let us assume a power factor of 0.8 lagging. This corresponds to $\varphi = -36.870$. The angle of the line impedance is 82.46°. Thus, $(\theta + \varphi) = (82.46° - 36.87°) = 45.59°$. Using Equation (5.26), the loadability of a mho relay for a 0.8 pf lagging load is

$$S_{1,mho} = 3 \times \frac{20^2 \times 100}{8.917 \times 288.6 \times \cos{45.59°}} = 66.63 \, \text{MVA}.$$

Of course, one must check the loadability of all the zones, but the zone 3 loadability, being the smallest, will usually be the deciding criterion.

5.12 Summary

In this chapter, we have examined the protection of a transmission line using distance relays. Distance relays, both single-phase and polyphase, are usually used for transmission line protection when changes in system configuration, or generating pattern, provide too wide a variation in fault current to allow reliable settings using only current as the determining factor. Distance relays are relatively insensitive to these effects. We have reviewed a number of characteristics that are available depending on the protection required. We have also discussed several common problems associated with nonpilot line protection, including the problem of excessive or unusual load, protection of multiterminal lines, parallel transmission lines, and lines with series or shunt compensation.

Problems

5.1 Determine the three zone settings for the relay R_{ab} in the system shown in Figure 5.27. The system nominal voltage is 138 kV, and the positive-sequence impedances for the various elements are given in the figure. The transformer impedance is given in ohms as viewed from the 138 kV side. Assume that the maximum load at the relay site is 120 MVA, and select a CT ratio accordingly. The available distance relay has zone 1 and zone 2 settings from 0.2 to 10 Ω, and zone 3 settings from 0.5 to 40 Ω, in increments of 0.1 Ω. The angle of maximum torque can be adjusted to 75° or 80°. Remember that the zone 3 of the relay must back up the line BC as well as the transformer.

5.2 Consider the system shown in Figure 5.13. Assume that F is at the end of the line. If the maximum load in the line is 10 MVA, what are the zone 1 and zone 2 settings for this line? You may assume that the available distance relay is the same as that in Problem 5.1.

5.3 For the system in Figure 5.27, and using the relay settings determined in Problem 5.1, plot the three zone settings on the *R–X* diagram. Also show the 120 MVA load on the *R–X* diagram. You may assume a load power factor of 0.8 (lagging). Plot the load impedance when the load is flowing into the line and when it is flowing into the bus. Is there any problem in setting a directional impedance relay with this load? If you perceive a problem, what solution do you suggest?

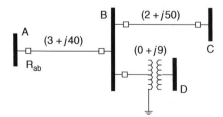

Figure 5.27 System for Problem 5.1.

5.4 Calculate the appropriate relaying voltages and currents for faults between phases a and b, and between phase b and the ground, for the system in Example 5.4. You may use the results obtained in Example 5.4 to simplify the work involved.

5.5 Repeat the calculations for the system in Example 5.4 for a b–c–g fault. Check the response of the b–c (phase) distance relay as well as the a–g and b–g (ground) distance relays.

5.6 For the a–b fault studied in Problem 5.4, what is the impedance seen by the b–c and c–a relays? What is the impedance seen by the a–g and b–g relays for the a–b fault?

5.7 Consider the multiterminal line in the system shown in Figure 5.28. Each of the buses C, D, G, H, and J has a source of power behind it. For a three-phase fault on bus B, the contributions from each of the sources are as follows:

Source	Current, *I*
J	600
C	200
D	300
G	800
H	400

You may assume that the fault current contributions from each of these sources remain unchanged as the fault is moved around throughout the system shown. Determine the zones 1, 2, and 3 settings for the distance relays at buses A and B. Remember to take into account the effect of the infeed for determining the zones 2 and 3 settings, while no infeeds are to be considered for the zone 1 settings.

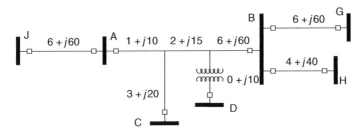

Figure 5.28 System diagram for Problem 5.7.

5.8 For relays R_d and R_g at buses D and G in Figure 5.18, determine the settings for the three zones of their distance relays. You may assume that the positive-sequence impedance of line AB is $(3 + j30)$ Ω. Also, you may assume that nothing beyond bus G needs backup. Assume $I_2/I_3 = 0.5$.

5.9 In the problem considered in Example 5.7, the current in the parallel circuit is not available to the relay R_1. What is the impedance seen by the ground distance relay if I_{02} is omitted from the relay? Is it necessary to set the ground distance relay differently to accommodate this error?

5.10 A zone of a distance relay is set at $10\,\Omega$ secondary. The CT ratio is 500:5 and the VT ratio is 20 000:69.3. What is the fault resistance that this relay tolerates for a fault at a distance of 80% of the zone boundary? Find the answers for a directional impedance relay and also for a mho relay. You may assume that the remote terminal of the line is out of service. The line impedance angle and the relay maximum torque angle are both equal to 80°.

5.11 For the system shown in Figure 5.19, a simultaneous fault occurs on phase b to ground on line 1, and phase c to ground on line 2. The fault is at the midpoint of the lines. You may assume that there is no source connected to the far end of the lines. (i) Calculate the phase-to-neutral voltages at the sending end of the lines and the phase currents in the two lines. (ii) What is the impedance seen by the phase (b–c) distance relays of the two lines?

References

1 Horowitz, S.H., Thorp, J.S., Phadke, A.G., and Beehler, J.E. Limits to impedance relays. *IEEE Trans. Power Appar. Syst.* **98** (1): 246–260.

2 IEEE Standard 100-1972 (1976). *ANSI C42.100-1976 IEEE Standard Dictionary of Electrical and Electronic Terms*. Piscataway, NJ.: IEEE Standards Association, IEEE.

3 Horowitz, S.H. and Phadke, A.G. (2006). Third zone revisited. *IEEE Trans. Power Deliv.* **21** (1): 23–29.

4 Blackburn, J.L. (1952). Ground relay polarization. *AIEE Trans. Power Appar. Syst.* **71**: 1088–1093.

5 Lewis, W.A. and Tippett, L.S. (1947). Fundamental basis for distance relaying on 3-phase systems. *AIEE Trans.* **66**: 694–708.

6 Mason, C.R. (1956). *The Art and Science of Protective Relaying*. New York: John Wiley & Sons, Inc.

7 Anderson, P.M., Henville, C., Rifaat, R. et al. (2022). *Power System Protection*, 2e. Wiley-IEEE Press.

8 Warrington, A.R.v.C. (1962). *Protective Relays, the Theory and Practice*, vol. **1**, Chapman and Hall, London. New York: John Wiley & Sons, Inc.

9 IEEE Std C37.113™-2015 (2015). *IEEE Guide for Protective Relay Applications to Transmission Lines*. New York, NY: IEEE.

10 IEEE PES PSRC Committee Report (2012) Performance Specification and Testing of Transmission Line Relays for Frequency Response available on line at http://www.pes-psrc.org/kb/published/reports/D22%20final%204-18-12.pdf

11 Westinghouse (1976). *Applied Protective Relaying*. Newark, NJ: Westinghouse Electric Corporation.

12 Phadke, A.G., Ibrahim, M., and Hlibka, T. (1977). Fundamental basis for distance relaying with symmetrical components. *IEEE Trans. Power Appar. Syst.* **96** (2): 635–646.

13 IEEE Power System Relaying Committee (1979) *Protection Aspects of Multi-terminal Lines*, IEEE special publication 79 TH0056-2-PWR.

14 Phadke, A.G. and Jihuang, L. (1985). A new computer based integrated distance relay for parallel transmission lines. *IEEE Trans. Power Appar. Syst.* **104** (2): 445–452.

6

Pilot Protection of Transmission Lines

6.1 Introduction

The protection principles described in Chapters 4 and 5, nonpilot protection using overcurrent, and distance relays contain a fundamental difficulty. It is not possible to instantaneously clear a fault from both ends of a transmission line if the fault is near one end of the line. This is due to the fact that, in detecting a fault using only information obtained at one end, faults near the remote end cannot be cleared without the introduction of some time delay. As discussed in Chapter 5, there is always an uncertainty at the limits of a protective zone. Referring to Figure 6.1, to avoid loss of coordination for a fault at F_2, the relays at terminal B trip instantaneously by the first zone and the relays at terminal A use a time delay for the second zone or backup tripping. This results in slow clearing for a fault at F_1. The ideal solution would be to use the differential principle described in Section 2.2.3 (Chapter 2). Until the late twentieth century, this solution was not practical for the majority of transmission lines because of the distances involved. For a three-phase line, six pilot conductors would be required, one for each phase, one for the neutral, and two for the DC-positive and -negative leads required to trip the circuit breaker(s). For distances beyond 5–10 miles, the cost of the cable alone would make this prohibitive. In addition, there could be "error current" introduced by current transformer (CT) saturation caused by heavy load, transmission line charging current, or voltage drops in the cable itself due to its length and the large secondary currents during a fault. The proximity of the control cable to the transmission line, and its exposure to lightning, would require very high cable insulation. All of these factors tended to rule out this method of protection except for very short lines (less than 2 miles). More recently, however, with the increasing use of broadband communications and digital relays, transmission line current differential protection has become popular. This is further described in Section 6.10.

Pilot protection is an adaptation of the principles of differential relaying that avoids the use of control cable between terminals. The term "pilot" refers to a communication channel between two or more ends of a transmission line to provide instantaneous clearing over 100% of the line. This form of protection is also known as "teleprotection." The communication channels generally used are

- Power line carrier (PLC).
- Microwave.

Power System Relaying, Fifth Edition. Stanley H. Horowitz, Arun G. Phadke and Charles F. Henville.
© 2023 John Wiley & Sons Ltd. Published 2023 by John Wiley & Sons Ltd.

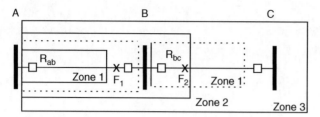

Figure 6.1 Coordination between the first and second zones.

- Fiber optics.
- Communication cable.

6.2 Communication channels

A comprehensive discussion of communication channels and equipment is beyond the scope of this book. However, a general understanding of this technology is essential in understanding and applying transmission line protection. Reference [1] provides in-depth discussion of the various communications technologies that may be applied for protection purposes. Two classes of technologies may be noted, analog and digital. Digital technologies allow the use of teleprotection equipment as an interface between a relay and a communications system, or direct communications between relays in a peer-to-peer mode. In the class of digital technologies, two subclasses may be identified.

1) Individual manufacturers have developed their own "vendor-specific" communications protocols for peer-to-peer communications between relays of the same manufacturers.
2) The International Electrotechnical Commission (IEC) has developed a suite of standards under the heading of "IEC 61850." This suite defines a protocol that allows digital peer-to-peer communications between relays from different manufacturers as long as they comply with the standard.

A significant benefit of digital peer-to-peer communications is that the digital protection devices themselves can monitor the health of the teleprotection and raise an alarm if the communications become impaired

6.2.1 Power line carrier

Figure 6.2 shows the basic one-line diagram of the main PLC components. PLC systems operate in an on–off mode by transmitting radio frequency signals in the 10–490 kHz band over transmission lines. They may also operate in frequency shift or digital modes by transmitting tones or digital code over the radio frequency carrier [1]. PLC systems with power outputs of 10 W are reliable up to about 100 miles and those with 100 W outputs are effective at over 150 miles. Coupling capacitors are used to couple the carrier equipment to the high-voltage (HV) transmission line. They are low-impedance paths to the high frequency of the carrier current, but high-impedance paths to the 60 Hz power frequency. In conjunction

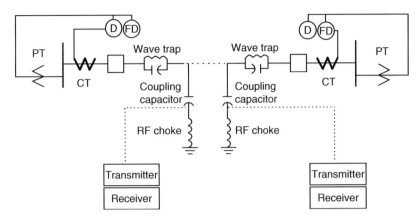

Figure 6.2 One-line diagram of a power line carrier system.

with the coupling capacitors, line tuners and wave traps are used, which present low impedance to the power frequency and high impedance to the radio frequency. The signal is thus trapped between the ends of the line. Normally, only one 4 kHz bandwidth channel is provided exclusively for protection. The transmission time is approximately 5 ms for on/off carrier, or 8–17 ms for frequency shift tones and digital signals. In the United States, the transmit/receive power and the number of available frequencies are limited by government regulations.

PLC is subject to high impulse noise associated with lightning, faults, switching, or other arcing phenomena. PLC is an extremely versatile communication link that can be applied to directional or phase comparison fault-detection schemes, to block or trip circuit breakers, or with on–off or frequency shift keying (FSK) modes of operation [1, 2]. These are discussed in detail later in this chapter. PLC has been the most common protective communication link used in the United States, but microwave and, more recently, fiber-optic links are also widely used. PLC is not used as much in Europe or Asia.

There are a number of ways of coupling the carrier frequency signal to one or more of the conductors of the transmission line. The simplest, and most common, on HV transmission lines is to use one phase of the power line with ground as the return path. This system, commonly called "line-to-ground," "phase-to-ground," or "single-phase" coupling, requires less coupling capacitors and tuners and wave traps. The return path is usually the overhead ground wire, although the ground itself can be used with accompanying higher attenuation. Coupling between any two phases of the transmission line is an alternative, referred to as "phase-to-phase" or "line-to-line" coupling when the coupling is on the same three-phase transmission line, or "interphase" coupling when the coupling is between adjacent lines on the same tower. Phase-to-phase coupling appears to have an advantage, in that a ground on one of the phases would not affect the communication signal as it would seem to do with phase-to-ground coupling. Actually, however, there is enough electrostatic and electromagnetic coupling between the conductors to transfer enough energy around the ground to maintain a useful signal. With the advent of extra-high voltage (EHV), however, several conditions have been introduced that argue in favor of coupling that is more reliable and provides greater channel capacity and less attenuation. The lines are longer, introducing more attenuation, and they require relaying, voice, and data communication reliability that is increasingly more important. In addition, the voltage source for the relays protecting EHV lines usually comes from line-side potential devices. These are the coupling capacitor voltage transformers (CCVTs) discussed in Section 3.7 (Chapter 3). Figure 6.2 shows the voltage source for the relays coming from bus potential transformers. This is the common application for HV systems. At the EHV level, however, the cost of potential transformers is high and CCVTs are used. Since three-phase voltages are required, three coupling capacitors are also required. The preferred method for EHV lines is to couple to all three phases. To analyze the high-frequency transmission signals, an analytical technique known as "modal analysis" is used. This is a procedure that is similar to symmetrical components for analyzing 60 Hz voltages and currents. During the development of this technology, however, a great deal of confusion unfortunately was introduced. The three modes of coupling, in decreasing order of attenuation and increasing order of reliability and cost, are phase-to-ground coupling using the center phase, phase-to-phase coupling between the outside phases, and coupling to all three phases. Early literature referred to these modes, respectively, as modes 1, 2, and 3. Later literature reversed these designations; so it is important to check the diagrams to be sure which coupling mode is involved. Additional special considerations in applying PLC for protective relaying are provided in Reference [3].

6.2.2 Microwave

Microwave operates at frequencies between 150 MHz and 20 GHz. This bandwidth can be put at the disposal of protection systems with many 4 kHz channels operating in parallel to provide high bandwidth. Analog microwave technology was originally used, but modern applications use digital microwave for greater channel capacity [1]. Protection, however, is usually a small part of the total use of a microwave system. The large bandwidth allows a wide variety of information to be sent, such as voice, metering, alarms, and so on. Microwave is not affected by problems on the transmission line, but is subject to atmospheric attenuation and distortion. Frequency diversity and/or space diversity is used to increase reliability in the case of unfavorable atmospheric conditions [1]. The transmission length is limited to a line-of-sight path between antennas, but can be increased through the use of repeaters for increased cost and decreased reliability [1].

6.2.3 Fiber-optic links

Optical cable is an often used option. The interface between the teleprotection equipment and the fiber-optic multiplexer is standardized in Reference [4]. Dedicated relay-to-relay fiber connection is also sometimes used.

Fiber-optic links have virtually unlimited channel capacity. Single fibers having as many as 8000 channels are available and this can be increased significantly using multiple fiber cables. Figure 6.3 shows a typical construction of a fiber cable. There can be any number of fibers in the cable depending upon the mode of operation and the application. Each glass fiber is protected by a plastic tube, all of the tubes are further protected by an aluminum-alloy tube, additional strength members are used for support, and the entire construction is embedded in galvanized steel. The fiber cable itself, being nonconducting, is immune to interference from electric or magnetic fields and provides excellent transmission quality. There is very little signal attenuation and, through the use of repeaters, the transmission

Figure 6.3 Typical construction of a fiber cable.

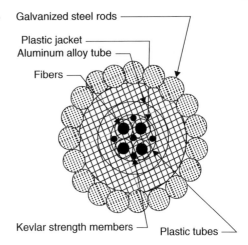

Galvanized steel rods

Plastic jacket
Aluminum alloy tube

Fibers

Kevlar strength members

Plastic tubes

length can be several hundreds of miles. The channel capacity is virtually unlimited, providing as many as 8000 4-kHz channels per fiber. The use of fiber cable, however, is rarely justified just for protection, but, with its large data transmission capacity, it is used for dispatching and telemetering. Once available, however, it makes an excellent communication channel for relaying. Many utilities are installing multiple paths of fiber-optic cable using sophisticated computer programs to monitor the integrity of the cable and to reroute the signals in the event of difficulty on any path. Figure 6.4 shows several methods of stringing fiber-optic cable. Two of the most common methods are to embed the fiber cable within the aluminum conductors of the overhead ground wire (Figure 6.4d) or to wrap the fiber cable around one of the phase or ground conductors.

6.2.4 Pilot wire

Telephone cable consists of shielded copper conductors, insulated up to 15 kV, and is the most popular form of pilot protection using a separate communication medium for short distances (up to 10–15 miles). Figure 6.5 shows a typical pilot wire construction. Usually, the conductors are #19AWG, twisted pairs to avoid cross-talk and shielded in either braided or corrugated copper or aluminum to minimize inductive interference from nearby power lines. The cable can be strung along a steel messenger, or can be self-contained in a figure-of-eight configuration with the steel messenger in the upper loop and the telephone cable in the lower loop. This type of communication channel has a bandwidth from 0 kHz (DC) to 4 kHz. Attenuation is a function of cable type, cable length, and frequency. Overhead cable is vulnerable to induced voltages by power line faults and lightning. Buried cable is subject to damage by digging or animals.

In the United States, pilot channels can be privately owned by the utility or industrial facility or can be rented from a telephone company. If rented, the utility loses a significant measure of control over a very important protective channel. For instance, although agreements are made between the telephone company and the utility regarding maintenance practices, it is extremely difficult for the telephone maintenance worker atop a pole always

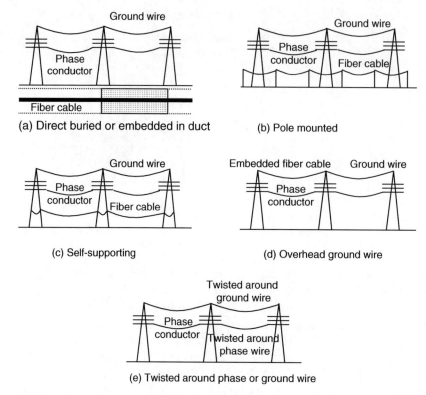

Figure 6.4 Methods of stringing fiber cable. (a) Directly buried or embedded in duct; (b) pole-mounted; (c) self-supporting; (d) overhead ground wire; and (e) twisted around phase or ground wire.

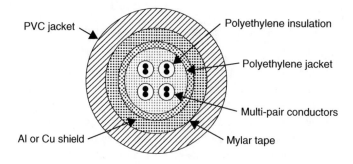

Figure 6.5 Typical pilot cable construction.

to recognize and isolate the utility's relay circuit. The worker can then inadvertently remove it from service or impress a test voltage across the pairs and cause a false trip. The utility can put a fault detector on to ensure that no trip occurs unless there is a fault, but this will restrict the sensitivity of the protection. In addition, when telephone companies convert their metallic wire circuits to microwave, this protection principle is not applicable. In other countries, the telephone system is owned by the government and use of this channel is provided by the appropriate agency [5].

6.3 Tripping versus blocking

The selection of a communication channel for protection is based upon a great many factors such as cost, reliability, the number of terminals and the distance between them, the number of channels required for all purposes, not only relaying, available frequencies, and the prevailing practices of the power company. In addition to these fundamental considerations, either as another factor to be included or as a result of the selection already agreed upon, a decision must be made whether to operate in a blocking or a tripping mode. A blocking mode is one in which the presence of a transmitted signal prevents tripping of a circuit breaker, and a tripping mode is one in which the signal initiates tripping a circuit breaker. There are different relay schemes that accommodate one mode or the other. These are discussed in detail. Basically, however, the criterion upon which the decision to use a block or trip signal is based on the relationship between the power line and the communication channel. The use of a blocking signal is preferred if the communication medium is an integral part of the protected line section, such as PLC. In this case, an internal fault may prevent or seriously attenuate the signal so that a trip signal would not be received. If a separate transmission medium such as microwave, fiber-optic cable, or a pilot cable is used, then the integrity of the power line during an internal fault will have no effect on the transmitted signal and a tripping scheme is a viable application. In EHV and many HV systems, two primary protection schemes are used, in which case one may be a tripping system and the other may be a blocking system, which will then provide diversity [6, 7].

In many modern applications, the teleprotection uses digitally coded signals. In most cases, digital teleprotection will not use PLC channels and will be unaffected by short circuits on the transmission line primary conductors. If digital teleprotection is applied with PLC, there is a concern that the fault itself will interfere with the communication and delay or interrupt the protection to make tripping schemes ineffective. The concern may be mitigated by using a blocking scheme instead of a tripping scheme.

6.4 Directional comparison blocking

The directional comparison blocking scheme, using PLC, is a widely used pilot relaying scheme in the United States [8]. The fundamental principle upon which this scheme is based utilizes the fact that, at a given terminal, the direction of a fault, either forward or backward, is easily determined. As discussed in Chapter 4, a directional relay can differentiate between an internal or an external fault. By transmitting this information to the remote end and by applying the appropriate logic, both ends can determine whether a fault is within the protected line or external to it. For phase faults, a directional or nondirectional distance relay can be used as a fault detector to transmit a blocking signal, that is, a signal that, if received, will prevent the circuit breaker from tripping. Tripping is allowed in the absence of the signal, plus other supervising relay action. As discussed in Section 6.3, a blocking signal is used, since the communication path uses the transmission line itself. A directional relay, usually an admittance-type (mho) or quadrilateral relay, is used to stop the transmitter. In the event of a ground fault, either a ground distance relay similar to the phase relay or a directional ground overcurrent relay is used. A trip is initiated if the directional relays at both ends have operated and neither end has received a blocking signal. Each receiver receives signals from both the local and the remote transmitters [9].

The directional comparison blocking scheme is shown in Figure 6.6. Figure 6.7a shows a DC circuit for electromechanical relay implementations. Figure 6.7b shows the logic for digital relay implementation. Both panels (a) and (b) of Figure 6.7 are simplified diagrams showing only the essential logic for the blocking scheme.

If we assume an internal fault at F_1, since the tripping relays D_{ab} and D_{ba} are directional and are set to see faults from the bus into and beyond the line, contact 21-1 (Figure 6.7) will close. If the fault detector starting relays FD_{ab} and FD_{ba} are directional, they are set with reversed reach, that is, to see behind the protected line, and they will not see the fault at

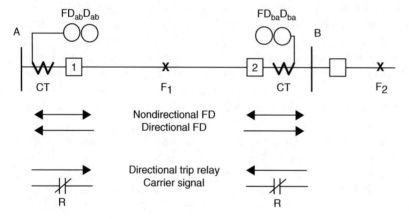

Figure 6.6 Directional comparison blocking scheme.

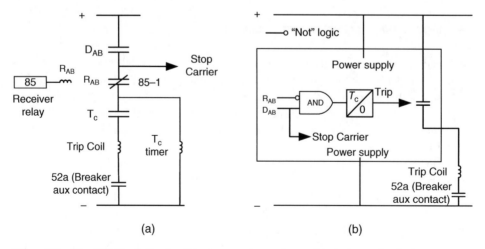

Figure 6.7 Simplified logic for directional comparison scheme: (a) electromechanical implementation; (b) digital implementation.

F_1 and will not send a blocking signal (contact 85-1 in Figure 6.7 will remain closed). If they are nondirectional, they will start transmission, but the directional relays will stop it. In either case, since the receiver relay contact 85-1 will be closed due to the absence of a carrier signal, and the directional relays have operated, both ends will trip their associated breakers. An external fault, F_2, will result in the operation of the directional relay D_{ab} at breaker 1, but if fault detector FD_{ab} is directional, it will not operate for the fault at F_2. If FD_{ab} is nondirectional, it will start transmitting a blocking signal, but D_{ab} will stop it. At breaker 2, however, FD_{ba} will operate whether it is directional or not, and the tripping relay D_{ba} will not operate to stop transmission. Breaker 1 will not trip, since it is receiving a blocking signal from terminal B, opening its receiver relay contact, and breaker 2 will not trip for two reasons: it is receiving its own blocking signal and relay D_{ba} did not operate.

Note that a short coordinating time delay (T_C in Figure 6.7) is normally added in the trip path to ensure that the local blocking signal can be asserted before the tripping relay operates for an external fault beyond the remote terminal. Fault location F_2 in Figure 6.6 is an example external fault. This time delay allows time for the blocking signal to be transmitted from the remote terminal and received locally plus any necessary additional time if the blocking function at the remote terminal is not as fast as the local tripping relay. Typically, a delay of about 8 ms may be used with on–off PLC, and up to 17–34 ms for slower communication channels.

6.4.1 Settings

A major advantage in applying distance relays in a directional comparison scheme is the relative ease and consistency of the settings. Since the distance relay operates on the ratio of V/I, load is not usually a concern. On internal faults, for most applications, the ratio of normal voltage and load current is considerably higher than that at the balance point of the relay. Even when the two ratios are close, the phase angles are not; load power factors are close to unity, and fault current phase angles are reactive. Except for very heavy loads or extremely large relay settings, the relay characteristic does not encompass the load impedance. Under stressed system conditions, however, as discussed in Chapters 10 and 11, the relay characteristic may expand and load then becomes a factor. The major criteria are that the blocking signal must be present for all external faults and the tripping relay must operate for all internal faults. The usual setting for the carrier trip relay is 175–200% of the protected line section. For the reversed mho carrier start relay, the carrier start at one terminal must overreach the carrier trip relay at the other terminal. If this relay is a directional, reversed mho-type relay, it is set at 125–160% of ($M_{trip} - Z_{line}$). If it is a nondirectional impedance-type relay, it is set at 150% beyond the longest tripping zone.

Example 6.1

a) Consider the transmission line shown in Figure 6.8. The line impedance is shown with a resistance component of $0\,\Omega$. This is acceptable, since the resistance is small compared to the inductance. However, the line angle is assumed equal to the relay angle of maximum torque to simplify the diagram, as if there were a resistance component. Both of these conventions are commonly used despite the seeming contradiction. Line section AB is protected by a directional comparison blocking scheme using, at each end, an

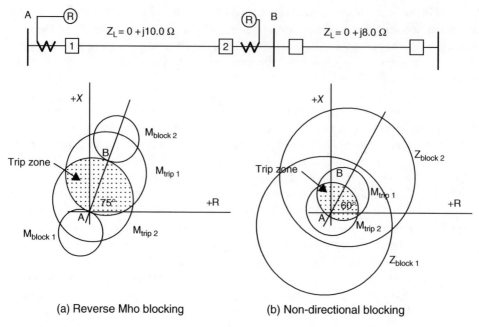

(a) Reverse Mho blocking (b) Non-directional blocking

Figure 6.8 One-line and R–X diagrams for Example 6.1. (a) Reverse mho blocking and (b) nondirectional blocking.

admittance-type tripping relay (M_{trip}) and a reverse admittance-type blocking relay (M_{block}) without any offset. Calculate the settings of the tripping relay at A and the blocking relay at B. Draw the relay characteristics on an R–X diagram. Assume that 75° is the line angle and also the angle of maximum torque of each relay.

b) Repeat (a) for a nondirectional impedance blocking relay. Assume that 60° is the line angle and also the angle of maximum torque of each relay.

a) Set M_{trip} at bus A between 175 and 200% of Z_{line} : $M_{trip} = 17.50\ \Omega$.

Set M_{block} at bus B between 125 and 160% of ($M_{trip} - Z_{line}$); so it will overreach M_{trip} at bus A.

If $M_{trip} = 17.5$: M_{block} at bus B = $1.25(17.5 - 10) = 9.375\ \Omega$

or M_{block} at bus B = $1.60\ (17.5 - 10) = 12.0\ \Omega$.

If $M_{trip} = 20$: M_{block} at bus B = $1.25(20 - 10) = 12.5\ \Omega$

or M_{block} at bus B = $1.6(20 - 10) = 16\ \Omega$.

All of these settings are acceptable. In general, the smallest setting would be chosen to avoid any problems with load, as discussed in Section 5.11 (Chapter 5).

b) In (a), the blocking relay at A must overreach the tripping relay at B, and vice versa. However, if the blocking relay is an impedance type, the blocking relays at both terminals must coordinate with both tripping relays as shown in Figure 6.8b. For the system shown in this example, the relays at both terminals are set the same.

M_{trip} can be 17.5 or 20.0 Ω. Use the smaller number.

$$Z_{block} = 1.5(17.5) = 6.25 \, \Omega.$$

These settings must be checked against loadability.

For the system shown in Figure 6.8, assume the same parameters as in Example 5.8 (Chapter 5), that is, $E = 20 \, kV$, $n_i = 100$, $n_v = 288.6$, and $(\theta + \varphi) = 45.49°$. Loadability is checked for the largest characteristic that might encompass the load at a reasonable load power factor.

$$M_{trip} = 20\Omega; \qquad S_{1mho} = 3 \times \frac{E^2 n_i}{Z_r \cos{(\theta + \varphi)} n_v} = 29.80 \, MVA.$$

$$Z_{block} = 26.25\Omega; \quad S_{1imp} = 3 \times \frac{E^2 n_i}{Z_r n_v} = 15.8 \, MVA.$$

These values must be compared against the maximum reasonable load before determining the final relay settings.

One of the advantages of using the directional mho relay for the carrier starting element compared to the nondirectional impedance relay is an increase in the dependability of the protection system. With a directional starting relay, it is not necessary to stop the transmitter for an internal fault, since it never starts. This eliminates the possibility of not tripping because the transmitter continues to send a blocking signal. In addition, as shown in Example 6.1, the nondirectional impedance characteristic may be so large that it may encompass the load impedance and result in a continuous carrier signal. For long lines, this concern of encompassing the impedance of heavy or emergency loads also exists with the tripping relays. Solid-state and digital relays can avoid this by shaping the characteristic closer to the line and thus avoiding operation for heavy load. As shown in Figure 6.9, the reverse mho electromechanical relay (M_{block1} and M_{block2}) is usually slightly offset from the origin to encompass the line side of the CT and the breaker. This avoids the low torque that would exist for a fault on the relay characteristic, biasing the operation toward blocking for a bus fault. This precaution is not necessary with solid-state or digital relays due to their more positive operating characteristics near the balance point.

Figure 6.9 Offset reverse mho blocking.

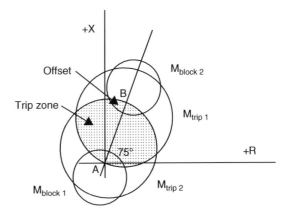

6.5 Directional comparison unblocking

The blocking signal in the directional comparison scheme is transmitted only when a fault occurs. When on–off PLC is used with a directional comparison blocking scheme, a malfunction can occur in the relay or communication equipment such that no signal is sent. The result would be a failure to block during an external fault when the directional tripping relay picks up and a false trip would occur (see the logic circuits of Figure 6.7a,b). To avoid this, a low-energy continuous carrier signal can be transmitted as a check on the communication link. The frequency shifts (is unblocked) when a fault occurs. A schematic and simplified DC circuit of the unblocking scheme is shown in Figure 6.10 [10].

Under normal conditions, a blocking signal is sent continuously—so there is no input to OR_1—therefore, AND_1 and AND_2 (through path "a") receive no input. During an internal fault, the tripping relay D at each end causes the transmitter to shift to the unblocking frequency, providing an input to OR_2, and then to the receiver relay R, allowing the circuit breaker to trip. The monitoring logic, that is, path "a," provides protection in the event there is a loss of a blocking signal without the transmission of an unblocking signal. With no blocking frequency, there is an input to OR_1 and to AND_2, and, with no unblocking signal, there is an input to OR_2 through AND_1 and the timer (until it times out). This allows a trip for the interval of the timer (X is usually 150 ms) after which the circuit is locked out. With the directional comparison blocking scheme, a failure of the communication link will go

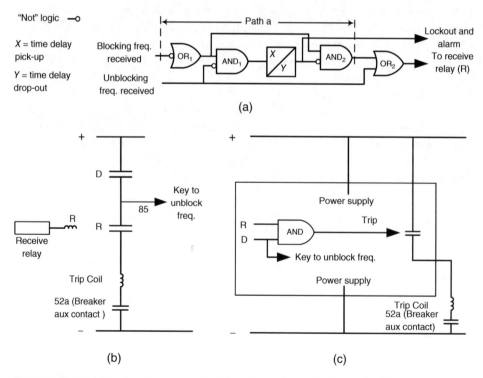

(a)

(b)   (c)

Figure 6.10 Directional comparison unblocking scheme: (a) unblocking logic; (b) electromechanical implementation of tripping logic; (c) digital implementation of tripping logic.

unnoticed until a false trip occurs during an external fault. With the unblocking scheme, this condition is immediately recognized and the relay is made inoperative. If a "hole" in the carrier should occur during a fault, both schemes are vulnerable to false trips.

6.6 Direct underreaching transfer trip (DUTT)

If the communication channel is independent of the power line, a tripping scheme is a viable protection system [8]. There are some slight advantages to tripping versus blocking. Tripping can generally be done faster, since there is no need to include coordination time. (Referring to Figure 6.7, the receiver relay must open before the directional relay closes and T_c times out to avoid an incorrect trip on an external fault.) In the blocking scheme, the blocking relay must be more sensitive than the tripping relay. This is not always easy to accomplish, particularly if there is a big difference between the fault contributions at each end. In making this setting, the relay must avoid operating on heavy load or system unbalance.

6.6.1 DUTT schemes using frequency shift tones

In many cases, the transfer trip signal is a frequency shift system as described in the following text. This provides a continuously transmitted guard signal that must shift to a trip frequency. It is this two-pronged action that will prevent a false trip due to random electrical noise; the guard frequency must stop, allowing its contact to close, and the trip frequency must be received, allowing its contact to close. The continuous guard frequency also monitors the communication channel similar to the directional comparison unblocking scheme.

Figure 6.11 shows a schematic and DC circuit of circuit breaker 1 of a DUTT scheme (circuit breaker 2 has a similar circuit). The potential for the relays is derived from CCVTs as discussed in Section 6.2. This is the simplest application of the tripping scheme and uses an underreaching relay at each terminal. A guard signal is transmitted continuously from each terminal, energizing guard relays of the two receivers of 85-1 and 85-2 and opening their

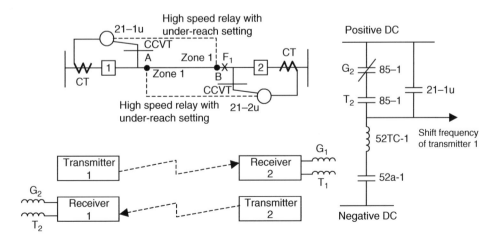

Figure 6.11 One-line and DC diagrams for direct underreaching transfer trip.

contacts (85-1 G_2 and 85-2 G_1) in the associated breaker trip circuits. If a fault occurs within the reach setting of the directional underreaching relay 21-1$_u$, it will trip circuit breaker 1 directly. This relay is set the same as a zone 1 instantaneous relay. In addition, 21-1$_u$ shifts the transmitted signal from the guard frequency to the trip frequency. If the fault occurs beyond the reach of 21-1$_u$, but within the reach of 21-2$_u$, that relay shifts to the trip frequency. A trip occurs when the guard relay contact 85-1 G_2 drops out and the trip relay contact 85-1 T_2 closes. Each terminal receives only the signal from the remote transmitter. The frequencies of this equipment are selected so that there is no interaction between channels. However, this scheme is not secure inasmuch as a trip can occur simply by having the receiver relay's tripping contact close. This can happen inadvertently during maintenance or calibration, or it can be caused by electrical noise accompanying switching in the substation or by control circuit transients during relay operations.

The frequencies are usually selected so that in normal operation, the receiver will "see" the guard frequency (allowing its contact to open) before it sees a trip frequency (when its contact closes). This cannot always be relied upon, however, in the presence of random noise. Therefore, to prevent a false trip, two transmitter/receiver sets are sometimes used as shown in Figure 6.12. These receivers operate on two different frequencies, and tripping requires both receivers to operate. Security is improved, since it is unlikely that the two frequencies would be present simultaneously due to switching or other electrical noise. There is, however, a decrease in dependability with this solution, since there are twice as many components involved. In addition, the capital and maintenance costs are greater.

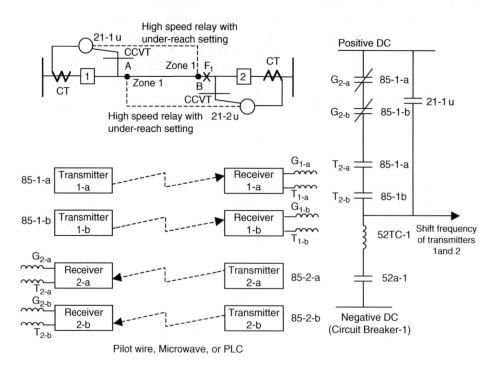

Figure 6.12 Direct underreaching transfer trip with dual transmitter/receiver sets.

The relays designated 21-1$_u$ and 21-2$_u$ in Figures 6.11 and 6.12 represent both phase and ground directional, high-speed, first-zone devices that are set to overlap each other but not to reach beyond the remote terminal. For a fault in this zone, that is, between A and B, the underreaching relays at both ends of the line operate and trip their respective breakers directly. At the same time, trip signals are sent from both terminals. Receipt of these signals will energize the trip coils of both breakers and trip them, if they have not already been tripped directly by their respective underreaching relays.

6.6.2 DUTT schemes using relay-to-relay communications

Modern digital communications systems can support direct relay-to-relay communications with transparent digital teleprotection facilities. The relay-to-relay communications may be accomplished by vendor-specific communications protocol, or by IEC 61850. The digital communications may be interrupted by noise, but since they are coded signals, they are not susceptible to giving an incorrect trip signal due to noise. Thus, the guard signals in transmitter/receiver sets shown in Figures 6.11 and 6.12 are neither necessary nor applied. The digital communications are continuously monitored for health, and an alarm raised if any problems are detected.

The underreaching zone 1 function trips the local breaker and keys a simple direct transfer trip to the remote terminal. The remote terminal trips directly upon receiving the transfer trip without further qualification.

Example 6.2   Consider the transmission line shown in Figure 6.13 protected by the DUTT scheme using admittance-type relays at each end. Calculate the settings of the underreaching relays. Draw the relay characteristics on an R–X diagram.

At bus A, R_u should be equal to 80–90% of Z_{line}. If we use 85%, then M_1 equals $0.85 \times (2 + j10)$ or $8.67 \angle 79\,^\circ\ \Omega$. At bus B, M_2 also equals $0.85 \times (2 + j10) = 8.67 \angle 79\,^\circ\ \Omega$.

6.6.3 DUTT scheme application issues

The DUTT scheme requires instantaneous zone 1 functions at both terminals with reaches that overlap each other for at least 60% of the middle portion of the line. If overlapping settings are not possible, such as on very short lines where an underreaching relay cannot be

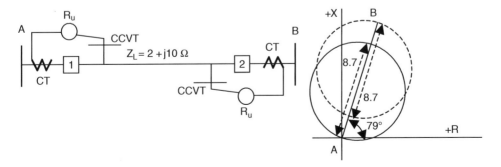

Figure 6.13   Calculation and diagram for DUTT setting of Example 6.2.

set reliably to distinguish between an end-of-line fault and an external fault, this scheme cannot be used. Also, in cases where the remote terminal is open or too weak for application of a zone 1 function, the scheme is not effective for faults near the weak or open terminal. Even if the overlapped region is acceptable, the sensitivity of the zone 1 function to resistive faults may be unacceptably limited in some applications. As noted in the following paragraph, if time-delayed clearing of resistive faults is acceptable, the sensitivity to resistive single-line-to-ground faults may be increased by the application of directional time-delayed ground overcurrent relays.

For this type of scheme, it is a common practice to use directional distance functions for phase faults and either directional overcurrent or ground distance functions for ground faults. On systems where the ground current can change significantly for different operating conditions, it may not be possible to set the overcurrent functions to overlap each other and ground distance functions must be used. In any case, the same functions that are used for the transfer trip protection could also be used with other units to provide backup protection that is independent of the communication channel. For instance, the zone 1 distance function plus the directional instantaneous ground overcurrent function could serve the underreaching relay functions. In addition, the three zones of the phase distance protection would provide three-step distance protection in the normal manner, while the directional instantaneous and time-delayed units ground overcurrent functions would provide two-step ground fault protection.

6.7 Permissive overreaching transfer trip

A more common, and less expensive, solution to increase the security and sensitivity of the DUTT is to provide an overreaching fault detector in a permissive overreaching transfer trip (POTT) scheme [8]. This is somewhat similar to the directional comparison blocking scheme in which the directional overreaching function serves as both the fault detector (similar to the nondirectional impedance function in the blocking scheme) and as a permissive interlock to prevent inadvertent trips due to noise. The received signal provides the tripping function. A trip signal, instead of a blocking signal, is used and each terminal will only respond to the remote transmitter's signal.

6.7.1 POTT scheme using frequency shift tones

Referring to Figure 6.14, directional overreaching relays $21\text{-}1_o$ and $21\text{-}2_o$ send tripping signals to the remote ends. Note that only an internal fault will cause both directional overreaching relays to operate; an external fault at either terminal will be seen by only one of the two directional overreaching relays. A trip, therefore, will be initiated if the local overreaching relay operates and a tripping signal is received from the remote terminal. Tripping is now dependent on both a transmitted signal from the remote end, which will cause 85-1 G_2 to drop out and 85-1 T_2 to pick up, and the overreaching fault detector $21\text{-}1_o$ picking up at the local terminal.

6.7.2 POTT scheme using digital peer-to-peer communications

Digital peer-to-peer communications eliminate the need for frequency shift tones provided by teleprotection equipment and associated logic. The self-monitoring features of such

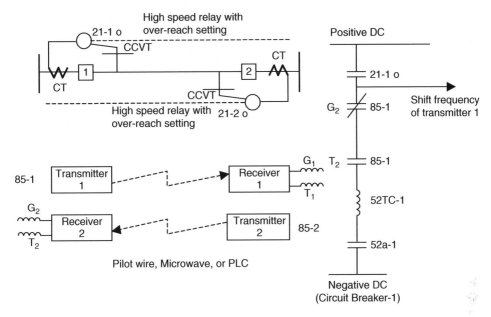

Figure 6.14 POTT using frequency shift tones.

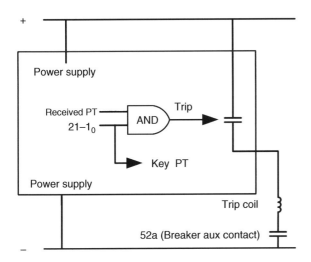

Figure 6.15 POTT logic using peer-to-peer communications.

communications provide an immediate alarm if the teleprotection system becomes impaired. The POTT logic becomes much simpler, as shown in Figure 6.15.

Example 6.3 Consider the transmission line shown in Figure 6.16 protected by a POTT scheme using admittance-type relays at both ends for the overreaching function. Calculate the settings and draw the relay characteristics on an R–X diagram.

At bus A, M_1 equals 150 % $\times Z_{\text{line}}$ or $1.5 \times (2 + j10) = 15.3 \angle 79^\circ \, \Omega$. Set relay at $15.0 \angle 60^\circ$.

Similarly, at bus B, $M_2 = 150 \% \times Z_{\text{line}}$ or $1.5 \times (2 + j10) = 15.3 \angle 79^\circ \, \Omega$. Set relay at $15.0 \angle 60^\circ$.

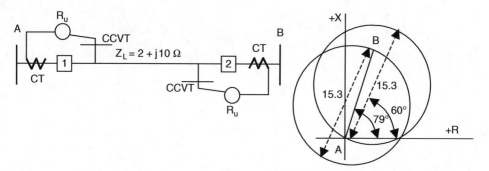

Figure 6.16 Calculation and diagram for POTT setting of Example 6.3.

6.7.3 Supplementary POTT logic

POTT logic may also be supplemented by additional features as noted here.

Hybrid logic adds a reverse looking blocking function to increase security. This logic prevents tripping and keying of permissive trip if a fault in the reverse direction is detected. A small time delay on reset of the blocking function prevents incorrect tripping during current reversals for faults on a parallel line. Current reversals have previously been discussed in Section 4.5.1 (Chapter 4). In the case of hybrid POTT (without the small reset delay), the current reversal could cause an incorrect trip due to a race between reset of the reverse-looking blocking function and the forward-looking tripping function.

Weak source echo logic allows the POTT logic to operate even if one terminal of the line is too weak to cause a forward-looking protection function to operate. This function increases sensitivity of the scheme. The echo function will echo a received permissive trip back to the sending terminal even if no forward-looking fault detector operates at the receiving terminal. The echo function is only applied if the reverse-looking hybrid logic is also applied to prevent an echo for a fault in the reverse direction. Special logic is required to reset the echo function to prevent ping pong from end-to-end or permanent lock up of the permissive trip signal.

Weak source tripping logic may also be added to increase the dependability of the logic in some applications. This logic is an optional supplement to the weak source echo logic. If one terminal is too weak for a distance or overcurrent fault detector to operate under all conditions, a voltage fault detector could also cause tripping if a permissive tripping signal is received and no reverse fault has been detected. The voltage fault detector has two parts:

- Phase-to-phase measuring undervoltage detectors between all three phase pairs or positive-sequence undervoltage detector to detect multiphase faults.
- Zero-sequence overvoltage detector to detect single-line-to-ground faults.

More details on these supplementary functions are provided in Reference [8].

6.8 Permissive underreaching transfer trip (PUTT)

The use of two functions, an underreaching and an overreaching relay, at each terminal, results in even greater security at the cost of reduced sensitivity. The supplementary hybrid blocking function that is often used to provide increased security in POTT schemes is not necessary in a PUTT scheme. The same sensitivity and short line application limitations exist in the PUTT scheme as noted in Section 6.6.3 for the DUTT scheme.

6.8.1 PUTT scheme using frequency shift tones

Figure 6.17 shows the PUTT scheme and logic using frequency shift tones. The underreaching functions $21\text{-}1_u$ and $21\text{-}2_u$ initiate the trip by shifting the transmitted frequency from guard to trip, and the overreaching functions $21\text{-}1_o$ and $21\text{-}2_o$ provide the permissive supervision. In addition, the underreaching functions can provide a zone 1, instantaneous, direct tripping function to the local breakers, and the overreaching functions, with a time delay, can provide backup second-zone protection. Again, each terminal only receives the remote transmitter's signal. Historically, when discrete electromechanical relays were used, this scheme had an advantage over the POTT scheme due to the availability of the two relays for backup protection.

6.8.2 PUTT schemes using relay-to-relay communications

The simplified logic scheme for a PUTT scheme using digital relays with peer-to-peer communications is shown in Figure 6.18. Note that the underreaching and overreaching function settings are the same as shown in Figure 6.17. Figure 6.18 also shows the logic of the communications independent backup instantaneous and time-delayed tripping functions.

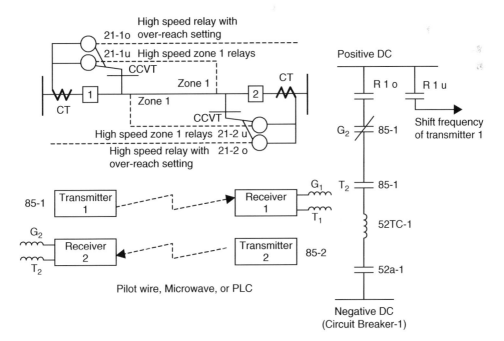

Figure 6.17 PUTT scheme using frequency shift tones.

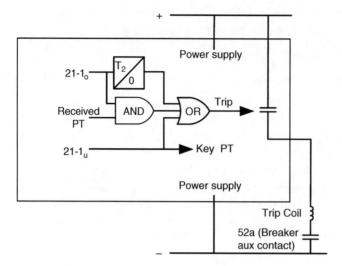

Figure 6.18 PUTT scheme using digital relay.

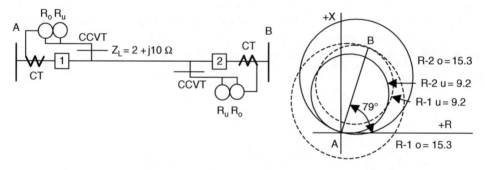

Figure 6.19 Calculation and diagram for PUTT setting of Example 6.4.

Example 6.4 Consider the transmission line shown in Figure 6.19 with section AB protected by a PUTT scheme using admittance-type relays at both ends for the underreaching and overreaching functions. Calculate the settings and draw the relay characteristics on an R–X diagram.

At bus A, R_o equals 150% of Z_{line} or 15.3 Ω, and R_u is 90 % × 10.0 or 9.2 Ω.

At bus B, the same settings could be used, that is, $R_o = 15.3$ Ω and $R_u = 9.2$ Ω.

6.9 Phase comparison relaying

Phase comparison relaying is a differential scheme that compares the phase angle between primary currents at the two ends of a line [8]. If the two currents are essentially in phase, there is no fault in the protected section. If these two currents are essentially 180° out of phase, there is a fault within the line section. Any communication link can be used.

Figure 6.20 shows the direction of the primary current in the protected line between circuit breakers A and B. For faults at F_1 or F_3, I_A and I_B will be in phase. For a fault at F_2, I_A and I_B are 180° out of phase. The outputs of the secondaries of the three CTs at each end of the line are fed into a composite sequence network that combines the positive-, negative-, and zero-sequence currents to produce a single-phase voltage output. This voltage is directly proportional to the phase angle and magnitude of the three-phase currents at each terminal. A squaring amplifier converts the single-phase voltage to a square wave. The square wave is used to key the channel transmitter to send a signal to the remote terminal. In a single-phase comparison scheme, the squaring amplifier creates a mark on the positive half-cycle and a space, that is, nothing, on the negative half-cycle. For an internal fault, the CT secondary current is in phase at both ends, as shown in Figure 6.20. As a result, both ends have a mark at the same time and each end transmits this mark to the remote end where a comparator sees the local mark and the remote mark at the same time and issues a trip signal. For an external fault, one end sees a positive half-cycle and creates a mark and the other end sees a

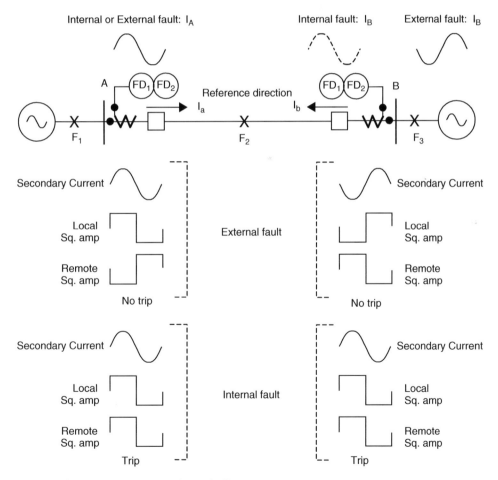

Figure 6.20 Phase comparison schematic diagram.

negative half-cycle at the same time and does not transmit any signal. Each local comparator, therefore, sees only one mark and one space simultaneously, and does not issue a trip signal. In a dual-phase comparison scheme, two frequencies are used, one in place of the mark and the other in place of the space. The comparators at each end then see either two signals of the same frequency at the same time and will trip, or two different frequencies and will not trip.

The overreaching fault detectors FD_1 at each terminal start the transmitter and, at the same time, provide an input to a comparator. The higher set overreaching element FD_2 at each end arms the system for tripping, depending upon the result of the comparison of the local and remote square waves.

The scheme is such that the positive and negative blocks of the square wave of the secondary current are 180° from the positive and negative half-cycles of the primary current wave. This is due to the relative location of the CT polarities, as shown in Figure 6.18. Since a single-phase comparison relay transmits only on the positive half-cycle, there is a possible tripping delay of one half-cycle. A dual-phase comparison relay transmits on both halves of the square wave and is therefore ready to trip on each half-cycle. It is important to recognize that a phase comparison relaying scheme provides protection only for faults within the protected line. It does not provide backup protection for faults on an adjacent line, nor does it have inherent backup protection for its protected line section. For this reason, additional protection must be used at each terminal for backup protection, and to allow maintenance on the communication equipment. In addition, since this system operates only on line current, it is not subject to tripping on power swings or out-of-step conditions. Another advantage is the fact that the phase comparison scheme does not require a voltage to establish directionality or distance measurements. Distance or directional relays are subject to errors caused by mutual coupling of parallel circuits, a common configuration at lower transmission voltages. Lower-voltage systems often are installed on double-circuit towers with the resulting induced zero-sequence voltages. Directional relays may operate for "backward" faults, and distance relays can either underreach or overreach depending upon the relative angle between the ground current and phase current [11, 12].

A commonly used scheme is a non-segregated phase comparison scheme in which identification of the faulted phase is lost due to the mixing transformers. Additional monitoring for postfault analysis may be required. Also used, however, is a segregated phase comparison scheme, which uses one relay per phase, each relay individually coupled to each phase. The segregated phase comparison scheme may also use a fourth phase comparison of neutral currents for more sensitive ground fault protection. In the case of electromechanical relays and on–off carrier or tone-based teleprotection, this is a more expensive scheme, but it provides faulted phase identification, which is useful in single-phase tripping, postfault analysis, and other operating strategies. In the case of digital multifunction protection relays and digital communications, there is no extra cost. This makes segregated phase comparison the preferred scheme when implemented by digital relays.

The specific settings of a phase comparison system are determined in very large measure by the filters and the timing coordination required by its blocking and tripping logic. This is unique to the various manufacturers, and their instruction books must be followed. In general, however, there are two fault detectors involved: FD_1, a low-set relay, that is, a relay with a low setting, which starts the carrier; and FD_2, a high-set relay, a relay with a high setting, which arms the tripping circuits, waiting for the comparison of the transmitted signals. FD_1 must be set above maximum load if possible, to avoid continuous transmission and it must operate for all internal faults. FD_2 must not operate on maximum load and is set between 125 and 200% of FD_1. FD_2 must be set high enough to reset itself in the presence of heavy loads following clearing an external fault. A distance relay is added if the minimum internal three-phase fault is less than twice the maximum load current. The same setting in primary amperes should be used at both ends to ensure coordination. This can be difficult if the contributions from the two ends are significantly different.

An important factor in the application of phase comparison relaying is accounting for the latency in transmitting the square waves from one terminal to the other. It is important that the current at the local end be delayed before comparison so that the correct phase relationship between the local and remote terminals is maintained. This delay is provided by a local delay timer which is usually calibrated during commissioning. The communications system should be carefully designed so that there is no jitter, drift, or change in the delay time. Some digital relays may periodically send a timing signal from one end to the other at which it is bounced back to the sending end using a "ping-pong" technique. The sending end will then measure the time difference between sending and receiving, and set the local time delay to half the total time. This technique is satisfactory as long as there is symmetry in the delays in each direction. Symmetry is the normal expectation in utility-owned and -operated teleprotection systems, but may not always be the case in synchronous optical networks (SONET) systems where asymmetry may cause difference in send and receive times.

Example 6.5 Consider that the transmission line shown in Figure 6.18 is a 138-kV, 100-mile line and its primary impedance is 80 Ω. Calculate the settings for the low-set and high-set relays.

Assume that the equivalent impedances at the end of the line are negligible compared to the line length. A three-phase fault at either end of the line is equal to $138\,000/(\sqrt{3} \times 80)$ or 1000 primary amperes. Assume that the minimum phase fault is 866 A and the maximum load is 300 A. Set FD_1 equal to $1.5 \times 300 = 450$ A and FD_2 equal to $1.25 \times 450 = 562.5$ A. The minimum fault, 866 A, is almost twice FD_1, which is acceptable, and is more than twice the maximum load; so no distance relay is required. The actual relay settings, of course, must be calculated in secondary ohms, that is, using the appropriate CT ratio.

Example 6.6 Assuming that the system behind one of the buses in Example 6.5 is very weak and has an equivalent primary impedance of 40 Ω, a three-phase fault at the remote end would be equal to $138\,000/(\sqrt{3} \times 120) = 664$ A, and the assumed minimum fault is 575 A. The same settings would now result in FD_2 being less than twice the maximum load of 300 A and a distance relay would be required to supervise the trip.

6.10 Current differential

In a current differential scheme, a true differential measurement is made [8, 13]. Ideally, the difference should be zero or equal to any tapped load on the line. In practice, this may not be practical due to CT errors, ratio mismatch, or any line charging currents. Information concerning both the phase and magnitude of the current at each terminal must be made available at all terminals in order to prevent operation on external faults. Thus, a communication medium with sufficient bandwidth must be provided, which is suitable for the transmission of these data. The same issues that were mentioned in Section 6.9 with synchronizing measurements at the various terminals (considering channel delay times) exist with current differential schemes. In cases where asymmetry in channel times may invalidate the "ping-pong" timing technique, discrete current measurements may be time-stamped by precision global positioning satellite (GPS) clocks as discussed in Chapter 13.

There are two main types of current differential relaying. Legacy implementations may combine the currents at each terminal into a composite signal and compare these composite signals through a communication channel. Modern digital implementations measure individual phase, neutral, and, in some cases, negative-sequence currents, and transmit these measurements between terminals [13].

Digital communications monitoring features are generally used to block differential protection trips and alarm in the event of communications failure. Alarms and blocking options are also provided to alert loss of one or more CT inputs. Phase current differential minimum pickup settings are sometimes set above load currents to provide security in the event of loss of one or more CT inputs. Negative- and zero-sequence current differential functions can provide more sensitive protection for unbalanced faults, since they are not sensitive to load currents.

Current differential schemes tend to be more sensitive than distance-type schemes, since they respond only to the current, that is, no potential device is required. This tends to be more dependable, but at a cost to security. Since no potential is involved in the basic differential principle, they are not affected by system swing conditions or blown fuse problems. Some relays include voltage inputs. These can be used to compensate for line charging currents in EHV lines or cables where such currents may be significant. Even though, there are no inherent backup capabilities in the differential principle, the voltage inputs in many modern implementations enable supplementary distance and ground overcurrent functions to provide backup in the case of communications failure.

As communication systems are becoming more sophisticated, transmission line current differential is now well established. It inherently accommodates multiterminal lines by the logic applied within the relay that can monitor all of the current inputs. In addition, the computer can detect second and fifth harmonics, allowing it to differentiate between transformer energization and magnetizing inrush, permitting its application to a line with a tapped transformer (refer to Section 8.4 in Chapter 8).

Communication media can be metallic pilot wires, audio tones, leased pilot wires, microwave, or fiber optics. PLC is not suitable due to the wide bandwidth required.

Three modern technologies are available for line current differential protection using digital communications.

6.10.1 Percentage-restrained line current differential

This type of protection uses the same principle as generator, transformer, and bus differential described in Chapters 7, 8, and 9, respectively. Similar to the case of any other differential protection, CT errors may cause incorrect apparent differential current. The percentage-restraint principle can mitigate the effect of CT errors by appropriate setting of the slope of the differential characteristic. As in the case of other differential protections, the use of multiple slopes can be helpful with providing sensitivity for low-magnitude internal faults while providing security for high-magnitude external faults which could cause partial CT saturation. Unlike the case with other differential protections, the percentage restraint applied to lines will include local delay timers to mitigate the effect of channel delay in receipt of current samples from the remote terminal.

Many North American substation arrangements provide two breakers at line terminals in a ring bus or breaker-and-a-half or other arrangement. Such an arrangement facilitates taking one of the breakers out of service for maintenance. As discussed for the case of three winding transformers in Section 8.6.1 (Chapter 8), it is helpful for security of transmission line differential protection if separate restraints are provided for each of the two breakers at a line terminal. Several types of modern line current differential protection systems include facilities for obtaining such separate restraint currents.

6.10.2 Alpha plane differential

The alpha plane differential protection technique calculates the complex ratio of currents entering each end of the protected line. If the current entering the local terminal is the phasor value I_L and entering the remote terminal is the phasor value I_R, the ratio I_R/I_L is measured. With no internal fault, or measuring errors, this ratio would be -1, as I_L would be equal and 180° out of phase with I_R. Figure 6.21 shows the operating characteristic of an alpha plane differential. The protection will trip for any ratio outside the restrain region. Depending on whether or not there is significant fault resistance, for most external faults the ratio will lie close to the real axis. Phase comparators measure α which (depending on the setting) will compensate for timing and CT errors. Magnitude comparators measure R which (depending on the setting) will compensate for CT error line charging currents and small tapped loads if significant. A small value of R will increase sensitivity (and reduce security) as will a large value of α. Typical values of α and R are 75° and 6, respectively.

6.10.3 Charge comparison differential

Charge comparison differential compares the charge or integral of each half-cycle of current (ampere seconds) at each end of a line [14]. If there is no fault, this sum should be close to zero. The main advantage of this technique is that communications requirements are not as stringent as is the case with percentage-restrained differential or alpha plane differential. Percentage or alpha plane differentials compare a number of samples (usually 8 or more) each cycle. The comparison of charges need only be twice in every cycle, so

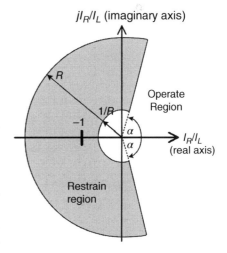

Figure 6.21 Alpha plane differential operating characteristic.

the error in setting the local time delay, or jitter or shifting in communication delay need only be less than ±4 ms. Further, only two values need to be sent every cycle, so the bandwidth requirements of communications for the charge comparison differential are significantly less than other differential techniques. It may be noted, however, that modern communications facilities have no difficulty in meeting the communications requirements for percentage or alpha plane differential systems.

6.11 Pilot wire relaying

Pilot wire relaying is a form of differential line protection similar to phase comparison, except that the phase currents are compared over a pair of metallic wires. As in the directional comparison versus transferred tripping schemes, there are two variations, a tripping pilot and a blocking pilot.

6.11.1 Tripping pilot

The operation of a tripping pilot wire relay scheme is illustrated in Figure 6.22. The secondary currents from the three CTs at each terminal are fed into a mixing network that produces a single voltage output. This is one of the disadvantages of the pilot wire scheme. As with the non-segregated phase comparison scheme, the faulted phase identification is lost. This output voltage is applied to the associated sensing unit, essentially composed of an operating and a restraining circuit, and on to the pilot wires through an insulating transformer. The operating circuit is considerably more sensitive than the restraining circuit. When the current entering bus B is equal to, and in phase with, the current leaving bus A, as in the case of load or an external fault at F_1, the voltages V_A and V_B are also equal and in phase. For this condition, there will be no current circulating in the pilot wires or in the operating circuits of the relays. The voltages at each end will provide restraining coil current and the relays will not operate.

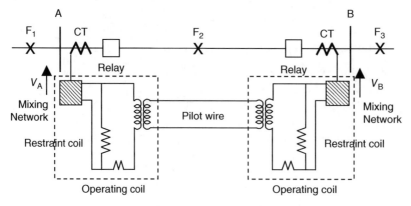

V_A and V_B shown for external fault

Figure 6.22 Schematic for tripping pilot system.

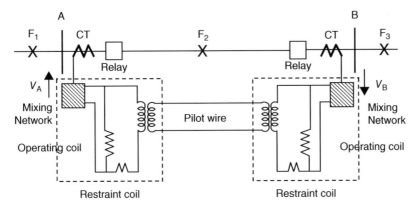

V_A and V_B shown for external fault

Figure 6.23 Schematic for blocking pilot system.

For an internal fault at F_2, the current entering bus A will be 180° out of phase with the current entering bus B, although not necessarily of equal magnitude. For this condition, V_A and V_B will also be 180° out of phase and, depending on the fault currents, probably unequal. This will result in a circulating current flowing in the pilot wires and in the operating circuits of the relays at both terminals. While some current will flow in the restraining circuit, the operating current will dominate and the relays will operate. Since this scheme requires a circulating current to trip, if the channel is open-circuited, no tripping can occur. However, if the pilot wires are shorted, false tripping can occur. To monitor for open circuits, a few milliamperes of a DC monitoring current can be continuously impressed across the pilot pair. A fault detector can be provided to prevent false tripping, although this reduces the sensitivity of the protection to that of the setting of the fault detector.

6.11.2 Blocking pilot

This scheme is shown in Figure 6.23. The difference between the tripping and the blocking schemes is in the relative location of the restraining and operating circuits and in the phase relation of the voltage output for internal and external faults. Following the same sequence of operation as described for the tripping pilot, it can be seen that the blocking scheme requires a circulating current in the pilot wires to prevent tripping. The effect of open and shorted pilot pairs is opposite, but the remedies are the same.

6.12 Multiterminal lines

In Chapter 5, we discussed the problems associated with infeeds and multiterminal lines and the difficulty in calculating the settings for overcurrent and distance relays that would adequately protect the lines for all system configurations. The same difficulties confront us with pilot relaying. As with two-terminal lines, pilot relaying is the only way to protect 100% of the lines instantaneously. If there is no outfeed from any of the terminals, directional comparison is preferred. If, due to the specific system configuration, the fault contribution from one terminal flows out of that terminal, the fault detector relays, either directional or nondirectional, will start the carrier, and the tripping relays will not stop it. Tripping is, therefore, prevented. One solution to this problem is to wait until the zone 1 relays at one or more terminals trip the associated breaker, eliminating the multiterminal configuration and normal operation will resume. Another solution is to use phase comparison relays. Although outfeed during load or an external fault does not affect phase comparison, the setting criteria in a three-terminal configuration are extremely stringent. In the best situation with equal contributions from all terminals for an external fault, one terminal will see twice as much fault current as either of the other two. During the half-cycle in which the terminal with the larger fault current is sending a blocking signal, there is no problem. However, during the other half-cycle, either of the other two terminals must send the blocking signal. If the fault is assumed to be just at the trip setting of the relay with the larger contribution, neither of the other terminals will have enough current to cause a blocking signal to be sent with normal settings. The optimum technical solution is line current differential protection. This protection measures the sum of all currents flowing into a line. As long as CT and timing errors are managed properly (as in a two-terminal application), this solution will elegantly handle inflows and outflows. The main obstacle to line current differential protection is the availability of communications systems with adequate bandwidth between all terminals.

Example 6.7   Consider a symmetrically configured three-terminal line, 1, 2, and 3, that is, all legs and all fault contributions are equal (Figure 6.24). Assume an external three-phase fault beyond terminal 3. Terminals 1 and 2 contribute equal fault current to the fault, but terminal 3 contributes twice their current. If the fault is just at the level of FD_h of terminal 3, say 125 A, then the FD_1 settings at terminals 1 and 2 would be 100 A, but their fault contributions would be 62.5 A. They would not send blocking signals to terminal 3.

The alternatives are (1) to halve the carrier start level or (2) to double the tripping level. Alternative (1) has the undesired effect of transmitting during normal load conditions, which monopolizes the communication channel, making it unavailable for other uses. Alternative (2) has the disadvantage of reducing the sensitivity. If possible, the trip setting must therefore be 250% of the start setting and the start setting must be three times the maximum load, conditions that cannot always be met. If there is a limited source of fault current from one terminal, the situation worsens [15–17].

With no outfeed, directional comparison schemes can be applied. A tapped load is not a concern, since the distance relay characteristics are relatively immune to load, as discussed

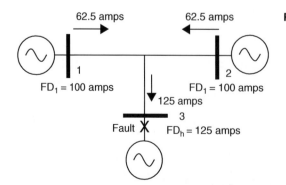

Figure 6.24   System for Example 6.7.

before. All carrier start relays must reach beyond all carrier trip relays with good margin. Recall the rules stated in Chapter 5 concerning infeed: set the carrier start relays without any infeed and set the trip relays with maximum infeed. If any terminal can experience out-feed, then the directional feature is fooled and a blocking signal is sent to all terminals. One remedy is to rely on zone 1 relays at one of the infeed terminals to open its breaker, thus (sequentially) allowing all other terminals to respond correctly. The best technical remedy is to use line current differential protection.

Example 6.8 Consider the system shown in Figure 6.25. We may assume that the relative magnitudes of I_1, I_2, and I_3 remain unchanged for any fault on the system between buses A through G. This is clearly an approximation, and in an actual study we must use appropriate short-circuit calculations for each of the faults. We are required to set the directional compar-ison carrier trip and carrier start relays at bus B. The system nominal voltage is 138 kV. Assume that the maximum load is 360 MVA. $I_{f1} = 360\,000/(\sqrt{3} \times 138) = 1506.2$ A. The CT ratio is therefore $1500/5 = 300:1$. The primary turns ratio is $138\,000/(1.73 \times 69.3) = 1150$. Assume that $I_2/I_1 = 0.5$.

Carrier Trip
This must be set equal to 175% of the longer of the two impedances between buses B and D, and between buses B and G. Infeed will be considered to be present, and will apply to the impedance segment C–D. Thus, the carrier trip setting is $1.75 \times [4 + j40 + 1.5 \times (2 + j20)] = 12.25 + j122.5$ or $123.11 \angle 84.2°$ primary ohms or $32.12 \angle 84.2°$ secondary ohms. Assume the distance relay has settings from 10 to 40 Ω and angle adjustments of 75° and 80°. Set the relay at 30 Ω and 80°.

Carrier Start
This must be set equal to 125% of the carrier trip setting minus the impedance between buses B and D without the infeed. Thus, the carrier start setting is equal to $\mathbf{1.25} \times [\mathbf{119} \angle \mathbf{80}°$ $- (\mathbf{6} + \mathbf{j60})] = \mathbf{59.04} \angle \mathbf{75.6}°$ primary ohms or $\mathbf{15.4} \angle \mathbf{75.6}°$ secondary ohms. Set the relay at 16 Ω and 80°.

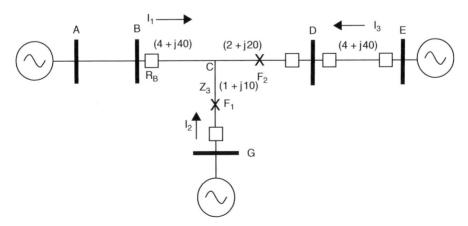

Figure 6.25 System with infeed.

6.13 The smart grid

The smart grid, also described as "distribution grid integration," has become one of the most innovative subjects of research topics today. "Smart grid" generally refers to a class of technology designed to modernize the existing utility distribution system to improve the efficiency on the power network and energy use in homes and businesses. The smart grid is made possible by two-way communication between the power generation from dispersed facilities and the consumers in homes and businesses. It includes sensing and measurements using phasor measurements streaming over high-speed channels, smart meters that record system parameters and report system conditions and quality to central dispatching centers, and advanced control and protection methods that include differential line relaying, adaptive settings, and various system-integrity protection schemes that rely on low-latency communication. Improved interfaces and decision support will utilize instantaneous measurements from phase measurement units (PMUs) and other sources to drive fast simulations and advanced visualization tools that can help system operators assess dynamic challenges to system stability.

The smart grid can introduce load information to the digital protection elements to assure coordination between load and fault. These features have not yet been fully implemented, but are clearly future possibilities. The increasing use of solar and wind-generating devices at the distribution system level has made the introduction of a smart grid almost a necessity [18].

6.14 Summary

In this chapter, we have examined the application of pilot channels to provide instantaneous protection over the entire length of a given line section. Two operating modes are possible: a blocking mode when the communication channel is an integral part of the power line and a tripping mode when the channel is independent of the line. This form of protection involves the transmission of relaying information from one terminal of the zone of protection to the other by means of an appropriate communication channel. These channels can be PLC, microwave, fiber optics, or pilot cable. The relaying schemes can be classified as directional comparison, phase comparison, current differential, or pilot wire depending on the type of sensing used, and are further described as blocking, unblocking, or transfer trip depending on how the transmitted signal is used. The transfer trip schemes are again divided into direct, permissive underreaching, and permissive overreaching. There are, of course, advantages and disadvantages associated with each scheme, and the specific application depends on all of the individual factors and conditions involved. Directional distance relaying is the most commonly used throughout the world, but it has application and setting problems when series capacitors are present. Phase comparison and current differential are immune to such problems, and only require current inputs, eliminating the need for potential sources. There is no inherent backup provided, however, and the coupling requirement may become expensive. If a microwave or fiber-optic channel is available, then a differential or tripping scheme is preferred. If PLC is used, then a blocking scheme is used. Pilot cable is only used if the distance between terminals is short. We have examined setting criteria and the specific application to multiterminal lines.

Problems

6.1 For each of the following pilot schemes, select a channel of communication, that is, PLC, microwave, or pilot wire, and select a mode of operation, that is, blocking or tripping, and explain why such a choice was made: (i) directional comparison blocking; (ii) PUTT; and (iii) phase comparison.

6.2 Relay R_{ab} in Figure 6.26 represents a directional comparison blocking relay using directional mho relays for tripping and reversed mho relays, without offset, for blocking. Determine the carrier start and carrier trip settings for the phase relays. The system nominal voltage is 138 kV, and the positive-sequence impedances for the various elements are given in the figure. Assume that the maximum load is 120 MVA and select a CT ratio accordingly. The available relays can be set from 0.2 to 10 Ω or from 0.5 to 40 Ω whichever is required. The angle of maximum torque can be adjusted to 75° or 80°.

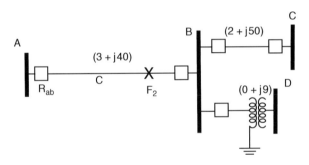

Figure 6.26 System for Problem 6.2.

6.3 Draw the R–X diagram for the relays of Problem 6.2.

6.4 Repeat Problem 6.2 for a POTT scheme.

6.5 Consider the multiterminal line in the system shown in Figure 6.27. Each of the buses C, D, G, H, and J has a source of power behind it. For a three-phase fault on bus B, the contributions are shown in the table in Figure 6.27. Assume that the fault current contributions from each of the sources remain unchanged as the fault is moved around the system. Determine the start and trip settings for the directional comparison blocking relays at buses A and B. Assume the same relays as in Problem 6.2.

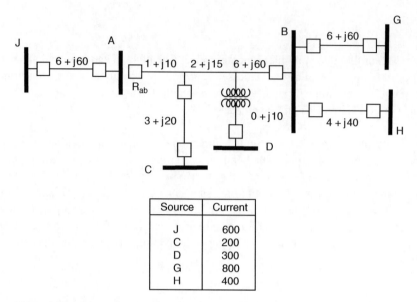

Source	Current
J	600
C	200
D	300
G	800
H	400

Figure 6.27 System for Problem 6.5.

6.6 Repeat Problem 6.2 for a phase comparison scheme between terminals A and B. Assume infinite equivalent sources behind terminals A, C, and D.

References

1 IEEE Power and Energy Society, Power System Relaying Committee Report (2013), *Communications Technology for Protection Systems.* On Line at http://www.pes-psrc.org/kb/published/reports/IEEE%20PSRC%20Subcommittee%20H9%20Understanding%20Comm%20Tech%20for%20Protection%20-20130113%20D8F.pdf

2 IEEE (2004). *Std 643-2004 (Revision of IEEE Std 643-1980) IEEE Guide for Power-Line Carrier Applications IEEE Standards Association.* New York, NY: IEEE.

3 IEEE Power and Energy Society, Power System Relaying and Control Committee Report, *Special Considerations in Applying Power Line Carrier for Protective Relaying* On line http://www.pes-psrc.org/kb/published/reports/SpecialConsiderationsPLC.pdf

4 IEEE Std C37.94-2017 (2017) (Revision of IEEE Std C37.94-2002). *IEEE Standard for N times 64 kbps Optical Fiber Interfaces between Teleprotection and Multiplexer Equipment.* New York, NY: IEEE Standards Association, IEEE.

5 IEEE Power System Relaying Committee (2004). *ANSI/IEEE standard C37.93-2004, (Revision of IEEE Std C37.93-1987), IEEE Guide for Power System Protective Relay Applications of Audio Tones Over Voice Grade Channels.* Piscataway, NJ: IEEE Standards Association, IEEE.

6 AIEE (1962). Evaluation of transfer trip relaying using power line carrier. Committee report. *AIEE Trans. Part III* **81**: 250–255.

7 Cheek, R.C. and Blackburn, J.L. (1952). Considerations in selecting a carrier relaying system. *AIEE Trans. PAS* **71**: 10–15.

8 IEEE (2015). *Standard C37.113-2015 IEEE Guide for Protective Relay Applications to Transmission Lines.* Piscataway: IEEE Standards Association, IEEE.

9 Cordray, R.E. and Warrington, A.R.v.C. (1944). The MHO carrier relaying scheme. *AIEE Trans.* **63**: 228–235.

10 Westinghouse (1976). *Applied Protective Relaying.* Newark, NJ: Westinghouse Electric Corporation.

11 Blackburn, J.L. (1962). Voltage induction in parallel circuits. *AIEE Trans.* **81**: 921–929.

12 Elmore, W.A. and Blackburn, J.L. (1962). Negative sequence directional ground relaying. *AIEE Trans.* **81**: 913–921.

13 IEEE (2015). *Std C37.243-2015 IEEE Guide for Application of Digital Line Current Differential Relays Using Digital Communication.* Piscataway: IEEE Standards Association, IEEE.

14 Ernst, J.L., Hinman, W.L., Quam, D., and Thorp, J.S. (1992). *Charge comparison protection of transmission lines-relaying concepts. IEEE Trans. Power Delivery* **7** (4): 1834–1852.

15 Horowitz, S.H., McConnell, A.J., and Seeley, H.T. (1960). Transistorized phase-comparison relaying: application and tests. *AIEE Trans.* **79**: 381–392.

16 AIEE Power System Relaying Committee (1961). Protection of multiterminal and tapped lines. Committee report. *AIEE Trans.* **80**: 55–65.

17 IEEE Power System Relaying Committee (1979). *Protection Aspects of Multi-Terminal Lines.* IEEE special publication 79-TH0056–2-PWR.

18 Horowitz, S., Phadke, A.G., and Renz, B. (2010). The future of power transmission. *IEEE Power Energy* **8** (2): 34–40.

7

Rotating Machinery Protection

7.1 Introduction

The protection of rotating equipment involves the consideration of more possible failures or abnormal operating conditions than any other system element. Although the frequency of failure, particularly for generators and large motors, is relatively low, the consequences in cost and system performance are often very serious. Paradoxically, despite many failure modes that are possible, the application principles of the protection are relatively simple. None of the complications require a pilot scheme. Those failures involving short circuits are usually detected by some type of differential or overcurrent protection. Many failures are mechanical in nature and use mechanical devices such as limit, pressure, or float switches, or depend upon the control circuits for removing the problem [1].

Some of the abnormal conditions that must be dealt with are the following:

1) Stator winding faults—phase and ground faults.
2) Overload.
3) Overspeed.
4) Abnormal voltages and frequencies.

For generators, we must consider the following:

5) Underexcitation.
6) Motoring and start-up.

For motors, we are concerned with the following:

7) Stalling (locked rotor).
8) Single phase.
9) Loss of excitation (synchronous motors).

There is, of course, some overlap in these areas, particularly in overloads versus faults, unbalanced currents, and single phasing, and so on. Thus, relays applied for one hazard may operate for others. Historically, when discrete electromechanical relays were applied, some relays would back each other up for the same hazard, and many such legacy systems are still in service. Modern practice is often to apply multifunction digital relays that provide a suite of protection functions that protect against almost all hazards. Such multifunction

Power System Relaying, Fifth Edition. Stanley H. Horowitz, Arun G. Phadke and Charles F. Henville.
© 2023 John Wiley & Sons Ltd. Published 2023 by John Wiley & Sons Ltd.

protection results in a possible dependability problem with a single relay failure disabling most or all the protection. For this reason, multifunction relays are usually applied with redundant protection systems or at least with separate backup for the major fault types. In the figures in this chapter, discrete protection relays are shown with various function numbers for various hazards. In practice, the various protection functions may be provided in a single multifunction device with a variety of voltage transformer (VT) and current transformer (CT) inputs.

Since the solution to a given failure or abnormality is not the same for all failures or abnormalities, care must be taken that the proper solution is applied to correct a specific problem. In some instances, tripping of the unit is required; in other cases, reduction in load or removing some specific equipment is the proper action. This will be discussed in greater detail as we examine each type of failure.

Several of these abnormal conditions do not require automatic tripping of the machine, as they may be corrected in a properly attended station while the machine remains in service. Hence, some protective devices only actuate alarms. Other conditions, such as short circuits, require fast removal of the machine from service. The decision, whether to trip or alarm, varies greatly among utilities and, in fact, between power plants of a given utility or between units in a single plant. The conflict arises because there is a justifiable reluctance to add more automatic tripping equipment than is absolutely necessary. Additional equipment means more maintenance and a greater possibility of incorrect operation. In today's systems, the loss of a generator may be more costly, in terms of overall system performance, than the delayed removal of a machine. On the other hand, failure to promptly clear a fault, or other abnormality, may cause extensive damage and result in a longer, more expensive outage. The decision is not obvious nor is it the same for all situations. It requires judgment and cooperation between the protection engineer and the appropriate operating and plant personnel.

7.2 Stator faults

7.2.1 Phase fault protection

For short circuits in a stator winding, it is standard practice to use differential protection on generators rated 1000 kVA or higher and on motors rated 1500 hp or larger or rated 5 kV and above. Rotating equipment provides a classic application of this form of protection, since the equipment and all of the associated peripherals such as CTs, breakers, and so on are usually in close proximity to each other, thereby minimizing the burden and possible error due to long cable runs. In addition, since there is only one voltage involved, the CT ratios and types can be the same, with matched characteristics. They should be dedicated circuits and should not be used with any other relays, meters, instruments, or auxiliary transformers without a careful check on the effect on CT performance.

The CTs used for the generator differential are almost invariably located in the buses and leads immediately adjacent to the generator winding. This is done to limit the zone of protection; so a fault in the generator is immediately identifiable for quick assessment of damage, repair, and restoration of service. The buses themselves are usually included in their own differential or in some overall differential scheme.

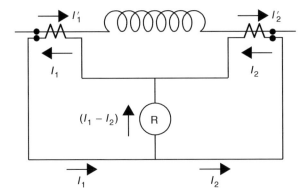

Figure 7.1 Generator differential connection with overcurrent relay.

In motor differential circuits, three CTs should be located within the switchgear in order to include the motor cables within the protection zone. The other three CTs are located in the neutral connection of the motor. Six leads must be brought out of the motor: three on the incoming cable side to connect to the switching device and three on the motor neutral to accommodate the CTs before the neutral connection is made (refer to Figure 7.15). The provision and connections for the CTs must be specified when the motor is purchased [2].

Figure 7.1 shows the basic differential connection using a simple overcurrent relay. This protection scheme is described in Section 2.2, and shown in Figure 2.5 (Chapter 2). For an external fault, the relay sees $I_1 - I_2$, which is zero or very small. For an internal fault, the relay will see $I_1 + I_2$, which can be very large. This big difference between the current in the relay for an internal fault compared to an external fault makes the setting very easy, that is, sufficiently above the external fault for security and enough below the internal fault for dependability. This precise distinction between the location of an internal and an external fault is what makes the differential circuit such an ideal protective principle.

Example 7.1 Consider the system shown in Figure 7.2 which represents a generator prior to being synchronized to the system. The generator is protected by an overcurrent relay, 87,[1] connected in a differential circuit as shown. The maximum load is $125\,000/(\sqrt{3} \times 15.5) =$ 4656.19 A. For this maximum load select a 5000:5 (1000:1) CT ratio. This results in a secondary current of 4.66 A at full load. Before the unit is synchronized, a three-phase fault at either F_1 or F_2 is $(V_{pu}/x''_d) \times I_{F1}$ or $(1.0/0.2) \times 4656.19 = 23\,280.95$ primary amperes or 23.28 secondary amperes.

For the external fault at F_2, 23.28 A flows through both CT secondary circuits and nothing flows in the overcurrent relay.

For the internal fault at F_1, 23.28 A flows through only one CT secondary and the operating coil of 87.

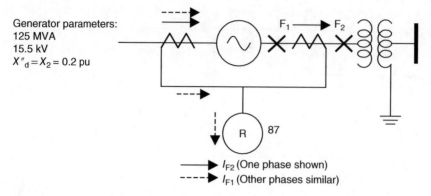

Generator parameters:
125 MVA
15.5 kV
$X''_d = X_2 = 0.2$ pu

I_{F2} (One phase shown)
I_{F1} (Other phases similar)

Figure 7.2   One-line diagram for Example 7.1.

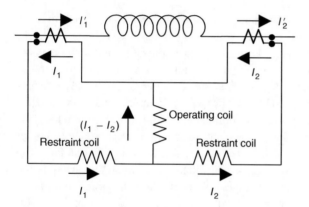

I'_1

I_1

I'_2

I_2

Operating coil

$(I_1 - I_2)$

Restraint coil

Restraint coil

I_1

I_2

Figure 7.3   Generator differential using percentage differential relay.

This arrangement would be ideal if the CTs always reproduced the primary currents accurately. Actually, however, the CTs will not always give the same secondary current for the same primary current, even if the CTs are commercially identical. The difference in secondary current, even under steady-state load conditions, can be caused by the variations in manufacturing tolerances and in the difference in secondary loading, that is, unequal lengths of leads to the relay, and unequal burdens of meters and instruments that may be connected in one or both of the secondaries. What is more likely, however, is the error current that can occur during short-circuit conditions. Not only is the current magnitude much greater, but there is the possibility of DC offset so that the transient response of the two CTs will not be the same. This difference in secondary current will flow in the relay. An overcurrent relay must then be set above the maximum error current that can flow during the external fault; yet it must be set significantly below the minimum fault current that can accompany a fault that is restricted due to winding or fault impedance.

The percentage differential relay solves this problem without sacrificing the sensitivity. The schematic arrangement is shown in Figure 7.3. This scheme has already been introduced in Section 2.2.1 (Chapter 2). Depending upon the specific design of the relay, the differential current required to operate this relay can be either a constant or a variable (or multi-slope) percentage of the current in the restraint windings. The constant percentage differential relay operates, as its name implies, on a constant percentage of the through or total restraint current. For instance, a relay with a 10% characteristic would require at least 2.0 A in the operating winding with 20 A through-current flowing in both the restraint windings. A variable (or multi-slope) percentage relay requires proportionately more operating current at the higher through-currents, as shown in Figure 7.4. Regardless of the specific design, however, conceptually, the contact-closing torque (for an electromechanical relay) or tripping action (in a solid-state or digital relay) caused by the current in the operating coil is proportional to the **phasor sum** of the secondary currents. By contrast, the contact-opening torque, or non-tripping action, caused by the through-current in the restraint coils is proportional to the **sum of the magnitudes** of the two currents, with the additional non-tripping requirement that there must be current in both the restraint windings. Note that in Figure 7.4, the restraining current is shown as $(|I_1| + |I_2|)/2$, but in some designs, a different restraining current may be used, such as $(|I_1| + |I_2|)$ or the maximum restraining current.

For external faults, the restraining windings receive the total secondary current and function to desensitize the operating winding, particularly at high currents. The effect of the restraint windings is negligible on internal faults, since the operating winding has more ampere-turns and it receives the total secondary current while the net ampere-turns of the restraint windings are decreased by virtue of the opposite direction of current flow in the windings during an internal fault. Note that in the description, for clarity and simplicity, we have referred to operate and restraint "coils" and "windings" which imply electromechanical relays. In modern implementations of solid-state analog, or digital percentage-restrained differential relays, it would be more correct to refer to operate and restraint circuits, since currents would be electronically manipulated and measured as required to perform the percentage-restrained function.

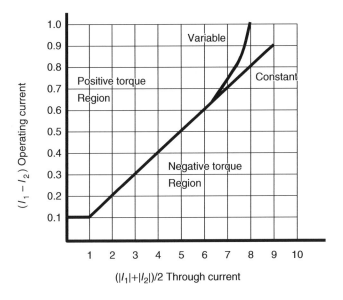

Figure 7.4 Percentage differential relay characteristics.

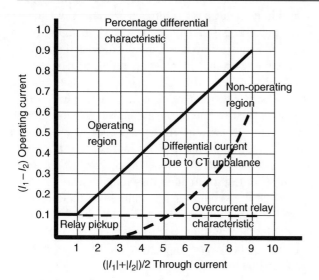

Figure 7.5 Comparison between percentage differential and overcurrent relay performance.

Example 7.2 Figure 7.5 shows the difference in the operating characteristics of the generator differential when using an overcurrent relay as shown in Figure 7.1 or a constant percentage differential relay as shown in Figure 7.3. Also shown is a typical plot of the error current due to CT unbalance caused by different burdens or saturation. Both relays are set for the same pickup of 0.1 A. It is clear that for the given plot of differential current due to CT unbalance, a through-current greater than 5 A will cause the errror current to exceed 0.1 A. This error current will flow in the operating coil and will trip the overcurrent relay incorrectly, whereas the entire error current plot lies in the nonoperating region of the differential relay, and there would be no tendency for the percentage differential relay to operate.

Example 7.3 Consider the system and the associated positive-sequence network shown in Figure 7.6 when the unit is synchronized to the system. The generator is protected by a percentage differential relay (87) set for a minimum pickup of 0.2 A. Full-load current is

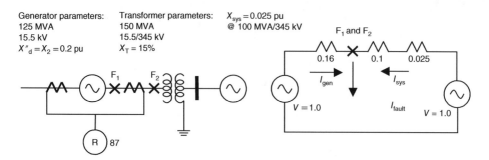

Figure 7.6 System and positive-sequence network for Example 7.3.

$125\,000/\left(\sqrt{3} \times 15.5\right) = 4656.19$ A. Select a CT ratio of 5000:5 (1000:1). The per unit reactances on a 100 000 kVA, 15.5 kV base are

$$x''_d = 0.2 \times \left(\frac{100}{125}\right) = 0.16 \text{ pu,}$$

$$x_t = 0.15 \times \left(\frac{100}{150}\right) = 0.1 \text{ pu,}$$

$$x_{sys} = 0.025 \text{ pu.}$$

Three-phase faults at F_1 and F_2 are

$$I_{1f} = \frac{1.0}{0.07} = 14.29,$$

$$I_{base} = \frac{100\,000}{\left(\sqrt{3} \times 15.5\right)} = 3724.9 \text{ A,}$$

$$I_f = 14.29 \times 3724.9 = 53228.82 \text{ A,}$$

$$I_{gen} = \frac{23\,346}{1000} = 23.35\text{A, secondary,}$$

$$I_{sys} = \frac{29\,883}{1000} = 29.89 \text{ A, secondary.}$$

For a fault at F_2, that is, outside the generator differential zone, the generator contribution flows through both sets of CTs. If both sets of CTs reproduce the primary current accurately, there will be no current in the operating winding.

For a fault at F_1, that is, within the differential zone of protection, the generator contribution flows through one set of CTs and the system contribution flows through the other set of CTs. Each restraint winding sees its associated current flowing in opposite directions (which decreases the net restraining torque) and the operating winding sees the sum of the two contributions, that is, 53.24 A.

The security of the percentage-restrained differential protection can be challenged during a black start if a circuit breaker is applied at the generator terminal voltage instead of at the high-voltage side of the generator step-up (GSU) transformer. The pros and cons of such an application are discussed in Section 7.10.6. When there is a breaker between the GSU transformer and the generator, the unit is normally synchronized to the system across this breaker. In the case of a breaker on the high-voltage side of the GSU transformer, the generator and the GSU transformer are both energized together when the generator excitation is applied during start-up, so there is never any energizing inrush current to the GSU transformer.

During a black start with an open breaker at the generator terminals, the GSU transformer will need to be energized when the generator is already energized, at nominal voltage. In such a case, the GSU transformer will experience energization inrush current. The initial inrush current will be several times rated current and will have a large DC offset that will persist for much longer than a normally cleared short circuit. This sustained large current with significant offset will often cause severe saturation of some of the differential

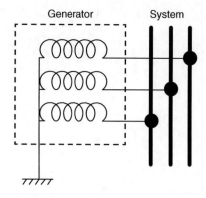

Figure 7.7 Direct-connected and solidly grounded generator.

protection CTs. If the generator is capable of energizing the GSU transformer while running at nominal voltage, the slope of the percentage differential protection may have to be set higher than necessary to prevent misoperation due to CT saturation during a normally cleared external short circuit.

7.2.2 Ground fault protection

The method of grounding affects the amount of protection that is provided by a differential protection. When the generator is solidly grounded, as in Figure 7.7, there is sufficient phase current for a phase-to-ground fault to operate almost any differential relay. If the generator has a neutral impedance to limit ground current, as shown in Figure 7.8, there are relay application problems that must be considered for the differential protections that are connected in each phase. The higher the grounding impedance, the less the fault current magnitude and the more difficult it is for the differential protection to detect low-magnitude ground faults.

If a CT and a relay are connected between ground and the neutral point of the circuit, as shown in Figure 7.9, sensitive protection will be provided for a phase-to-ground fault, since the neutral relay (51N)[2] sees all of the ground current and can be set without regard for load current. As the grounding impedance increases, the fault current decreases and it becomes more difficult to set a current-type protection. The lower the protection pickup, the more difficult it is to distinguish between ground faults and normal third harmonic unbalance.

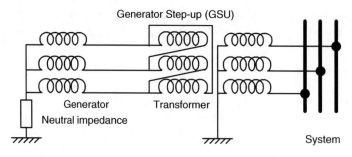

Figure 7.8 Neutral impedance grounding.

---

2 Suffix "N" is generally used in preference to "G" for devices that are connected in the secondary of a CT whose primary winding is located in the neutral of a machine or power transformer. "N" may also be used for devices connected in the residual circuit of the three secondary windings of the CTs connected in the primary as shown in Figure 7.12. However, if there are relays in both the neutral connection to ground and the residual circuit of the three-phase CTs, it is common practice to differentiate between these two sets of relays by designating one "N" and the other "G." In the case of transmission line relaying, the suffix "G" is used for those relays that operate on ground faults, regardless of the location of the CTs.

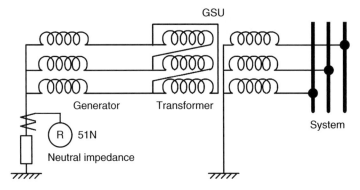

Figure 7.9 Neutral impedance grounding with neutral CT and relay.

This unbalanced current that flows in the neutral can be as much as 10–15% of the rated current. Other spurious ground current may flow due to unbalances in the primary system. The total false ground current flows through the neutral CT and relay. However, only the difference between the secondary currents will flow through the generator differential CTs. Since the spurious ground current is small, there should be no effect on the accuracy of the CTs.

If the machine is solidly (or low-impedance) grounded, and protected with a neutral CT and overcurrent protection 51N as shown in Figure 7.9, an instantaneous overcurrent function is applicable. In high-impedance grounding schemes, with the same protection, although the fault current is low and the potential damage is reduced, a time-delay overcurrent function is preferred, since it can be set lower than an instantaneous function to accommodate the lower ground current and it would be set with sufficient time delay, for example, 5–10 cycles, to override any false ground current that could be caused by switching or other system transients.

Example 7.4 Referring to the system diagram and the sequence networks shown in Figure 7.10, the phase current for the differential function (87) and the neutral overcurrent protection (51N) for various values of grounding impedances are as follows:

a) Solidly grounded: $R_n = 0$,

$$I_1 = I_2 = I_0 = \frac{j1.0}{j(0.2 + 0.2 + 0.03)} = 2.33 \text{ pu} \times 4656 = 10\,828.35 \text{ A},$$

$I_g = I_a = 3 \times I_0 = 32\,485$ A primary. The secondary current in the generator differential function will be 32.5 A. The typical minimum pickup of this type of protection is 0.2–0.4 A; so 32.5 A is sufficient to reliably operate on ground faults even with additional resistance or within the generator winding.

b) Moderately grounded: $R_n = 1.0\,\Omega$,

$$1.0\,\Omega = \frac{[(1) \times (125\,000)]}{[(1000) \times (15.5)^2]} = 0.52 \text{ per unit},$$

$$I_1 = I_2 = I_0 = \frac{j1.0}{[3(0.52) + j0.43]} = 0.617\angle 74.60° \text{ per unit},$$

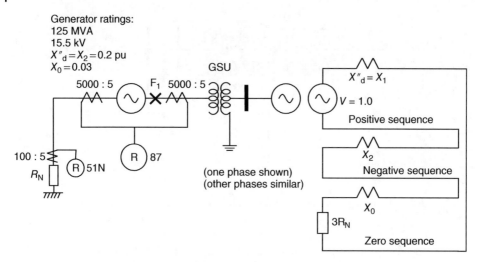

Figure 7.10 System and sequence networks for Example 7.4.

$I_g = I_a = 3 \times I_0 = 8618$ A primary. The generator differential CT secondary will be 8.62 A, which is still adequate. However, if a neutral CT connected to an overcurrent function were installed, a CT ratio as low as 100:5 would result in a neutral CT secondary current of 431 A. In practice, if a 1.0 Ω neutral grounding resistance was used, a higher neutral CT ratio would be applied to keep the maximum secondary current during a ground fault below 100 A.

c) High-impedance grounding: $R_n = 10\,\Omega$,

$10\,\Omega = 5.2$ pu,

$$I_1 = I_2 = I_0 = \frac{j1.0}{[3(5.2) + j0.43]} = 0.06\angle1.5°,$$

$I_g = 3I_0 = 837$ A primary. The generator differential secondary current (0.84 A) is above the pickup but does not allow for additional fault resistance. An even lower neutral CT ratio of (say) 100:5 could be used. Then the neutral CT secondary current (41.85 A) is sufficient to make a good neutral overcurrent protection setting.

Example 7.4 shows that, for a solidly grounded generator, there is enough secondary current to operate the generator differential (32.5 A). As the impedance increases, the differential current falls to just above differential pickup (0.29 A). Any higher neutral impedance or fault resistance and the differential protection will not operate. On the other hand, since the neutral CT does not have to consider load, it can have any ratio that is determined by the available fault current. In case (c), a ratio of 100:5 will result in a neutral secondary current of 41.85 A, enough to operate an overcurrent protection function set at, say, 5.0 A.

If the machine is not grounded, then the first ground fault does not result in any current flow. This situation does not require immediate tripping, since there is no fault current to cause any damage. However, a second ground fault will result in a phase-to-phase or turn-to-turn fault. This condition can result in heavy current or magnetic unbalance and does require immediate tripping. It is, therefore, essential to detect the first fault and start to take appropriate action. The usual ground fault detector is zero-sequence overvoltage protection. The zero-sequence voltage may be provided by a potential transformer with the primary winding connected in a grounded wye configuration and the secondary winding connected in broken delta. This results in $3E_0$ across the broken delta, as shown in Figure 7.11. Alternatively, the zero-sequence voltage may be determined by a solid-state relay measuring the sum of the three phase voltages, which is $3E_0$.

If the neutral is not accessible, or there is no neutral CT, an alternative protection scheme to the neutral relay is a residually connected relay, as shown in Figure 7.12. This is a relay that is connected in the residual of the CT secondary circuit so that it sees the vector sum of the secondary current of all three phases. This is particularly applicable in motor installations where each phase is energized at a different point of the voltage wave, resulting in different DC offsets and inrush currents. Because of these differences, there is a difference in the CT secondary output, and this difference flows in the common residual circuit.

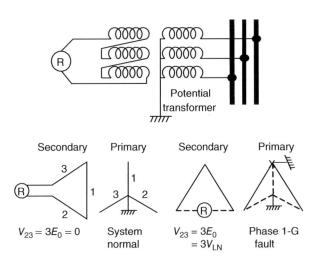

Figure 7.11 Broken-delta ground detector.

Figure 7.12 Residually connected ground relay.

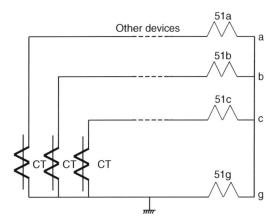

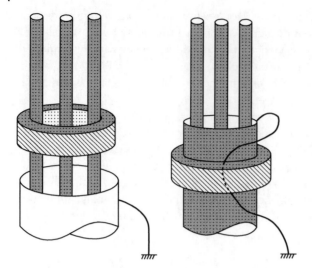

Figure 7.13 Toroidal CT.

Therefore, instantaneous relays cannot be used and a very-inverse or short-time induction relay, or its solid-state or digital equivalent, is required. The time delay prevents this type of relay from tripping falsely during start-up. Typical pickup settings of the time-delay relay are one-fifth to one-third of the minimum fault current with some time delay. In power plants, to avoid incorrect trips due to vibration, the time delay is usually not set at the lowest setting, although virtually any time delay will be longer than the starting current error [3].

An alternative to the residually connected ground relay in motor applications is the toroidal CT, shown in Figure 7.13 and discussed in Section 3.4 (Chapter 3). This CT encircles all three-phase conductors and thereby allows all positive- and negative-sequence currents to be canceled out, so only zero-sequence current appears in the relay. Care must be taken to keep all ground wires and cable shields out of the toroid. If they conduct current during a ground fault, the net magnetic effect will be zero and no current will be seen by the relay [3]. Because the secondary current is a true reflection of the total three-phase primary current, there is no CT error due to any unbalanced primary current. The CT ratio can be any standard value that will provide the relay current from the available ground current for adequate pickup. Since there will be no error current, the relay can be an instantaneous relay set at a low value. Typical relay calculations are shown in the following examples.

Example 7.5   Consider the 2000 hp motor installation shown in Figure 7.14. The CT ratio is selected to provide some margin above the trip setting, so meters will not read off-scale. Normally, overcurrent relays are set at 125% of full load and the CT ratio should allow less than 5.0 A for this condition. If the motor is vital to the operation of the plant, advantage is taken of the motor service factor which is 115%. This results in a maximum load of 245 A × 1.15 = 282 A and a relay pickup setting of 1.25 × 282 = 352.5 A. Select a CT ratio of 400:5 (80:1).

The time-delay overcurrent relay 51 sees 352.5/80 or 4.4 secondary amperes. To set the relay, the manufacturer's instruction book and characteristic curves must be used. For our purposes, however, assume there is a 5.0 A pickup setting and the characteristics

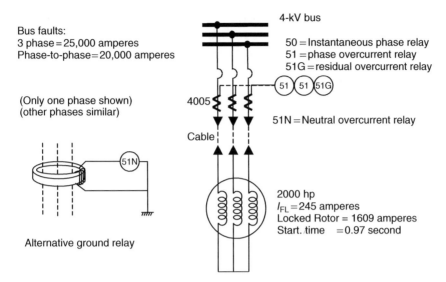

Figure 7.14 Protection for 2000 hp motor.

of Figure 4.7 (Chapter 4) apply. The time delay must be set longer than the starting time of the motor. This assumes that the starting current is equal to the locked rotor current and lasts for the full starting time. This is not strictly true. The starting current starts to decrease at about 90% of the starting time. However, this is a conservative setting that is often used to cover any erratic motor behavior and to avoid false tripping during starting. The relay pickup during starting is $1609/(80 \times 5) = 4 \times$ pu; the time delay must be at least 0.97 s, which results in a time dial setting of 2.8. In electromechanical relay implementations, two over-current relays are usually used, one in phase 1 and the other in phase 3, on the assumption that an overload is a balanced load condition. An ammeter (not shown) is usually connected in phase 2. All three phase currents are measured in digital multifunction relays.

The instantaneous relay 50 must be set above the asymmetrical value of the locked rotor current, that is, $1.7 \times 1609 = 2735$ primary amperes or 34.19 secondary amperes. Set at 35 A. Check for pickup at minimum 4 kV bus fault; $20\,000/(35 \times 80) = 7.14 \times$ pu. In electrome-chanical relay implementations, three relays were used, one per phase, to provide redundancy for all phase faults. With multifunction solid-state relay implementations, redundant relays may be applied if other protections cannot provide independent backup.

The residual overcurrent relay 51G is set at one-third of the minimum ground fault. If the auxiliary system has a neutral resistor to limit the ground fault to 1200 A,[3] the residual CT current will be $1200/80 = 15$ A. Set the relay at 5.0 A or less to ensure dependability. Set the time dial at a low setting. There are no criteria for these settings except what is the usual practice at a given plant.

If the alternative scheme using a toroidal CT is used, the CT ratio can be 1200:5 (240:1) and an instantaneous relay set at 1.0 A. This will give five times pickup at the maximum

---

3 See ground fault protection discussion and Figure 7.22.

ground fault current and provide sufficient margin above any false ground currents to prevent false tripping and still allow for reduced ground fault current due to fault resistance.

Example 7.6 Figure 7.15 shows a 7500 hp motor connected to the same auxiliary bus as the motor in Example 7.5. The time-delay overcurrent relays follow the same setting rules as for the 2000 hp motor.

The pickup of the phase overcurrent protection 51 is equal to $1.15 \times 1.25 \times 918$ or 1320 primary amperes.

Select a CT ratio of 1500:5 (300:1). The relay pickup setting is therefore $1320/300 = 4.4$; choose a 5.0 A setting as the nearest integral above. Pickup during starting is $5512/(5 \times 300) = 3.7 \times$ pu and the time delay must exceed 3 s. From Figure 4.7 (Chapter 4), use #8 dial. The residual overcurrent relay 51G is set at one-third of the limited ground current ($1200/300 = 4$) or 1.33 A. Use the 1.0 A pickup setting and the same dial setting as in Example 7.5. If a toroidal CT is used, the setting rules used in Example 7.5 apply.

The setting of the instantaneous protection, however, introduces a problem. Using the criteria of Example 7.5, we should set the relays at 1.7×5512 or 9370 A. Since the minimum 4 kV bus fault is 20000 A, we would only have $2.1 \times$ pu. This is not enough of a margin to ensure fast tripping. For a motor of this size, we would use differential protection (87). There is no setting required, since the sensitivity of the differential protection is independent of the starting current.

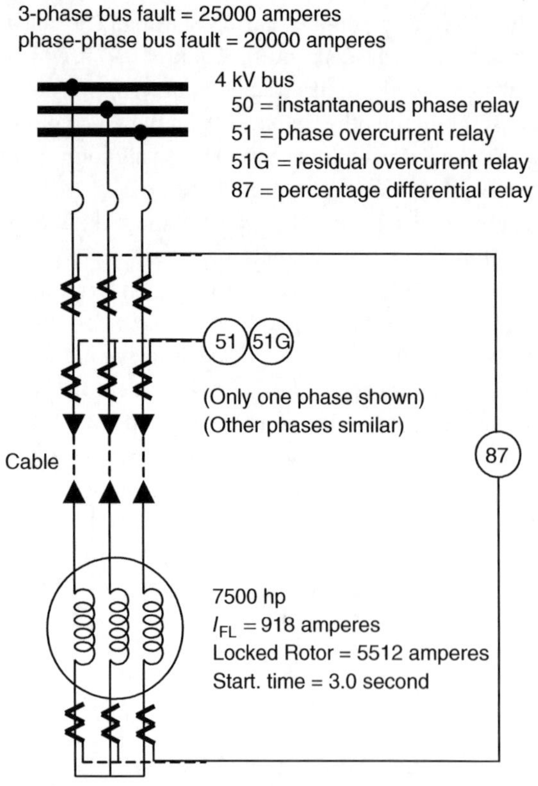

3-phase bus fault = 25000 amperes
phase-phase bus fault = 20000 amperes

Figure 7.15 Protection for 7500 hp motor.

4 kV bus
50 = instantaneous phase relay
51 = phase overcurrent relay
51G = residual overcurrent relay
87 = percentage differential relay

51 51G

(Only one phase shown)
(Other phases similar)

Cable

87

7500 hp
$I_{FL} = 918$ amperes
Locked Rotor = 5512 amperes
Start. time = 3.0 second

If a generator is connected directly to a grounded transmission system, as shown in Figure 7.7, the generator ground relay may operate for ground faults on the system. It is therefore necessary for the generator ground relay to coordinate with any other relays that see the same fault. If the generator is connected to the system through a wye–delta transformer as shown in Figure 7.16, zero-sequence current cannot flow in the generator bus beyond the delta connection of the GSU transformer. Faults on the wye side will, therefore, not operate ground relays on the delta side.

The most common configuration for large generators today uses the generator and its GSU transformer as a single unit, that is, failure in the boiler, turbine, generator, GSU transformer, or any of the associated auxiliary buses will result in tripping the entire unit-connected system. The generator is grounded through some resistance to limit the fault current, yet provides enough current or voltage to operate relays. A primary resistor or reactor can be used to limit the ground fault current, but for economic reasons the most popular arrangement uses a distribution transformer and resistance combination, as shown in Figure 7.16. The primary voltage rating of the distribution transformer must be equal to or greater than the line-to-neutral voltage rating of the generator, usually with a secondary rating of 120, 240, or 480 V. The distribution transformer should have sufficient overvoltage capability so that it does not saturate at 105% rated voltage. A secondary resistor is selected so that, for a single line-to-ground fault at the terminals of the generator, the power dissipated in the resistor is equal to the reactive power that is dissipated in the zero-sequence capacitance of the generator windings, leads, surge arresters, and transformer windings. The secondary resistor is chosen to limit the primary fault current to 10–35 A. As we have seen, this is not enough to operate the generator differential relay. However, the voltage across the secondary resistor for a full line-to-ground fault, that is, at the phase terminals of the generator, is equal to the full secondary voltage of the distribution transformer (120, 240, or 480 V) and is more than enough to operate a voltage relay.

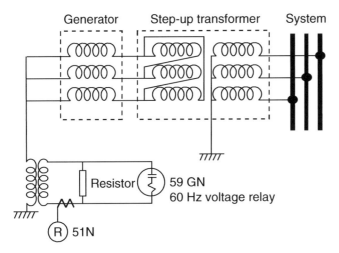

Figure 7.16 Generator with transformer ground.

The impedance reflected in the primary circuit is the secondary impedance times the square of the voltage ratio. For example, given a 26 kV generator (15 kV line-to-ground) and a 15 000 V/240 V distribution transformer with a 1 Ω secondary resistor, the resistance reflected in the primary will be $(15000)^2/(240)^2$ times 1 Ω, or 3906 Ω. The resulting generator ground current is 15 000/3906 or 3.8 A for a full line-to-ground fault.

Ground protection is provided by a voltage relay, 59GN, connected across the secondary resistor. Electromechanical relays must discriminate between 60 Hz and third harmonic voltages. Digital protection normally includes a fundamental frequency filter that significantly desensitizes it to harmonics but must still be set above noise. Typical pickup settings are in the range of 2–10% of rated secondary line to neutral voltage, and usually have a time delay. There is adequate security in this scheme, since the voltage relay can easily be set between normal and abnormal operations. Dependability, however, should be improved, since we are depending upon a single relay to protect against the most common fault. A backup to this scheme could be a CT in the secondary of the distribution transformer, as shown in Figure 7.16. The CT ratio is selected to give approximately the same relay current as that flowing in the generator neutral for a ground fault. In our example provided in the earlier text, the secondary current is 3.8 A times 15 000/240, or 237.5 A. A 250:5 CT ratio would be adequate with the 51N relay set at 0.5 or 1.0 A and the same time delay as 59GN. Another backup scheme that is commonly used is to connect a potential transformer ground detector as shown in Figure 7.11 between the generator and the generator GSU.

The generator ground protection should be slower than the ground fault protection on the secondary of the generator voltage transformers. The generator neutral ground impedance will limit the single line-to-ground fault current on the VT secondary if the secondary neutral is grounded [1, 4]. The time delay of 59GN should be long enough to ensure that the VT secondary protection will operate first for a secondary single line-to-ground fault, which is the most common type of fault. Some applications will have the VT secondary windings ground on one phase with the neutral floating. All VT secondary loads would be connected phase to phase and the exposure of the neutral point to ground faults would be small. This arrangement will provide high currents and fast fault clearing for a single line-to-ground fault on either of the two ungrounded phases. The fast fault clearing of VT secondary faults will allow faster operation of the generator ground fault protection.

Example 7.7   Consider the system shown in Figure 7.17. The distributed generator capacitance to ground = 0.22 μF/phase; the distributed leads and transformer capacitance to ground = 0.10 μF/phase; and the surge arrester capacitance = 0.25 μF/phase. Therefore, the total capacitance = 0.57 μF/phase:

$$X_c = \frac{10^6}{2\pi f C} = \frac{10^6}{377 \times 0.57} = 4650 \ \Omega \text{ at 60 Hz,}$$

$$I_c = \frac{3V_{1-g}}{X_c} = \frac{3V_{1-1}}{\sqrt{3} \times X_c} = 5.77 \text{ A.}$$

Total capacitive kilovolt amperes $= 5.77 \times \dfrac{15.5 \text{ kV}}{\sqrt{3}} = 52 \text{ kVA.}$

To prevent ferroresonance, $I_{neut} > I_c$ and $kW_{loss} > kVA_{capac}$. Choose $I_{neut} = 10$ A:

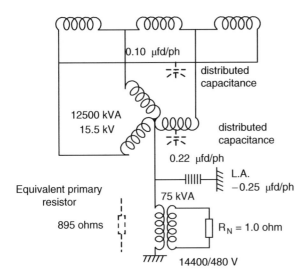

Figure 7.17 System for Example 7.7.

$$R_{\text{neut}} = \frac{15\,500}{\sqrt{3} \times 10} = 895\,\Omega \text{ primary or } 895 \times \frac{(480)^2}{(1440)^2} = 1.0\,\Omega \text{ secondary},$$

$$I_{\text{sec}} = 10 \times \frac{14\,400}{480} = 300\,\text{A},$$

Resistor loss $= (300)^2 \times 1.0 = 90\,\text{kW} > 52\,\text{kVA}.$

Assume third harmonic voltage is 3% of normal line-to-neutral voltage:

$$V_3 = 0.03 \times \frac{15\,500}{\sqrt{3}} = 268\,\text{V}.$$

Reactance to third harmonic is $1/3X_c = 4650/3 = 1550\,\Omega$ on a per-phase basis or $1550/3 = 517\,\Omega$ on a three-phase basis:

$$Z_3 = R - jX_3 = 895 - j517 = 1035\angle 30°,$$

$$I_3 = \frac{268}{1035} = 0.259\,\text{A},$$

$$V_{\text{pri}} = 0.259 \times 895 = 232\,\text{V},$$

$$V_{\text{sec}} = 232 \times \frac{480}{14\,400} = 7.74\,\text{V}.$$

An electromechanical relay should be set at about twice this value to ensure security. Assume the relay has a 16 V tap; that would be its setting. A digital relay with filtering could be set a little more sensitively. Choose a time setting to coordinate with VT secondary ground fault protection.

For ground faults at the phase terminal of the generator, the voltage across the relay and resistor is

$$\frac{15\,500 \times 480}{\sqrt{3} \times 14\,400} = 298 \text{ V},$$

$$\frac{298}{16} = 18.6 \times \text{pickup}.$$

Primary pickup voltage is 16 V times 14 400/480 primary, which is the lowest voltage the relay will see. This results in an unprotected part of the winding equal to $480/\left(15\,500/\sqrt{3}\right)$ or 5% of the total winding.

Since the entire winding is not protected, faults near the neutral end will not be detected by 59G or 51N. Although ground faults near the neutral are rare, they are possible, and there is also a possibility of the neutral grounding impedance being accidentally left short-circuited after maintenance activity in which a worker protection ground at the neutral was not removed. If a solid ground near the neutral end is left on continuously while the generator is in operation, there is risk of severe damage if a second ground fault should occur further away from the neutral. Several supplementary or alternative protective schemes have been developed which will protect 100% of the stator winding. They are covered extensively in the literature [4, 5]. These are called "100% stator ground fault protection" schemes.

Another method of providing high-impedance grounding to a unit-connected generator is the use of a reactor connected in the neutral of the generator and is referred to as "resonant grounding" or a "Petersen coil." The reactor is tuned to the total system capacitance, so the only impedance in the circuit is the resistance of the conductors. As a result there is very little fault current for a line-to-ground fault [6, 7].

7.3 Rotor faults

The field circuits of modern motors and generators are operated ungrounded. Therefore, a single ground on the field of a synchronous machine produces no immediate damaging effect. However, the existence of a ground fault stresses other portions of the field winding, and the occurrence of a second ground will cause severe unbalance, rotor iron heating, and vibration. Response to a rotor ground fault varies from case to case, depending on company operating practices and the size and importance of the generator. Many operating companies alarm on the indication of the first ground fault and prepare to remove the unit in an orderly shutdown at the first opportunity. Others allow the protection to automatically reduce the generator load to zero to prevent overspeed, then trip the unit and shut it down.

Two commonly applied field ground detection schemes are as shown in Figure 7.18. The ground in the detecting circuit is permanently connected through the very high impedance of the relay and associated circuitry. If a ground should occur in the field winding or the buses and circuit breakers external to the rotor, the relay will pick up and actuate an alarm. A low-frequency square-wave current injection protection may also be used in some applications [1]. This allows continuous monitoring of field winding insulation resistance and different sensitivities of alarm and trip functions.

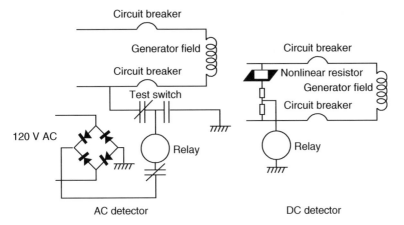

Figure 7.18 Generator field ground detectors.

For a brushless-type machine, access is not normally available to a stationary part of the machine field circuit and continuous monitoring is more difficult. One method is to use a monitoring system that is powered by, and rotates with, the field and illuminates an LED transmitter when the insulation is healthy. The receiver is mounted on the exciter housing. If a field ground occurs, the transmitter LED is extinguished and the receiver will initiate an alarm or trip [1]. Alternatively, pilot brushes can be provided that may be periodically lowered to the generator. If a voltage is read between ground and the brush, which is connected to one side of the generator field, then a ground exists. Alternatively, resistance measurements during maintenance can be used to evaluate the integrity of the field winding.

The primary concern with rotors in squirrel-cage induction motor construction or insulated windings in wound-rotor induction or synchronous motor construction involves rotor heating. In almost all cases, this is the result of unbalanced operation or a stalled condition. Protection is therefore provided against these situations rather than attempt to detect the rotor heating directly.

7.4 Unbalanced currents

Unsymmetrical faults may produce more severe heating in machines than symmetrical faults or balanced three-phase operation. The negative-sequence currents that flow during these unbalanced faults induce 120 Hz rotor currents, which tend to flow on the surface of the rotor forging and in the nonmagnetic rotor wedges and retaining rings. The resulting I^2R loss quickly raises the temperature. If the fault persists, the metal will melt, damaging the rotor structure.

Industry standards have been established which determine the permissible unbalance to which a generator is designed. The general form of the allowable negative-sequence current is $I_2^2 t = k$ (I_2 is the per unit negative-sequence current; t is the time in seconds). For directly cooled cylindrical rotors up to 800 MVA, the capability is 10. Above 800 MVA, the capability is determined by the expression $[10 - (0.00625)(MVA - 800)]$. For example, a 1000 MVA generator would have an $I_2^2 t = 8.75$. For salient pole generators with damper windings, the value of k is up to 40. Generators also are rated to operate continuously for small negative-sequence currents in the range of 5–10%. No standards have been established for motors, although $k = 40$ is usually regarded as a conservative value.

A basic question concerns the cause of the system unbalance. For generators, such operation is very often the failure of the protection or equipment external to the machine. For large motors, the unbalance can be caused by the supply equipment, for example, fused disconnects. Typical conditions that can give rise to the unbalanced generator currents are:

- Accidental single phasing of the generator due to open leads or buswork.
- Unbalanced generator step-up transformers.
- Unbalanced system fault conditions and a failure of the relays or breakers.
- Planned single-phase tripping without rapid reclosing.

When such an unbalance occurs, it is not uncommon to apply negative-sequence protection (46) on the generator to alarm first, alerting the operator to the abnormal situation and allowing corrective action to be taken before removing the machine from service. The protection itself consists of an inverse time-delay overcurrent function operating from the output of a negative-sequence filter. On a log–log scale, the time characteristic is a straight line of the form $I_2^2 t = k$ and can be set to closely match the machine characteristic. Legacy negative-sequence electromechanical relays in general do not respond to small values of negative-sequence currents which could damage generators if they persist for too long. Modern digital relay characteristics include a definite time function which puts an upper limit on the length of time, during which small negative-sequence currents in the range of 5–10% may persist before initiating an alarm or trip.

In the case of motors, such protection is generally reserved for the larger motors. On smaller motors, it is more common to use a phase-balance relay. For example, in an electromechanical relay, two induction disk units are used: one disc responding to $I_a + I_b$, the other to $I_b + I_c$. When the currents become sufficiently unbalanced, torque is produced in one or both units to close their contacts and trip the appropriate breaker. Solid-state and digital relays can perform in a similar manner by incorporating the appropriate logic elements or algorithms in their design.

7.5 Overload

Protective practices are different for generators and motors. In the case of generators, overload protection, if applied at all, is used primarily to provide backup protection for bus or feeder faults rather than to protect the machine directly. The use of an overcurrent relay alone is difficult because the generator's synchronous impedance limits the fault current of sustained faults to about the same or less than the maximum or rated load current. Typical three-phase 60 Hz generator synchronous impedance varies between 0.95 and 1.45 per unit. Using the unit in Example 7.1, this would result in a sustained fault current between 3211 and 4901 A; this is not enough to distinguish between a fault and full load of 4656 A. To overcome this problem, a voltage-controlled overcurrent function or an impedance function can be used. With this function, the current setting can be less than the rated current of the generator, but the function will not operate until the voltage is reduced by the fault. One hazard of all functions that rely on voltage is the inadvertent loss of voltage and consequent incorrect trip of the machine. This should be recognized and proper precautions taken through good design and adequate maintenance of the voltage supply. Modern digital protection relays also include logic (called "loss of potential" or "LOP") to optionally block this type of protection if the voltage supply fails.

Overload protection is always applied to motors to protect them against overheating. Fractional horsepower motors usually use thermal heating elements such as bimetallic strips purchased with the motor starter. Integral horsepower motors use time-delay overcurrent relays, as shown in Examples 7.5 and 7.6. However, heating curves are difficult to obtain and vary considerably with motor size and design. Further, these curves are an approximate average of an imprecise thermal zone, where varying degrees of damage or shortened insulation life may occur. It is difficult, then, for any relay design to approximate these variable curves adequately over the range from light, sustained overloads to severe locked rotor overload.

Thermal overload relays offer good protection for light and medium (long-duration) overloads, but may not be good for heavy overloads (Figure 7.19a). A long-time induction overcurrent relay offers good protection for heavy overloads, but overprotects for light and medium overloads (Figure 7.19b). A combination of two devices can provide better thermal protection, as in Figure 7.19c, but the complication in settings, testing, and so on weighs heavily against it and such an application is rarely used. Today, digital relays for motor protection are widely used to overcome the shortcomings of solid-state or electromechanical designs that use current as an indication of temperature or a thermal replica circuit that does not have the mass necessary to reproduce the thermal inertia of a motor. Digital relays take advantage of the ability to model the rotor and the stator mathematically and use algorithms that calculate the conductor temperature resulting from operating current, add the effect of ambient temperature, and calculate the heat transfer and the heat decay. They are therefore responsive to the effects of multiple starts, the major disadvantage of using only current as an indication of temperature. In addition, a digital device can record actual operating parameters such as ambient temperature and starting and running current, and adjust the algorithms accordingly [2, 8].

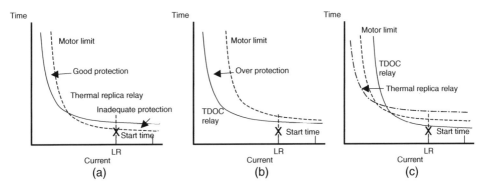

Figure 7.19 Motor overload protection (time-delay overcurrent [TDOC]). (a) Light-load overload protection; (b) heavy-load overload protection; and (c) combination overload protection.

The National Electrical Code requires that an overload protective device be used in each phase of a motor "unless protected by other approved means." This requirement is necessary because single phasing (opening one supply lead) in the primary of a delta–wye transformer that supplies a motor will produce three-phase motor currents in a 2:1:1 relationship. If the two units of current appeared in the phase with no overload device, the motor would be unprotected.

A motor that is rotating dissipates more heat than a motor at standstill, since the cooling medium flows more efficiently. When full voltage is applied, a motor with a locked rotor is particularly vulnerable to damage because of the large amount of heat generated. Failure of a motor to accelerate when the stator windings are energized may be caused by many things. Mechanical failures of motor or bearings, low supply voltage, or an open phase of a three-phase supply voltage are just a few of the abnormal conditions that can occur. If the motor fails to accelerate, stator currents may typically range from 3 to 7 or more times full-load value depending on motor design and supply system impedance. In addition, the heat loss in the stator winding is 10–50 times normal when the winding is without the benefit of the ventilation normally produced by rotation of the rotor.

Overtemperature from a locked rotor cannot reliably be detected by sensing the line current magnitude. Since motors can stand high current for a short time during starting, some time delay must be incorporated in the current-sensing device or provision must be made to sense motor winding temperatures as well as line current magnitude. Digital relays are particularly suited to this type of logic combined with temperature sensing. Another possible protective scheme is to shunt out the current-sensing device during starting.

Some larger motors are designed to have a maximum allowable locked rotor current time less than the starting time of the drive. This is permissible, since, during a normal start, much of the active power input during starting is utilized as shaft load, while on locked rotor all of the active power input is dissipated as heat. Therefore, a time delay sufficient to allow the motor to start would have too much delay to protect against locked rotor.

Two approaches are possible to solve this dilemma.

1) Use a motor zero-speed switch that supervises an additional overload relay set for locked rotor protection.
2) Use a relay that incorporates temperature change and discriminates between the sudden increase during locked rotor and the gradual increase during load increases.

7.6 Overspeed

Overspeed protection for generators is usually provided on the prime mover. Older machines use a centrifugal device operating from the shaft. More modern designs employ very sophisticated electrohydraulic or electronic equipment to accomplish the same function. It must be recognized that, in practical situations, overspeed cannot occur unless the unit is disconnected from the system. When still connected to the system, the system frequency forces the unit to stay at synchronous speed. The worst case overspeed will result from sudden load shedding which occurs when the unit breaker is opened while the generator is at maximum load. During overspeed, the turbine presents a greater danger than the generator. Hydraulic turbines are often subjected to much higher overspeed (up to 200% of rated) than combustion or steam turbines because slow closing of input gates (over tens of seconds) may be required to avoid excessive pressure in water penstocks. Overspeed is not a problem with motors, since their speed is limited by the system frequency.

7.7 Abnormal voltages and frequencies

In order to understand and appreciate the protection that should be provided against abnormal voltages and frequencies, the source of the abnormality must be examined [9].

7.7.1 Overvoltage

The voltage at the terminals of a generator is a function of the excitation and speed. There are two possible hazards resulting from overvoltage.

Overvoltage at rated frequency or nominal voltage at lower than rated frequency is known as **overexcitation**. Overexcitation may result in thermal damage to cores due to excessive high flux in the magnetic circuits. Excess flux saturates the core steel and flows into the adjacent structures causing high eddy current losses in the core and adjacent conductor material. Severe overexcitation can cause rapid damage and equipment failure. Protection sensing volts as a function of frequency (volts per hertz) is usually used to prevent damage due to excessive flux.

Since flux is directly proportional to voltage and inversely proportional to frequency, the unit of measure for excitation is defined as per unit voltage divided by per unit frequency ($V Hz^{-1}$). Overvoltage exists whenever the per unit $V Hz^{-1}$ exceeds the design limits. For example, the usual turbine generator design is for 105% of rated $V Hz^{-1}$. Overvoltage exists at 105% of rated voltage and per unit frequency or per unit voltage and 95% frequency. Transformers are designed to withstand 110% of rated voltage at no load and 105% at rated load with 80% power factor [10]. Overvoltage of a steam turbine generator set is not usually a problem. The excitation and regulator controls generally have inherent overvoltage limits and alarms, but separate overexcitation protection may be applied to back up the voltage regulator control. In a unit-connected generator transformer set, the transformer may be more prone to failure from this condition, and overexcitation protection settings may coordinate with transformer overexcitation capability as well as generator capability. Transformer overexcitation is discussed further in Chapter 8.

Overvoltage at higher than rated frequency may also occur due to sudden load rejection when the speed increases suddenly with full excitation on the field. In this case, overfluxing is not a concern, but **excessive voltage** may damage the dielectric properties of the insulation. The voltage regulator will normally quickly reduce the voltage to within rated limits, but overvoltage protection may be applied to generators to protect the insulation in case of load rejection while the excitation system is under manual control. Overvoltage protection is more commonly applied to hydrogenerators which may be subjected to much higher overspeeds (up to 100% above normal speed) than steam or combustion turbine-driven generators. Protection to prevent dielectric damage due to overvoltage is not volts per hertz, but true voltage measurement. Since the generator speed may be well above rated speed during dielectric stress, the overvoltage protection must not be tuned to fundamental frequency, but should have a flat frequency response up to the maximum possible generator speed. Overvoltage protection may include a time–delayed function and an instantaneous function. The time-delayed function typically starts at 110% of rated with sufficient time to allow the automatic voltage regulator to reduce the voltage before protection action. The instantaneous function is typically set in the range of 130–150% [1].

7.7.2 Undervoltage

Undervoltage presents a problem to the generator only, as it affects the auxiliary system, which will be discussed later. Low voltage prevents motors from reaching rated speed on starting or causes them to lose speed and draw heavy overloads. While the overload relays will eventually detect this condition, in many installations the low voltage may jeopardize production or affect electronic or digital controls, in which case the motor should be quickly disconnected. Protection from low line voltage is a standard feature of AC motor controllers. The contactor will drop out instantaneously when the voltage drops below the holding voltage of the contactor coil. If immediate loss of the motor is not acceptable, for example, in a manufacturing plant, the contactor must have a DC (or AC rectified) coil and a time-delay undervoltage relay can then be used [2].

7.7.3 Overfrequency

Overfrequency is related to the speed of the unit and is protected by the overspeed device. It is possible to use an overfrequency relay as backup to mechanical devices. Again, if the unit is connected to a stable system, the generator cannot operate above the system frequency. However, if the system is dynamically unstable, with severe frequency excursions, overfrequency relays can alert the operator. In general, the governing devices will protect the unit from overspeed, but the system conditions must be addressed.

7.7.4 Underfrequency

While no standards have been established for temporary abnormal frequency operation of generators, it is recognized that reduced frequency results in reduced ventilation; therefore, operation at reduced frequency should be at reduced kVA. Operating precautions should be taken to stay within the short-time thermal ratings of the generator rotor and stator [11]. Underfrequency is a system condition that affects the turbine more than the generator. The turbine is more susceptible because of the mechanical resonant stresses which develop as a result of deviations from synchronous speed [9].

System load shedding is considered the primary turbine underfrequency protection and is examined in detail in Sections 10.9 (Chapter 10), 11.3 (Chapter 11), and 11.4 (Chapter 11). Appropriate load shedding will cause the system frequency to return to normal before the turbine trouble-free limit is reached [12, 13]. The amount of load shed varies with coordinating regions and individual utilities, but varies from 25% to 75% of system load. Since the load shed program can be relied upon only to the extent that the original design assumptions are correct, additional protection is required to prevent steam turbine damage. In order to have the unit available for restart, it is desirable to trip the turbine to prevent damage. This action in itself is considered as a last line of defense and is sure to cause an area blackout. It will, however, allow the unit to be ready to restore the system. Turbine manufacturers have published curves of frequency versus time which can be used as a guide for operators. The question is to trip or not to trip. The problem is loss of life and it is not clear that the best interests of the system are served by tripping the unit too quickly. From the protective relay point of view, a simple frequency relay can be used. However, the loss of life of the turbine blades is a cumulative deterioration every time the turbine passes through a low-frequency operating zone. Computer monitoring of the history of frequency can be applied [9].

7.8   Loss of excitation

When a synchronous generator loses excitation, it operates as an induction generator running above synchronous speed with the system providing the necessary reactive support. Round-rotor generators are not suited for such operation because they do not have amortisseur (damper) windings and will quickly overheat from the induced currents in the rotor iron. The heating occurs in the end-iron region where the rotor bars leave one slot and enter another. Salient-pole generators, which are commonly used with hydro machines, have such damper windings and do not have the problem. However, in addition to overheating, both salient-pole and round-rotor synchronous machines require a minimum level of excitation to remain stable throughout their load range. The typical generator capability curve, shown in Figure 7.20, shows the various limits associated with overexcitation and underexcitation. The generator manufacturer supplies all of the temperature characteristics shown in Figure 7.20. The user must provide the steady-state stability limit.

Figure 7.20   Generator capability curve.

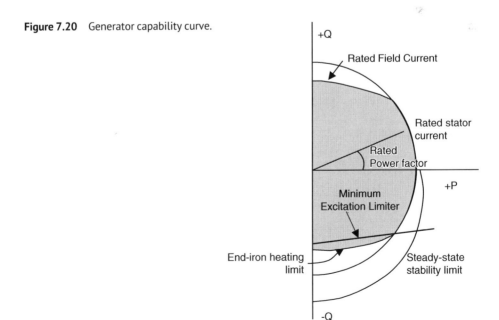

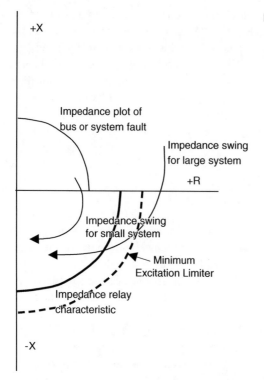

Figure 7.21 Typical impedance variations.

There are several methods of detecting underexcitation. Small units can use power factor or reverse-power relays. Manufacturers can provide current detectors in the excitation circuit. The most popular scheme, however, uses an impedance relay as the measuring element. This application is based on the behavior of the system impedance as seen from the generator terminals for various underexcited conditions. This behavior is explained in detail in Section 10.2.2 (Chapter 10). Figure 7.21 shows how the impedance varies with loss of excitation for different system sizes. Despite the complexity of the phenomenon and the variation in conditions, the end result is surprisingly simple. Since the final impedance lies in the fourth quadrant of the R–X diagram, any relay characteristic that will initiate an action in this quadrant is applicable as long as it coordinates with exciter control characteristics. Modern exciters include minimum or underexcitation limiters that allow the generator to operate in the underexcited region if it remains within the capability diagram shown in Figure 7.20. There are circumstances (such as high system voltages) where it may be necessary or desirable to operate the generator in the underexcited region. In this region, the impedance presented to a protection system will be in the third quadrant.

If the excitation system and its limiters are working properly, the generator should not be tripped off line by loss of excitation protection. Reliability regulators require that generator controls operate to keep the generator within acceptable limits before the protection operates to take the unit off line. When choosing a characteristic for the loss of excitation protection, the characteristics of the minimum excitation limiter should be plotted together with the generator capability characteristic on the P–Q diagram (like Figure 7.20) and on the R–X diagram (like Figure 7.21) [14]. The coordination process will check that the minimum excitation limiter (MEL) characteristic lies within the generator capability curve on the *P–Q* diagram and outside the loss of excitation protection characteristic which is plotted on an *R–X* diagram (like Figure 7.21). This coordination should be checked at maximum and minimum generator voltages, since the characteristics change with voltage when converting between the *P–Q* and *R–X* diagrams.

Various implementations are preferred by different relay manufacturers, but the concept is the same [1, 15, 16]. Once again, the question of whether to trip or to alarm for this condition must be addressed. In almost every case, an alarm is provided early in the locus of the impedance swing, so that the operator can take the appropriate corrective action. Whether this is followed by a trip after a time delay or further advance in the swing path is a utility's decision. Sometimes the protection may be arranged to trip faster (or automatically) if the terminal voltage is low, as this indicates that the excitation problem is adversely affecting the system voltage.

7.9 Loss of synchronism

The primary difference in the protection requirements between induction motors and synchronous motors is the effect of the excitation system. Loss of synchronism of a synchronous motor is the result of low excitation exactly as with the synchronous generator. For large synchronous motors or condensers, out-of-step protection is applied to detect pullout by counting the power reversals that occur as the poles slip. Small synchronous motors with brush-type exciters are often protected by operation of an AC voltage relay connected in the field. No AC voltage is present when the motor is operating synchronously. This scheme is not applicable to motors having a brushless excitation system. For such a system, a power factor relay is used [2].

7.10 Power plant auxiliary system

7.10.1 Auxiliary system design

The combination of motors, transformers, and other electrically driven devices that form an auxiliary system for a power plant presents a protection problem that is, in effect, a microcosm of power system relaying and deserves special mention. In addition to the protection of each of the elements of the auxiliary system, there is the overall system which must be considered. The following comments are also applicable to an industrial complex where the auxiliary system is required to sustain the main production facilities. Our interest here does not involve the protection, per se, of the motors, transformers, or other devices but the coordination of the protection of each of these devices from the point of view of the normal and emergency operation of the entire plant. Faulted equipment must be removed from service as fast as possible. For many faults or abnormal events within the plant, this may require that the generator be removed from the system, the excitation system tripped, the turbine valves closed, and the boiler fires extinguished. However, it is necessary that vital services such as bearing oil pumps, instrument air compressors, exhaust, and purging fans be maintained even though the unit has been tripped and is in the process of being shut down. In addition, the auxiliary system must be configured to allow the unit to return to service as soon as possible.

A portion of a typical auxiliary system of a unit-connected generator is shown in Figure 7.22. The 4-kV auxiliary bus is fed directly from the 20 kV generator leads or from the start-up transformer and is the source for the major motors. As unit sizes increase, the auxiliary load increases proportionately, requiring higher-rated transformers and higher-rated, higher-voltage motors. This has resulted in higher bus voltages, such as 6.9 and 13 kV. Phase fault currents are also increased, requiring switchgear with higher interrupting capacity. In sizing the switchgear, there are two contradictory factors that must be considered. The impedance of standard transformers increases as their ratings increase

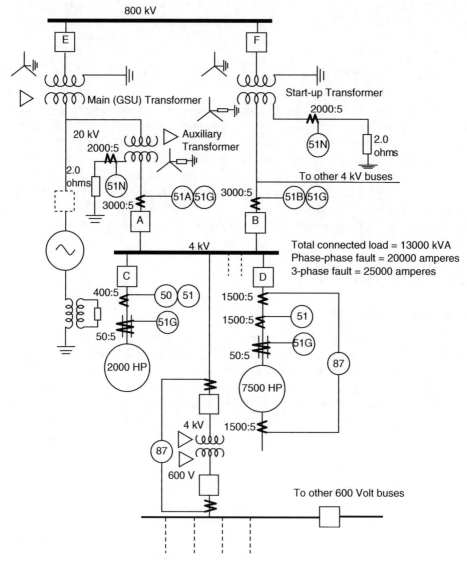

Figure 7.22 Typical power plant auxiliary system.

(Appendix C). Since the normal and short-circuit currents are also increasing, there is a greater voltage drop between the auxiliary bus and the motor. Normal design practice is to maintain at least 85% voltage at the motor terminals during motor starting. If the standard transformer impedance is specified to be at a lower value to reduce the voltage drop and maintain the 85% voltage criterion, the interrupting current will increase requiring larger rated switchgear. If the transformer impedance is raised to reduce the fault current, and hence the interrupting capacity requirement of the switchgear, the voltage drop will be too high. The art of designing the auxiliary system must take all of these factors into account. Transformers can be specified with special impedances at a greater cost. The auxiliary system can be designed with several bus sections, thus reducing the transformer rating for each section. Current-limiting reactors can be used either as separate devices or incorporated in the switchgear.

In addition to the 4 kV (or higher) bus, a lower-voltage auxiliary bus system is used to feed the dozens or hundreds of smaller motors, heating and lighting loads that are present in the plant. The nominal voltage rating of this lower-voltage bus system can be 600 or 240 V. Note that this is the voltage class as defined by its insulation rating. The actual operating voltage can be any standard voltage such as 600, 550, or 220 V depending on the practice and preference of the user. The lower-voltage buses are energized from the higher voltage bus as shown in Figure 7.22. Automatic throw-over schemes between the several bus sections or between the GSU and start-up transformer are used in the event of a 4 kV bus fault or failure of a 20 kV/4 kV or 4 kV/600 V transformer. In addition, manual throw-over provides flexibility for maintenance without removing the generator from service. The circuit breakers used on the lower-voltage buses are included in the metal-enclosed switchgear and are covered in American National Standards Institute (ANSI) standards C37.20-1 and C37.20-3. They may or may not be draw-out type, do not have CTs, and may be mounted in motor control centers. They may be air circuit breakers or molded case breakers with limited interrupting capacity. Protection is provided by series trip coils or thermal elements [2].

7.10.2   Circuit breaker application

As discussed in Section 1.5.2 (Chapter 1), there are many circuit breaker designs depending upon the particular application. Oil circuit breakers use the oil as both the insulating and the arc-extinguishing medium. The energy in the arc causes the oil to expand, enlarging, and cooling the arc. Air circuit breakers extinguish the arc by moving and stretching it into an insulating arcing chamber or arc chute. Vacuum circuit breakers extinguish the arc in a gap of less than 13 mm (0.5 in.) because there are no constituents in the vacuum that can be ionized to support the arc. Sulfur hexafluoride (SF_6) circuit breakers extinguish the arc using one of two methods: the puffer design blows the arc out with a small amount of gas blasted in a restricted arc space; the rotating arc design uses the electromagnetic effect to rotate the arc through SF_6 that cools and extinguishes it. Buses rated above 2400 V use metal-clad switchgear as defined in ANSI standard C37.20-2. The heart of the switchgear is the circuit breaker, and until the mid-1970s the use of air circuit breakers predominated. Nowadays, vacuum and SF_6 circuit breakers are more commonly used. These circuit breakers are draw-out types allowing the breaker to be removed for maintenance. The switchgear compartment contains the CTs, auxiliary contacts, and, usually, the relays and meters.

7.10.3 Phase fault protection

The phase overcurrent relays (51A and 51B) on the secondary of the unit auxiliary and start-up transformers provide bus protection and backup relaying for individual motor protection and switchgear. Figure 7.22 indicates the general arrangement of the buses and loads and shows the protection of the 2000 hp motor and the 7500 hp motor as discussed in Examples 7.5 and 7.6. Ideally, the 4 kV bus overcurrent relays 51A and 51B should have pickup settings greater than the load plus the highest motor starting current, and time delays longer than the longest starting time. These settings may be so high, or the times so long, that the protection is not acceptable and modifications or compromises are required as discussed in the following text. If the relays are also the primary bus protective relays, the settings may be so high that there may not be enough bus fault current to provide sufficient margin to ensure pickup for the minimum bus fault. Even if coordination is theoretically possible, the required time delay may be too long to be acceptable. Some compromises are possible. Since the largest motors will probably have differential protection, the backup function could consider coordinating with the overcurrent relays of the smaller motors with an associated reduction in pickup. Assuming that the differential relays are always operative, coordination with the larger motors is not a problem, since the differential protection is instantaneous. Coordination would be lost if the differential relays fail to clear a fault and the time-delay overcurrent relays must do it; this is usually an acceptable risk. A bus differential relay could be used to provide primary protection and the overcurrent relays provide backup protection for motor relay or switchgear failures. The time delay may then be acceptable. The pickup setting must still recognize the magnitude of starting current of the largest motor. If it cannot be set above this value, an interlock must be provided which will block the backup relay. Typically, an auxiliary switch on the motor circuit breaker is used to cut out the bus overcurrent protection, and a voltage relay is used to unblock this protection should a fault occur.

Example 7.8   Consider the auxiliary system shown in Figure 7.22. The 2000 hp motor is protected and set as described in Example 7.5. The protection of the 7500 hp motor is described in Example 7.6. The total bus load is $13\,000/(\sqrt{3} \times 4) = 1876$ A. Choose a 3000:5 (600:1) CT for both the main and reserve breakers. If the transformer overcurrent relays, 51A and 51B, are used primarily for bus protection, they are set at $20000/3 = 6666$ primary amperes. Choose a setting of 11 secondary amperes or 6600 primary amperes.

Check this setting against the starting current of the 2000 hp motor while the bus is fully loaded:

$$2000 \text{ hp at } 0.9 \text{ efficiency } = 1657 \text{ kVA},$$
$$\text{Total connected kVA less } 1657 \text{ kVA } = 11\,343 \text{ kVA}.$$

Assume that the start-up of this motor drops the bus voltage to 0.85 pu:

$$I_{bus} = \frac{11\,343}{(\sqrt{3} \times 0.85 \times 4)} = 1926 \text{ A},$$
$$I_{start} = 1609 \text{ A},$$
$$I_{rel} = 1926 + 1609 = 3535 \text{ A}.$$

This is below the 6600 A pickup of 51A and 51B, so that the setting is acceptable. Since the relays will not pick up during start-up of the motor, they can be set as fast as we want, for example, the #1 dial.

Check the bus overcurrent setting against start-up of the 7500 hp motor:

7500 hp motor at 0.9 efficiency = 6216 kVA.

The total connected load less the motor is 13 000 kVA − 6216 kVA or 6784 kVA, which is equal to 1153 A during motor start-up.

Motor starting current is 5512 A.

Therefore, total current through the bus overcurrent relay is 6665 A.

This is just about the same as the 6600 A pickup of the relay, and to ensure security, it must be controlled by some interlock as discussed in the earlier text.

7.10.4 Ground fault protection

The importance of ground fault protection cannot be overemphasized [17]. Ground is considered to be involved in 75–85% of all faults. In addition, phase overcurrent may often reflect a temporary process overloading, while ground current is almost invariably an indication of a fault. Auxiliary systems may be either delta- or wye-connected. A delta system is normally operated ungrounded and is allowed to remain in service when the first ground indication appears. It is generally assumed that the first ground can be isolated and corrected before a second ground occurs. It is not uncommon for systems of 600 V and less to be delta-connected. Medium-voltage systems (601 V–15 kV) are generally operated in wye, with a neutral resistor to limit the ground current to some definite value. The resistor has a time-related capability, for example, 10 s, at the maximum ground current and it is a function of the ground protective system to remove all faults within this time constraint. In Figure 7.22, ground faults on the 4 kV system are limited by the 2.0 Ω neutral resistors in the auxiliary and start-up transformers. The magnitude of the maximum fault current is the line-to-ground voltage divided by the 2.0 Ω resistor. The nominal voltage of the bus is 4 kV, but its normal operating voltage is 4160 V. Therefore, the maximum ground current is $4160/(\sqrt{3} \times 2)$ or 1200 A. Coordination must, of course, begin at the load. If the motor ground overcurrent protection is provided by the toroidal CT shown in Figure 7.13, there is no coordination problem. These can have a ratio of 50:5 resulting in a relay current of 120 A. Set an instantaneous relay at 5.0 A. If a residual ground relay is used as shown in Figure 7.12, the maximum ground fault through the CTs on breakers A and B is 1200/600 = 2.0 A. Set the time-delay ground overcurrent relays at 0.5 A and 15–30 cycles. The motor relays trip the associated feeder breaker, 51A and 51B trip the 4 kV main breakers, and the neutral relays 51N trip their associated primary breakers.

7.10.5 Bus transfer schemes

It is common practice to provide a bus transfer scheme to transfer the auxiliary bus to an alternative source in the event of the loss of the primary source. In power plants, the purpose of this alternative source is not to maintain normal operation but to provide a start-up source, to act as a spare in the event an auxiliary transformer fails, and to provide for orderly and safe shutdown. In industrial plants, the alternative source might have a different purpose, such as to provide flexibility in production or supply some facilities from the utility and others from a local generator. The transfer scheme must consider several factors. A manual, live transfer is performed by the operator while both the normal and start-up sources are still energized. If the two sources can be out of synchronism, it will be necessary to include synchronizing equipment. A dead transfer refers to the condition where the auxiliary bus has been disconnected from the generator. The speed of the transfer can be fast or slow depending upon the switchgear and the requirements of the process. It is common to check the outgoing breaker, by monitoring a breaker auxiliary contact, to be sure that the primary source is disconnected. There will always be an inrush current through the incoming source breaker and, in all probability, through the motor breakers, depending upon the residual voltage of the auxiliary bus at the instant of resynchronizing. Some schemes monitor this residual voltage and allow closing to the alternative source only after this voltage has been significantly reduced or is in synchronism.

7.10.6 Generator breaker

Figure 7.22 shows a generator breaker as an alternative facility. This is common for generators that are connected to a common bus, such as in a hydro plant. With the advent of the unit system, however, this configuration has not been used as often. The unit system requires that the boiler, turbine, generator, and GSU transformer be operated as a single entity and the loss of any one element requires that all of them be removed from service. The generator breaker is then unnecessary. In addition, as the unit sizes increased, the interrupting capability of a generator breaker became technically difficult. A 1300 MW generator can contribute as much as 100 000 A to a fault at the generator voltage level, for example, on the bus feeding the auxiliary transformers. Not only is such a breaker extremely costly, it must be placed between the generator and GSU transformer, which adds considerable length to the building. This introduces costs to every segment of the construction and installation. Nevertheless, the generator breaker can be extremely useful. Its most important advantage is the fact that, for a fault on the generator or auxiliary buses, without a generator breaker to remove the generator contribution from the fault, the generator will continue to feed the fault until the generator field decays. This can take as much as 7–10 s. During this time, the energy in the fault will result in extensive physical damage to all of the connected equipment and greatly increases the possibility of fire. With a generator breaker, the generator contribution is removed in three to five cycles; this is approximately the same time that the system contribution is removed by tripping the high-voltage breakers. A further advantage lies in the reduction in switching required when transferring the auxiliary bus. Referring to Figure 7.22, without a generator breaker, start-up is accomplished by energizing the auxiliary buses through the 800 kV breaker F, the start-up transformer, and 4 kV breaker B. Synchronizing is done through 800 kV breaker E. In the event of a unit trip, the unit is removed from the system by opening breaker E and the auxiliary bus is transferred to the start-up transformer by opening 4 kV breaker A and closing breaker B. Breaker F is operated normally closed. If the start-up transformer is connected to some other system, then breaker B must be closed with synchronizing relays. If a generator breaker is provided, at start-up the generator breaker is open and the auxiliary buses are fed through the GSU transformer and 4 kV breaker A. Synchronizing is done through the generator breaker. When removing the unit, only the generator breaker has to be opened; the auxiliary bus continues to be fed through the GSU transformer. There is no need for automatic or manual throw-over schemes. In fact, there is no need for the start-up transformer unless it is needed to provide an in-place spare for one of the auxiliary transformers. If the start-up transformer is used, it becomes a second source of start-up or shutdown power; a source that can be used to satisfy reliability requirements associated with nuclear units.

7.11 Winding connections

So far, we have been concerned with the protection principles associated with generator and motor short circuits and overloads and the appropriate relays that should be applied. The specific implementation of these principles, particularly with differential relays, varies also with the particular winding connections involved. Most machines have star (wye) connections. So three relays that are connected to star-connected CTs as shown in Figure 7.23 provide both phase and ground protection. With delta-connected windings, there is no connection to ground, and the phase currents differ from the winding currents by $\sqrt{3}$ and a phase shift of 30°. Care must be taken to obtain correct current flow, as shown in Figure 7.24. Note, however, with respect to Figure 7.24, that if digital differential relays are used, the phase shift and magnitude correction may often be accomplished within the relay (by setting adjustment) and with the lower set of CTs connections in star arrangement. This has an advantage of allowing the relay to record the actual winding current for

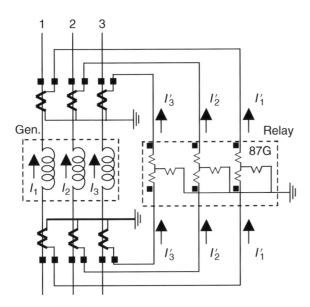

Figure 7.23   Connection for star-connected generator.

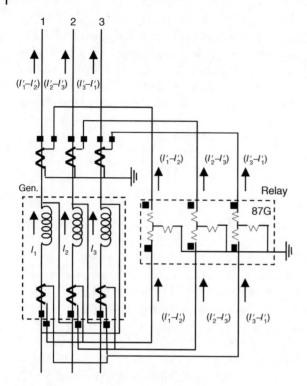

Figure 7.24 Connection for delta winding.

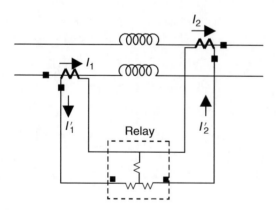

Figure 7.25 Connection for two windings per phase.

easier disturbance analysis. Split-phase windings can be protected, as shown in Figure 7.25. If the neutral connection is made inside the machine and only the neutral lead is brought out, differential relays can only be provided for ground faults, as shown in Figure 7.26. It must be noted that except in the case of split-phase windings, turn-to-turn faults cannot be detected by a differential relay, since there is no difference in the currents at the ends of the winding. In the case of split-phase windings, the differential protection connection shown

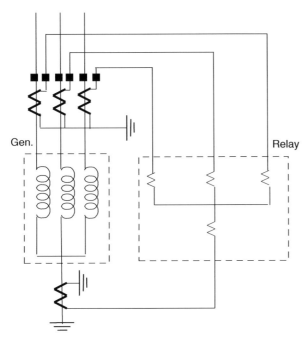

Figure 7.26 Relay connection with generator neutral CT.

in Figure 7.23 would detect the difference in phase current caused by a turn-to-turn fault in one of the windings. For windings that are not split-phase, turn-to turn faults may be detected by a zero-sequence voltage unbalance relay connected between the generator side of the neutral grounding impedance and the generator terminals [1], but this scheme is not frequently used in North America. Otherwise, a turn-to-turn fault would have to burn through to ground or to another phase before it would be detected.

7.12 Start-up and motoring

The synchronous speed of a four-pole generator is 1800 and 3600 rpm for a two-pole machine. A cross-compound turbine-generator unit consists of two shafts. Each shaft has its own steam turbine, generator, and exciter. Either shaft could have a synchronous speed of 1800 or 3600 rpm. When the unit is ready to be synchronized to the system, the two shafts must be at their respective synchronous speeds. However, the units are rolled off turning gear by admitting steam into the high-pressure turbine. The steam flow goes from the steam generator through the high-pressure turbine, back to the steam generator, and then to the low- or intermediate-pressure turbine. There is therefore a finite time before steam is admitted into the low- or intermediate-pressure turbine. If the speed of the two shafts were controlled only by the steam, the two shafts could never maintain the same speed ratio as they came up to synchronous speed. They therefore could not be synchronized to the system. To correct this problem, a cross-compound machine must have its excitation applied to each generator while on turning gear (the turning gears are designed to

Figure 7.27 Start-up protection.

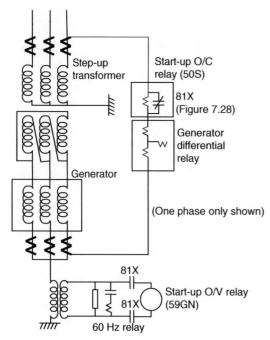

have the same speed ratio as the respective synchronous speeds of each shaft). By applying the field to the two machines, they act as a motor-generator set with the high-pressure turbine generator driving the low-pressure turbine generator at the proper speed ratio from turning gear up to synchronous speed. However, since both field and rotation are present, a voltage is generated during this start-up period. A fault can therefore result in short-circuit current, even at low voltage and low frequency. Since the magnitude of the short-circuit current will be low, and, since most differential relays are relatively insensitive at frequencies below 60 Hz, it is common practice to add an instantaneous overcurrent relay in the differential circuit and an instantaneous overvoltage relay in the grounding circuit, as shown in Figure 7.27. Electromechanical relays, such as the plunger or clapper type, are insensitive to frequency.

These relays can be set as low as necessary, provided they are removed from service prior to synchronizing the unit to the system. One circuit to accomplish this is shown in Figure 7.28. The start-up relays, 50S and 59GN, are connected to the breaker trip coil through time-delay dropout, auxiliary relay 81X, which takes both the operating coils and the tripping contacts out of service. The auxiliary relay is normally energized. When the system frequency goes above 55 Hz or the circuit breaker closes, the relay will drop out after a small time delay, usually 15 cycles. The time delay is necessary to give the start-up relays a chance to operate in the event the generator is inadvertently energized. This is a situation that will be discussed in the following text. Without the time delay, when the circuit breaker closes, there would be a race between the start-up relays, 59GN and/or 50S, operating and auxiliary relay 81X removing them from service. If the differential relays are solid-state or digital, their response at low frequencies must be determined and the need for start-up protection evaluated.

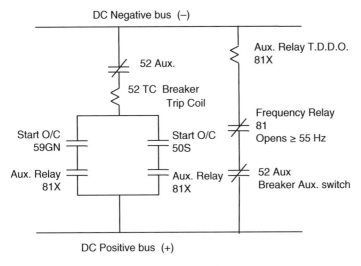

Figure 7.28 DC circuit for start-up protection.

7.13 Inadvertent energization

A common, catastrophic misoperation that has been reported many times involves the inadvertent closing of high-voltage breakers or switches while a unit is on turning gear or at some speed less than synchronous speed [1, 18]. When energized in this fashion, if field has been applied, the generator behaves as a synchronous motor or generator that has been badly synchronized. The result can destroy the shaft or other rotating element. There are several causes for this incorrect switching. Operating errors have increased dramatically as the complexity of stations and circuits increased. Stations are designed to have switching flexibility to allow a breaker or other switching device to be removed from service while still maintaining the generator in service. This has the opposite effect when the unit is offline. There are now several switching elements that can accidentally energize the generator. A flashover of breaker contacts is another possible cause, particularly when the unit is coming up to speed with field applied. As the unit rotates, the voltage increases and assumes a constantly rotating phasor not in synchronism with the system. The voltage difference, particularly when the generator and system phasors are 180° apart, can approach twice normal. If the pressure of the breaker insulating medium decreases, the breaker can flash over, connecting the unit to the system. The start-up protection described in Section 7.12 can usually act quickly enough to avoid or minimize the damage. Tandem machines, that is, turbine generators on one shaft, do not need start-up protection, since there is no need to apply field before the unit reaches synchronous speed. However, inadvertent energization is still a concern, since the machine will still behave as an induction motor when it is connected to the system before field is applied. The same protection provided for start-up can be used in this case. Some utilities use dedicated protective circuits that are activated when the unit is taken out of service.

7.14 Torsional vibration

The potential for shaft damage can occur from a variety of electrical system events. In addition to short circuits or bad synchronizing, studies have indicated that subsynchronous resonance or automatic reclosing, particularly high-speed reclosing, can produce torque oscillations leading to fatigue and eventual damage. Subsynchronous resonance is a phenomenon associated with series capacitors and results from a resonant condition that is caused by the series capacitor and the line inductance. This circuit oscillates at less than 60 Hz, resulting in extremely high voltages that are reflected back into the machine. There have been at least two well-documented incidents in the Western United States and several in Europe that initiated investigations into this problem. Specific protection packages have been designed to detect the onset of the oscillations and to remove the series capacitor and reconfigure the primary system so it will not have a series-resonant condition at subsynchronous frequencies [19]. The effect of high-speed reclosing is less certain and no events involving damage have been specifically reported at this time, although there are several monitoring studies in progress throughout the world [20, 21]. Nevertheless, criteria have been proposed, such as a 50% change in power flow following a switching event, that could serve as a warning to investigate further. The usual solution is to delay or remove high-speed reclosing or to prevent high-speed reclosing after a multiphase fault. This, of course, removes many of the advantages of high-speed reclosing and the total effect on the integrity of the system must be considered.

7.15 Sequential tripping

The purpose of sequentially tripping a synchronous generator is to minimize the possibility of damaging the turbine as a result of an overspeed condition occurring following the opening of the generator breakers [1]. With the breakers open, the unit is isolated from the system and the speed is determined by the energy delivered by the prime mover. If a valve fails to close completely after being given a trip signal, there is enough residual energy in prime mover to drive the turbine to dangerous overspeeds. Sequential tripping is accomplished by reducing the power delivered by the prime mover before tripping the generator and field breakers. Reverse-power relays, pressure switches, and/or valve limit switches are used to determine that the power input has been removed and then to complete the trip sequence. Sequential tripping is essential because overspeeding the turbine is a more damaging operating condition than motoring. There are recorded instances where overspeed resulted in throwing turbine blades through the turbine casing, resulting in injury and death to personnel in the area and, of course, extensive and costly damage to the unit. Motoring a steam turbine, in which the system supplies the rotational energy with little or no steam input, will result in heating the last-stage turbine blades, a situation that can be controlled by attemperator sprays and which allows enough time for the operator to take corrective action.

Simultaneous tripping, that is, tripping the boiler, closing all of the steam valves, and opening the generator and field breakers at the same time, is required in the event of an electrical failure. Sequential tripping is the proper action in the event of a mechanical failure. When the unit is manually tripped, it is commonly done sequentially.

7.16 Summary

In this chapter, we have examined the problems that can occur with AC generators and motors and the protective devices that can be used to correct them. The most common electrical failure involves short circuits in the stator winding. For phase faults, differential protections with percentage differential relays are almost invariably used. For ground faults, the protection depends upon the method of grounding. We have examined high-, medium-, and low-impedance grounding methods and the associated protection schemes. High-impedance grounding with the resistor on the secondary side of a transformer is the most common method for large unit-connected generators. Resonant grounding is also used, more in Europe than in the United States. Generator rotors are almost always ungrounded, so the only problem is to detect the ground and take some action before a second ground occurs. Two common ground detection methods are shown and others are referenced. For large generators, some owners choose to alarm and allow the operator to decide if a trip is warranted.

Almost all integral horsepower motors are protected with time-delay overcurrent relays to avoid overheating due to overloads, low or unbalanced voltages, or other abnormal operating conditions. Instantaneous or differential relays are used to protect against phase faults. Ground relays depend upon the method of grounding and the application of phase or toroidal CTs. Time-delay overcurrent relays are used if the CTs are connected in the residual or neutral circuit, and instantaneous relays are used with toroidal CTs.

Unbalanced voltages and currents are usually caused by system problems, but the harmful effects are felt by the rotating elements on the system. Detecting abnormal voltages, current, and frequency is not difficult. Volts/hertz, overvoltage or undervoltage, or negative-sequence current are parameters that are easily relayed. The problem arises as to the appropriate action to take. Very often, immediate tripping is not required, although if the abnormality continues damage will result and the unit must be removed from the system. Abnormal frequency is another system condition that can harm a turbine-generator set. Low frequency will seriously stress steam turbine blades and again, although the detection is simple, the remedy requires judgment.

The protection of power plant auxiliary motors has been studied, both from the point of view of the motor itself and as a system problem to ensure coordination. To examine the overall auxiliary system, we have introduced the various circuit breaker operating and interrupting mechanisms and bus transfer schemes. We have also examined a variety of operating or maintenance situations that can cause extensive damage to the turbine or the generator. Start-up of the generator, motoring, inadvertent energization, and torsional vibration are all potential hazards for which protection in the form of relays or logic circuits must be provided. The sequence of tripping the unit from the system is determined by the type of fault; an electrical fault, usually a short circuit in the generator or auxiliary bus, requires simultaneous tripping of the turbine and the generator and field breakers to remove the source of electrical energy and minimize damage. A mechanical failure, usually a boiler tube leak or turbine or pump problem, should initiate a sequential trip of the turbine, that is, extinguish the fire and close the steam valves, followed by opening the generator and field breakers when there is no danger of overspeed.

Problems

7.1 Consider the power system shown in Figure 7.29 which represents a unit-connected generator prior to being synchronized to the system and protected with an overcurrent relay connected as a differential relay. Determine the maximum load, select a CT ratio for the generator differential, calculate the relay operating currents for a three-phase fault at F_1 and F_2, and set the relay. Assume there is no CT error and the relay has the CO-11 time–current characteristics shown in Appendix D (Section D.2).

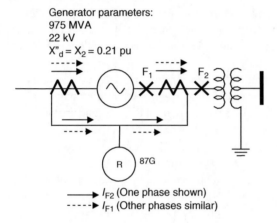

Generator parameters:
975 MVA
22 kV
$X''_d = X_2 = 0.21$ pu

I_{F2} (One phase shown)
I_{F1} (Other phases similar)

Figure 7.29 One-line diagram for Problem 7.1.

7.2 Repeat Problem 7.1 assuming that the line-side CT has an error of 1% of its secondary current. Set the overcurrent relay so it will not operate for an external fault.

7.3 Repeat Problem 7.1 for a phase-to-phase fault at F_1.

7.4 Figure 7.30 shows a percentage differential relay applied for the protection of a generator winding. The relay has a 0.1 A minimum pickup and a 10% slope. A high-resistance ground fault has occurred as shown near the grounded neutral end of the generator winding while it is carrying load with the currents flowing at each end of the generator as shown. Assume that the CT ratios are as shown in the figure and they have no error. Will the relay operate to trip the generator under this condition? Would the relay operate if the generator were carrying no load with its breaker open? Draw the relay-operating characteristic and the points that represent the operating and restraining currents in the relay for the two conditions.

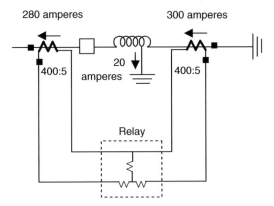

Figure 7.30 System for Problem 7.4.

7.5 Consider the system shown in Figure 7.31 with the generator, transformer, and system parameters as shown. Calculate three-phase and phase-to-phase currents due to faults at F_1 and F_2 and determine the restraining and operating currents in the percentage differential relay for the four conditions.

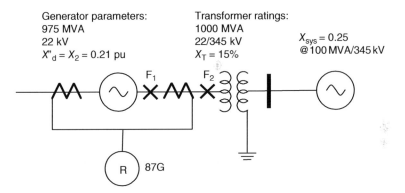

Figure 7.31 System for Problem 7.5.

7.6 For the system shown in Figure 7.32, draw the operating characteristics of an overcurrent and a percentage differential relay and show the tripping points for a fault at F_1 if R_N is, respectively, 0.5, 5, and 50 Ω.

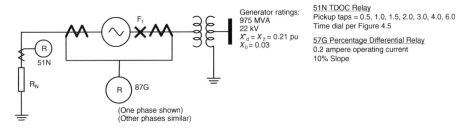

Figure 7.32 System for Problem 7.6.

7.7 Draw the one-line diagram showing a 200 hp motor connected to a 4 kV bus. Assume the following bus and motor parameters:

phase-to-phase bus fault = 15 000 A,
three-phase bus fault = 25 000 A,
maximum ground fault = 1500 A,
motor full-load current = 25 A,
motor locked rotor current = 150 A,
motor starting time = 1.5 s.

Select and set the phase and ground relays using the time–current characteristic of the three relays shown in Appendix D.

7.8 Repeat Problem 7.7 for a 1500 hp, 6.9 kV motor with the same bus fault parameters and motor full-load current of 110 A, locked rotor current of 650 A, and a starting time of 3 s.

7.9 For the distribution transformer, unit-connected generator shown in Figure 7.16, and the parameters given in Example 7.7, determine the value of the secondary resistor that will protect 85% of the winding. You may assume that a part winding voltage and leakage reactance are proportional to its length.

References

1 IEEE Std C37.102-2006 (2006). *(Revision of IEEE Std C37.106-1995), Guide for AC Generator Protection*. New York, NY: IEEE Standards Association, IEEE.

2 ANSI/IEEE standard C37.96-2012 (2013). *(Revision of IEEE Std C37.96-2000), Guide for AC Motor Protection*. New York, NY: IEEE Standards Association, IEEE.

3 Neff, W.F., Horowitz, S.H., and Squires, R.B. (1956). Relay protection of motors in steam power stations with 4 kV grounded neutral system. *AIEE Trans. PAS* **74**: 573–576.

4 IEEE Std C37.101-2006 (2007). *(Revision of IEEE Std C37.101–1993/Incorporates IEEE Std C37.101-2006/Cor1:2007), Guide for Generator Ground Protection*. New York, NY: IEEE Standards Association, IEEE.

5 Pope, J.W. (1984). Comparison of 100% stator ground fault protection schemes for generator stator windings. *IEEE Trans. PAS* **103**: 832–840.

6 Khunkhun, K.J.S., Koepfinger, J.L., and Hadda, M.V. (1977). Resonant grounding of a unit connected generator. *IEEE Trans. PAS* **96**: 550–559.

7 Gulachenski, E.M. and Courville, E.W. (1991). New England Electric's 39 years of experience with resonant neutral ground of unit-connected generators. *IEEE Trans. Power Deliv.* **6**: 1016–1024.

8 Zocholl, S.E. (1990). Motor analysis and thermal protection. *IEEE Trans.* **5** (3): 1075–1080.

9 ANSI/IEEE standard C37. 106-2003 (2004). *(Revision of ANSI/IEEE C37.106-1987), Guide for Abnormal Frequency Protection for Power Generating Plants*. New York, NY: IEEE.

10 ANSI/IEEE standard C57.12.01-2020, (n.d.) General requirements for dry-type distribution and power transformers.

11 ANSI/IEEE standard C50.13-2014 (n.d.). *(Revision of IEEE Std C50.13-2005), IEEE Standard for Cylindrical-Rotor 50 Hz and 60 Hz Synchronous Generators Rated 10 MVA and Above.* New York, NY: IEEE.

12 Maliszewski, R.M., Dunlop, R.D., and Wilson, G.L. (1971). Frequency actuated load shedding and restoration: I. philosophy. *IEEE Trans. PAS* **90**: 1452–1459.

13 Horowitz, S.H., Politis, A.P., and Gabrielle, A.F. (1971). Frequency actuated load shedding and restoration: II. Implementation. *IEEE Trans. PAS* **90**: 1460–1468.

14 IEEE Power System Relaying Committee Report (2004). Performance of generator protection during major system disturbances. *IEEE Trans. Power Deliv.* **19** (4): 1650–1662.

15 Mason, C.R. (1949). A new loss-of-excitation relay for synchronous generators. *AIEE Trans. Part II* **68**: 1240–1245.

16 Tremaine, R.L. and Blackburn, J.L. (1954). Loss-of-field protection for synchronous machines. *AIEE Trans. PAS* **73**: 765–772.

17 Love, D.J. (1978). Ground fault protection for electric utility generating station medium voltage auxiliary power systems. *IEEE Trans. PAS* **97** (2): 583–586.

18 IEEE Power System Relaying Committee (1989). Inadvertent energization protection. Report. *IEEE Trans. Power Deliv.* **4** (2): 965–974.

19 Committee, I.E.E.E. (1976). A bibliography for the study of subsynchronous resonance between rotating machines and power systems. Report. *IEEE Trans. PAS* **95** (1): 216–218.

20 Dunlop, R.D., Horowitz, S.H., Joyce, J.S., and Lambrecht, D. (1980) *Torsional oscillations and fatigue of steam turbine-generator shafts caused by system disturbances and switching events,* CIGRE 1980, paper 11–06.

21 Dunlop, R.D., Horowitz, S.H., Parikh, A.C. et al. (1979). Turbine generator shaft torques and fatigue: II. Impact of system disturbance and high speed reclosure. *IEEE Trans. PAS* **98** (6): 2308–2328.

8

Transformer Protection

8.1 Introduction

The inherent characteristics of power transformers introduce a number of unique problems that are not present in the protection of transmission lines, generators, motors, or other power system apparatus. Transformer faults—that is, short circuits—are the result of internal electrical faults, the most common one being the phase-to-ground fault. Somewhat less common are the turn-to-turn faults. Unlike a transmission line, the physical extent of a transformer is limited to within a substation, and consequently differential relaying, the most desirable form of protection available, can be used to protect transformers. In general, a transformer may be protected by fuses, overcurrent relays, differential relays, and pressure relays, and can be monitored for incipient trouble with the help of winding temperature measurements, and chemical analysis of the gas above the insulating oil. Which of these will be used in a given instance depends upon several factors as discussed in the following text.

- Transformer size: Transformers with a capacity less than 2500 kVA are usually protected by fuses. With ratings between 2500 and 5000 kVA, the transformer may be protected with fuses, but instantaneous and time delay overcurrent relays may be more desirable from the standpoint of sensitivity and coordination with protective relays on the high and low sides of the transformer. Between 5000 and 10 000 kVA, a time overcurrent relay connected in a differential configuration has often been applied, but differential protection similar to larger transformers is also often applied now. For transformers above 10 MVA, a harmonic-restrained percentage differential relay is recommended. Pressure and temperature relays are also usually applied with this size of transformer.
- Location and function: In addition to the size of the transformer, the decision regarding the specific protection application is significantly affected by consideration of the importance of the transformer within the power network. If the transformer is an integral part of the bulk power system, it will probably require more sophisticated relays in terms of design and redundancy. If it is a distribution station step-down transformer, a single differential relay and overcurrent backup may suffice. If the transformer is near a generation

Power System Relaying, Fifth Edition. Stanley H. Horowitz, Arun G. Phadke and Charles F. Henville.
© 2023 John Wiley & Sons Ltd. Published 2023 by John Wiley & Sons Ltd.

source, the high X/R ratio of the fault path will require harmonic-restrained relays to accommodate the higher magnetic inrush currents.

- Voltage: Generally, higher voltages demand more sophisticated and costly protective devices, due to the deleterious effect of a delayed fault clearing on the system performance, and the high cost of transformer repair.
- Connection and design: The protection schemes will vary considerably between autotransformers, and two- or three-winding transformers. The winding connection of a three-phase transformer—whether delta or wye—will make a difference to the protection scheme chosen. Also, important are the presence of tertiary windings, type of grounding used, tap changers, or phase-shifting windings.

As we develop protection ideas for transformers in this chapter, we also comment on the influence of these and other factors on the relaying systems of choice.

8.2   Overcurrent protection

As in all protection applications with overcurrent relays, the external faults or steady-state load currents must be distinguished from the currents produced by the internal faults. The effects of external faults that are not cleared promptly, or steady-state heavy loads, are overheating of the transformer windings and degrading of the insulation. This will make the transformer vulnerable to internal flashovers. Protection of transformers against internal faults can be provided by time delay overcurrent relays. The effects of a sustained internal fault are arcing, possible fire, and magnetic and mechanical forces that result in structural damage to the windings, the tank, or the bushings with subsequent danger to personnel or surrounding equipment. Protection for transformers can be provided by high-side fuses, instantaneous and time delay overcurrent relays, or differential relays.

8.2.1   Protection with fuses

As mentioned earlier, fuses are not usually used to protect transformers with ratings above 2.5 MVA. The basic philosophy used in the selection of fuses for the high-voltage side of a power transformer is similar to that used in other applications of fuses. Clearly, the fuse-interrupting capability must exceed the maximum short-circuit current that the fuse will be called upon to interrupt. The continuous rating of the fuse must exceed the maximum transformer load. Typically, the fuse rating should be greater than 150% of the maximum load. The minimum melt characteristic of a fuse indicates that the fuse will be damaged if conditions to the right of (greater than) the characteristic are obtained. The minimum melt characteristic of the fuse must coordinate with (i.e., should be well separated from) the protective devices on the low side of the power transformer. In considering the coordination, the ambient temperature, prior loading, and reclosing adjustment factors should be taken into account. All of these factors influence the prior heating of the fuse, and cause it to melt at different times than the specifications for a "cold" fuse would indicate. Example 8.1 provides further explanation of this procedure. It is also clear that the transformer magnetizing current should not cause damage to the fuse. This calls for a longer duration current which must be lower than the minimum melt characteristic. The "speed ratio" of the fuse is defined as the ratio between the minimum melt current values at two widely separated times: for example, 0.1 and 100 s. A smaller speed ratio would mean a more steeply sloped characteristic, and for proper degree of coordination with downstream protection, such a fuse would have to be set with a smaller sensitivity. It is desirable to have a fuse with as high a speed ratio as possible. Finally, if a current-limiting fuse is used, the lightning arresters on the line side of the fuse should have a rating equal to, or higher than, the overvoltages that the fuse may create, in order to avoid the arrester flashing over due to the operation of the fuse.

Example 8.1 Consider the transformer shown in Figure 8.1. It has a high-side delta winding and a low-side wye winding. We will consider the selection of the fuse on the high side. It must coordinate with the low-side feeder protection, which is assumed to be a fuse, with the characteristic "B" as shown. All currents are assumed to be given in terms of the secondary side, even though one fuse will be on the high side, and the actual currents in that fuse will be smaller by the turns ratio factor of the transformer.

The characteristic "A" has variability, due to prior loading, ambient temperature, and the timing of the reclosers, as shown by the dotted line in its group of characteristics. The downstream fuse, with characteristic "B," must have a sufficient coordinating margin with the dotted characteristic in the "A" group. The high-side fuse must also coordinate with the short-circuit capability curve of the transformer it is trying to protect. It should be noted that the transformer capability is not a simple matter to define, and often the capability curve is a composite of several segments obtained from different criteria.

Figure 8.1 Protection of a transformer with a fuse: coordination principles.

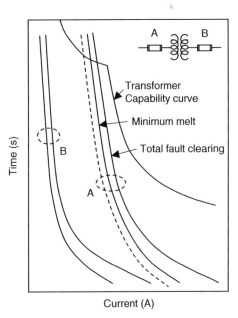

8.2.2 Time delay overcurrent relays

Protection against excessive overload, or persisting external fault, is provided by time delay overcurrent relays. The pickup setting is usually 115% of the maximum overload acceptable. This margin covers the uncertainty in the current transformers (CTs), relays, and their calibration. The time delay overcurrent relays must coordinate with the low-side protective devices. These may include low-voltage bus overcurrent relays for phase-to-phase faults, phase-directional relays on parallel transformers, and the breaker failure relay timers on the low-voltage breakers. These considerations are illustrated by the following example.

Example 8.2   Consider the transformer shown in Figure 8.2. It is rated at 2.5 MVA, with primary and secondary voltages of 13.8 and 2.4 kV, respectively. Thus, the full-load current for the primary is 104.6 A, and 600 A for the secondary. Let us assume an overload capability of 1.2 pu or 720 A in the secondary winding. The CT ratio should be selected to produce close to 5 A when the secondary current is 600 A. Select a CT ratio of 600:5 or 120:1. The pickup of the relay should be set at 115% of 720 A, or at 828 A primary, which is 828/120 = 6.9 A secondary. If an electromechanical relay with discrete taps is applied, select the nearest tap, say 7.0 A. As shown in Figure 8.2, the time dial selected for the relay should coordinate with overcurrent relays R_{ab} and R_{ac}, which protect the feeders on the low-voltage side. If a breaker failure relay is used for breaker B_{ab} or B_{ac}, the relay should also coordinate with those time delays.

8.2.3 Instantaneous relays

There are several constraints imposed upon the use of instantaneous relays; some of them depend upon the design of the relay. In all cases, of course, the relay must not operate on inrush, or for low-side faults. Peak magnetizing inrush current in a transformer can be as high as 8–10 times peak full-load current. Since the relay will see low-side faults, one must consider these faults when they are fully offset. Some relay designs—for example, electromechanical plunger-type relays—respond to the actual instantaneous value of the current, which includes the DC offset. Such a relay must be set above the low-side fault currents with full DC offset. Disk-type relays respond only to the AC portion of the current wave. Solid-state or computer-based relays may or may not respond to the DC offset, depending upon their design.

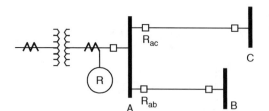

Figure 8.2   Coordination of transformer overcurrent relay with feeder protection on low side.

8.3   Percentage differential protection

Consider the single-phase, two-winding power transformer shown in Figure 8.3. During the normal operation of the transformer, the algebraic sum of the ampere-turns of the primary and the secondary windings must be equal to the magnetomotive force (MMF) required to set up the working flux in the transformer core. Because of the very small air gap in the transformer core, the MMF is negligible (often less than 0.5% of the MMF produced by the load current), and hence for a normal power transformer,

$$N_1 i_{1p} = N_2 i_{2p}. \tag{8.1}$$

If we use CTs having turns ratios of $1{:}n_1$ and $1{:}n_2$ on the primary and the secondary sides, respectively, under normal conditions, the currents in the secondary windings of the CTs are related by

$$N_1 n_1 i_{1s} = N_2 n_2 i_{2s}. \tag{8.2}$$

If we select the CTs appropriately, we may make $N_1 n_1 = N_2 n_2$, and then, for a normal transformer, $i_{1s} = i_{2s}$. However, if an internal fault develops, this condition is no longer satisfied, and the difference of i_{1s} and i_{2s} becomes much larger; in fact, it is proportional to the fault current. The differential current

$$I_d = i_{1s} - i_{2s} \tag{8.3}$$

provides a highly sensitive measure of the fault current. If an overcurrent relay is connected as shown in Figure 8.3, it could provide excellent protection for the power transformer except for practical issues as noted in the following.

Several practical issues must be considered before a workable differential relay can be implemented [1, 2]. First, it may not be possible to obtain the CT ratios on the primary and the secondary side which will satisfy the condition $N_1 n_1 = N_2 n_2$, as we must select CTs with standard ratios. Second, the errors of transformation of the two CTs may differ from each other, thus leading to significant differential current when there is normal load flow, or an external fault. Finally, if the power transformer is equipped with a tap changer, it will introduce a main transformer ratio change when the taps are changed. These three

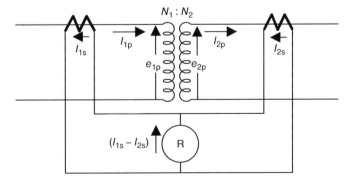

Figure 8.3   Differential overcurrent relay connections.

effects cause a differential current to flow in the overcurrent relay, and the relay design must accommodate these differential currents without causing a trip. Since each of these causes leads to a differential current which depends on the actual current flowing in the transformer primary and secondary windings, a percentage differential relay provides an excellent solution to this problem. In a single-slope percentage differential relay, the differential current must exceed a fixed percentage of the "through" current in the transformer. The through current is defined in different ways in different models of relays. One type of restraint is the average of the magnitudes of the primary and the secondary currents:

$$i_r = \frac{i_{1s} + i_{2s}}{2}. \tag{8.4}$$

Other types of restraint include the sum of the magnitudes of primary and secondary currents or the greater of those magnitudes. The current i_r is known as the restraint current—a name that comes from the electromechanical relay design, where this current produced a restraint torque on the moving disk, while the differential current produced the operating torque. The relay operates when

$$i_d \geq K\, i_r, \tag{8.5}$$

where K is the slope of the percentage differential characteristic. K is generally expressed as a percent value: typically 10%, 20%, and 40%. Clearly, a relay with a slope of 10% is far more sensitive than a relay with a slope of 40%. Although we have used lowercase symbols for the currents, signifying the instantaneous values, it should be clear that the corresponding relations exist between the root mean square (rms) or phasor values of the currents as well. Thus, Equations (8.1)–(8.5) also hold for the rms and phasor currents.

The inputs to a percentage restrained differential relay include special circuits to produce the restraint currents. In an electromechanical relay, these circuits would be coils to produce the restraint torque as shown in Figure 8.4. In solid-state implementations, the restraint inputs would be electronic circuits.

In electromechanical implementations, the restraint input coils will have taps that allow CT ratio mismatch to be minimized to very small levels. In solid-state implementations, the

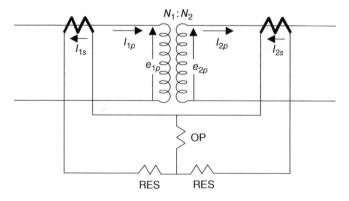

Figure 8.4 Percentage restrained differential relay connections.

Figure 8.5 Single-slope percentage differential relay characteristic.

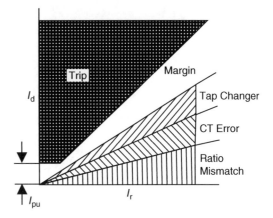

restraint circuits will be capable of compensating for CT mismatch to a level that makes the error negligible.

A practical single-slope percentage differential characteristic is shown in Figure 8.5. The relay slope determines the trip zone. The three sources of differential current during normal transformer operation are shown, as is the margin of safety used in arriving at the slope. The relay has a small pickup current setting, that is, the relay does not operate unless the differential current is above this pickup value. The pickup setting is usually set very low: typical values are 0.25 A secondary. This accounts for any residual CT errors at low values of transformer load current.

Example 8.3 A single-phase transformer is rated at 69/110 kV, 20 MVA. It is to be protected by a differential relay, with input taps of 3.0, 4.0, 4.5, 4.8, 4.9, 5.0, 5.1, 5.2, 5.5 A secondary. The transformer has an under load tap changer (ULTC) with a turns ratio of −5% to +5% in steps of 5/88%. Specify the CTs, the pickup setting, and the percentage differential slope for the relay. The available slopes are 10, 20, and 40%. What is the level of fault current for an unloaded transformer, for which the differential relay will not operate?

The currents in the primary and the secondary for the rated load are 289.8 and 181.8 A, respectively. We may select CT ratios of 300:5 and 200:5 for the two sides. These will produce $289.8 \times 5/300 = 4.83$ A, and $181.8 \times 5/200 = 4.54$ A in the two CT secondaries. In order to reduce a mismatch between these currents, we may use the relay tap of 4.8 for the CT on the primary side and the relay tap of 4.5 for the CT on the secondary side. This will give us a value of 4.83/4.8, or 1.0062×5 A, and 4.54/4.5, or 1.009×5 A in the relay coils. Thus, the differential current in the relay due to CT ratio mismatch would amount to $(1.009 - 1.006)$ pu, or about 0.3%. The tap changer will change the main transformer ratio by 5%, when it is in its extreme tap position. Thus, a total differential current of 5.3% would result from these two causes. If no information on unequal CT errors is available, we must make appropriate assumptions, so that we may select a proper percentage slope for the relay characteristic. Assuming the CTs to be of the 10CXXX type, we may expect a maximum error of under 10% in each of the CTs. It is therefore reasonable to assume that the errors in the two CTs will not differ from each other by more than 10% under all fault conditions. This gives a net differential current of 15.3% for the largest external fault, while the tap changer is at its

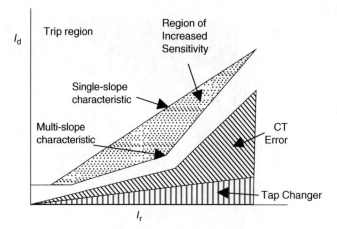

Figure 8.6 Multi-slope percentage differential relay characteristic.

farthest position. With about a 5% margin of safety, we may therefore select a 20% differential slope for the relay.

For the pickup setting, we may select the lightest available setting. A typical available value is 0.25 A. If there were a tertiary winding with no CTs available in the main transformer, we would find the contribution to the differential current from its rated load current, as it would not be available to the relay, and then select a pickup setting above that current.

With a 0.25 A pickup setting, the primary current on the 69 kV side will be $(300/5) \times (4.8/5) \times 0.25$, or 14.4 A. A fault on the 69 kV side producing currents smaller than this will not be seen by this relay.

It may be noticed in Example 8.3 that the slope setting is negligibly influenced by CT ratio mismatch, and CT error was assumed to be only 10%. In many applications, CT error will likely be much higher than this during high-magnitude external faults. Therefore, most percentage restrained differential relays will not have a single fixed slope. Electromechanical relays will have a variable slope characteristic as shown in Figure 7.4 (Chapter 7). Solid-state relays will usually have two different slope settings, a low slope for low levels of through current and a high slope for high-magnitude external fault when the probability of severe CT saturation is greater as discussed further in Section 8.4.6. The multiple or variable slopes as shown in Figure 8.6 allow greater sensitivity for low-magnitude internal faults, while retaining security for high-magnitude external faults.

8.4 Causes of false differential currents

The percentage differential relay described in the earlier text takes care of the differential currents which flow in the relay during normal load flow conditions, or during an external fault. However, certain other phenomena cause a substantial differential current to flow, when there is no fault, and these false differential currents are generally sufficient to cause a percentage differential relay to trip, unless some special precautions are taken. All such phenomena can be traced to the nonlinearities in the transformer core, or in the CT core or in both. We will now consider the effects of these nonlinearities.

8.4.1 Magnetizing inrush current during energization

Consider the energization of an unloaded power transformer as shown in Figure 8.7. As the switch is closed, the source voltage is applied to the transformer, and a magnetizing current is drawn from the source [1–3]. Let the source voltage be

$$e(t) = E_{max} \cos(\omega t - \varphi). \tag{8.6}$$

If we neglect the resistance, and the source and the leakage inductance in the circuit, the flux linkages of the transformer core are given by

$$\Lambda(t) = \Lambda_{max} \sin(\omega t - \varphi) + \Lambda_{max} \sin \varphi. \tag{8.7}$$

Note that the flux linkages—and hence the magnetizing current—are continuous at $t = 0$, when the switch is closed. Equation (8.7) assumes that there is no remnant flux in the core. Any remnant flux must be added to the right-hand side of this equation.

In reality, the magnetizing inductance of the transformer is nonlinear. Consider the two-slope approximation of a magnetizing characteristic shown in Figure 8.8a. As the flux linkages go above the saturation knee point, a much larger current is drawn from the source. The magnitude of this current is determined by the slope of the magnetizing characteristic in the saturated region, and by the leakage inductance of the transformer. It is obvious that magnetizing inrush currents of the order of fault currents are possible. Because of the losses in the circuit, the magnetizing current will decay to its nominal small value as shown in Figure 8.8b. Time constants of several seconds are common in most modern power transformers.

It should be clear that in most modern transformers very large inrush currents are possible, depending upon the instant of energization, and the remnant flux in the transformer core. Since the inrush current flows only in the primary and not in the secondary winding of

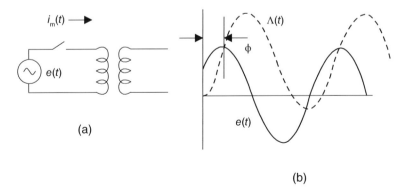

(a)

(b)

Figure 8.7 (a) and (b) Transformer core flux linkages upon energization at angle φ from maximum.

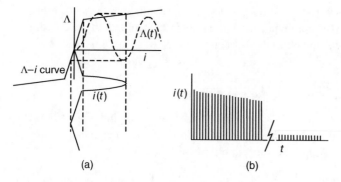

Figure 8.8 (a) and (b) Magnetizing current during energization of a transformer.

the transformer, it is clear that it produces a differential current which is 200% of the average restraining current, and would cause a false operation.

8.4.2 Harmonic content of the inrush current

As we see shortly, the false operation of a percentage differential relay for a transformer is prevented by taking advantage of the fact that the inrush current is rich in harmonic components, while the fault current is a pure fundamental frequency component (except for a possible decaying DC component). Let us calculate the harmonic components of a typical inrush current waveform. We will assume a simplified waveform for the inrush current. Let the magnetizing characteristic be a vertical line in the Λ–i plane, and be a straight line with a finite slope in the saturated region. This makes the current waveform of Figure 8.8a acquire the shape shown in Figure 8.9. The flux in the core is above the saturation knee point for a total angular span of 2α radians, and the corresponding current is a portion of a sine wave. For the remainder of the period, the current is zero. Although this is an approximation, it is quite close to an actual magnetizing current waveform. We may use Fourier series analysis to calculate the harmonics of this current. Consider the origin to be at the center of a current pulse, as shown in Figure 8.9. Then, the expression for the current waveform is

$$i(\theta) = I_m(\cos\theta - \cos\alpha) \quad 0 \le \theta \le \alpha, \ (2\pi - \alpha) \le \theta \le 2\pi$$
$$= 0 \quad \alpha \le \theta \le (2\pi - \alpha)$$
(8.8)

Since this choice of the origin gives a symmetric waveform about $\theta = 0$, we may use the cosine Fourier series for the current. The nth harmonic is given by

$$a_n = \frac{1}{\pi}\int_0^{2\pi} i(\theta)\cos n\theta \, d\theta = \frac{2}{\pi}\int_0^{\alpha} i_m(\cos\theta\cos n\theta - \cos\alpha\cos n\theta)d\theta$$

$$= \frac{I_m}{\pi}\left[\frac{1}{n+1}\sin(n+1)\alpha + \frac{1}{n-1}\sin(n-1)\alpha - 2\cos\alpha\frac{1}{n}\sin n\alpha\right]$$
(8.9)

The peak of the current wave is $I_m(1 - \cos\alpha)$, and the fundamental frequency component a_1 is given by

Figure 8.9 Idealized inrush current waveform.

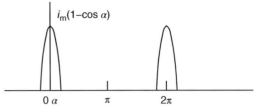

Table 8.1 Harmonics of the magnetizing inrush current.

Harmonic	a_n/a_1		
	$\alpha = 60°$	$\alpha = 90°$	$\alpha = 120°$
2	0.705	0.424	0.171
3	0.352	0.000	0.086
4	0.070	0.085	0.017
5	0.070	0.000	0.017
6	0.080	0.036	0.019
7	0.025	0.000	0.006
8	0.025	0.029	0.006
9	0.035	0.000	0.008
10	0.013	0.013	0.003
11	0.013	0.000	0.003
12	0.020	0.009	0.005
13	0.008	0.000	0.002

$$a_1 = \frac{I_m}{\pi}\left[\alpha - \frac{1}{2}\sin 2\alpha\right]. \tag{8.10}$$

The relative magnitude of various harmonic components with respect to the fundamental frequency component, as calculated from Equations (8.9) and (8.10), is tabulated in Table 8.1 up to the 13th harmonic, and for saturation angles of 60°, 90°, and 120°. It should be noted that when the saturation angle is 90° there are no odd harmonics present. As the angle of saturation increases, the harmonic content decreases: indeed, if α becomes π there will be no harmonics at all. However, in most cases, α is much less than π, and a significant amount of harmonics are present in the magnetizing inrush current. Of all the harmonic components, the second is by far the strongest.

8.4.3 Magnetizing inrush during fault removal

When a fault external to, but near, the transformer is removed by the appropriate circuit breaker, the conditions inside the transformer core are quite similar to those during magnetization of the transformer [2, 4]. As the voltage applied to the transformer windings jumps from a low prefault value to the normal (or larger) postfault value, the flux linkages in the transformer core are once again forced to change from a low prefault value to a value close to normal. Depending upon the instant at which the fault is removed, the transition may force a DC offset on the flux linkages, and primary current waveforms similar to those encountered during energization would result. It should be noted that as there is no remnant flux in the core during this process, the inrush is in general smaller than that during the transformer energization.

8.4.4 Sympathetic inrush

This is a somewhat unusual event, but occurs often enough, and deserves a mention. Consider a generator, connected to a bus through a transmission line having a resistance R and an inductance L, as shown in Figure 8.10a. The transformer T_1 is energized, and the transformer T_2 is being energized by closing the breaker B. As the breaker B closes, an inrush current is established in the primary winding of the transformer T_2, and is supplied by the generator through the impedance of the transmission line. The inrush current has a DC component, which decays with a somewhat long time constant as explained in the earlier text. This decaying DC component produces a voltage drop in the "resistance" of the transmission line.

The DC voltage drop is of a polarity as shown in Figure 8.10a for an assumed positive DC component flowing in the direction shown. Since the generator output is purely AC, and cannot be affected by this voltage drop, it is clear that the voltage of bus A acquires a negative DC component. This results in a negative change in the flux linkages (integral of the voltage change) of the two transformer cores. As the transformer T_2 was assumed to have a

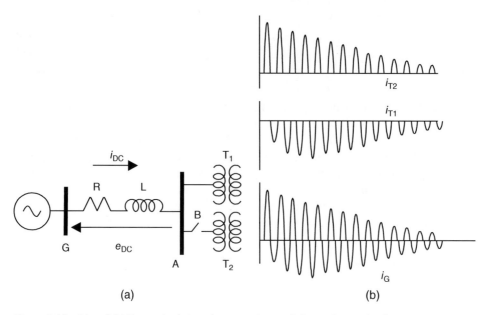

(a)   (b)

Figure 8.10 (a) and (b) Sympathetic inrush current in parallel transformer banks.

saturating flux in the positive direction (which led to the inrush current as shown), the effect of this flux change is to take T_2 out of saturation, and cause a possible saturation of T_1 in the negative direction. Consequently, the inrush current in T_2 decreases in time, and the inrush in T_1 increases in the opposite direction. When the DC components in the two inrush currents become equal to each other, there is no DC component in the current in the transmission line, but there may be DC components in both transformer magnetizing currents. The decay of this "trapped" DC component may be quite slow. The phenomenon which causes inrush to flow in a previously energized transformer, when a parallel bank is energized, is known as the sympathetic inrush. The current waveforms of a typical sympathetic inrush phenomenon are illustrated in Figure 8.10b. The remedy for the protection problem posed by sympathetic inrush currents is discussed in Section 8.9.

8.4.5 Transformer overexcitation

During load rejection and certain other operating conditions, a transformer may be subjected to a steady-state overvoltage at its nominal frequency. During overexcitation, the transformer flux remains symmetric, but goes into saturation for equal periods in the positive and the negative half-periods of the waveform. This condition is illustrated in Figure 8.11. The expression for the current in this case is given by

$$
\left.
\begin{aligned}
i(\theta) &= I_\text{m}(\cos\theta - \cos\alpha) & 0 \leq \theta \leq \alpha, (2\pi - \alpha) \leq \theta \leq 2\pi \\
&= 0 & \alpha \leq \theta \leq (\pi - \alpha), (\pi + \alpha) \leq \theta \leq (2\pi - \alpha) \\
&= I_\text{m}(\cos\theta + \cos\alpha) & (\pi - \alpha) \leq \theta \leq (\pi + \alpha)
\end{aligned}
\right\}. \tag{8.11}
$$

When the cosine Fourier series for this current is calculated, it turns out that all even harmonics are identically zero. The odd harmonics are twice those given by Equation (8.9), due to the contributions from the negative half-cycle of the current wave. The angles of saturation in an overexcited transformer are of the order of $10°$–$30°$, which corresponds to an angle of $5°$–$15°$ in Equation (8.9). For such small saturation angles, the harmonic proportions are somewhat different from those shown in Table 8.1. For example, for $\alpha = 5°$, the third and the fifth harmonics are almost as strong as the fundamental frequency component. And, of course, there is no second harmonic present in this waveform. It should be remembered that we have made significant approximations in computing these harmonic components: for example, we have assumed a two-straight-line approximation to the magnetizing characteristic, and also we have assumed that the unsaturated characteristic has an infinite slope. Under practical conditions, the harmonic magnitudes will tend to be somewhat different, although the general trend will be as determined here.

Transformer overexcitation protection (which is primarily a thermal problem) was discussed in Section 7.7.1 (Chapter 7), but it should be noted here that some solid-state transformer differential protection systems

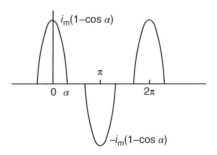

Figure 8.11 Magnetizing current during overexcitation.

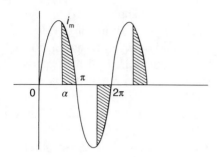

Figure 8.12 Current waveform during CT saturation.

may include fifth harmonic restraint to prevent instantaneous tripping due to overexcitation.

8.4.6 CT saturation

For certain external faults, where the fault currents are large, it is likely that one of the CTs may saturate. (We will disregard the possibility of DC offset of CT remnant flux for the initial discussion.) The resulting current waveform of that CT secondary winding is shown in Figure 8.12. The differential current in the relay will then equal the shaded area, which is the difference between the unsaturated current waveform and the saturated current waveform. The equation for the shaded current waveform is

$$
\left.
\begin{aligned}
i(\theta) &= 0 & 0 \leq \theta \leq (\pi - \alpha), \pi \leq \theta \leq (2\pi - \alpha) \\
&= I_\mathrm{m} \sin \theta & (\pi - \alpha) \leq \theta \leq \pi, (2\pi - \alpha) \leq \theta \leq 2\pi
\end{aligned}
\right\}. \tag{8.12}
$$

This being an odd function of θ, a sine Fourier series would be appropriate to determine its harmonic content. It is left as an exercise for the reader to show that this waveform contains no even harmonics, and that there is a significant third harmonic component in the current.

Note that when DC offset is present in the fault current, or if remnant flux was in the CT core before the fault, there could be a significant amount of even harmonics.

8.5 Supervised differential relays

The inrush current phenomenon in large power transformers presented an insurmountable obstacle to the percentage differential relay design when it was introduced. Measures were introduced for counteracting the tendency of the relay to trip for inrush conditions. One of the earliest ideas was to desensitize the differential relay when the transformer is energized. Thus, the less sensitive relay would not see the inrush current, thus avoiding a false trip. However, desensitizing (or disabling) the differential relay during energization is a poor practice, as it is precisely during the initial energization of the transformer, when the transformer is first energized, or some repair work on the transformer may have been completed, that the transformer is in need of protection. This is to ensure that the repair work has been successfully completed, and no maintenance tools inadvertently left inside or around the transformer.

The next significant step was the introduction of the concept of voltage supervision [2]. It may be expected that during the inrush conditions, the transformer voltage would be close to normal, while during faults, the voltage would be much less. Thus, an undervoltage relay may be used to supervise the differential relay. If the undervoltage relay can be set to distinguish between a normal transformer and a faulted transformer, then it could be used to block the differential relay when it detects a voltage above its setting. In general, this type of voltage supervision is not preferred, as the undervoltage relay tends to be slow, and consequently the entire protection becomes slower. More importantly, this type of protection requires a voltage source for the transformer relay, which is an added expense, and may not be justifiable in many cases.

The method currently in use on large transformers is based upon using the harmonic characterization of the inrush and overexcitation currents. The differential current is almost purely sinusoidal when the transformer has an internal fault, whereas it is full of harmonics when the magnetizing inrush current is present, or when the transformer is overexcited. Thus, the differential current is filtered with filters tuned to an appropriate set of harmonics, and the output of the filters is used to restrain the differential relay. We may consider a generic circuit for a harmonic-restrained percentage differential relay, although actual relay designs would depend upon the particular manufacturer's preference and experience. Consider the circuit shown in Figure 8.13. The restraint current is obtained from the through-CT T_1, while the operating current is obtained through the differential-CT T_2. The differential current is filtered through a fundamental frequency band-pass filter to provide the normal operating current. The differential current is passed through a fundamental-frequency-blocking filter, which lets through all but the fundamental frequency, and this signal is used to provide an additional restraint force. A typical adjustment is to use a restraint setting of 15% in the harmonic circuit; that is, when the harmonic current is 15% or greater than the fundamental frequency component of the differential current, the relay is on the verge of operation. As can be seen from Table 8.1, this setting is sufficient to prevent operation of the differential unit for all inrush conditions considered in that table. With more

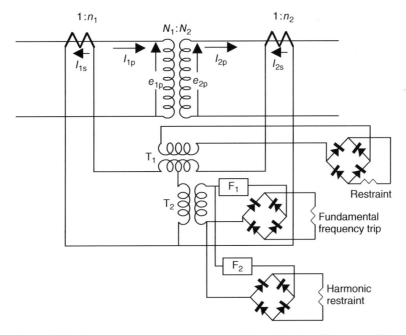

Figure 8.13 Harmonic-restrained percentage differential relay. F1 is a pass filter and F2 is a block filter for the fundamental frequency.

modern steels being used in modern transformers, it has been found that often the harmonic content of the inrush current may be as low as 7% in some phases. In such cases, the relay using 15% harmonic restraint would fail to prevent tripping under inrush conditions. For such cases, an even lower inrush setting is desirable. Alternatively, since the harmonic content in at least one phase is always over 15%, some relays may have a setting that allows harmonic content in one phase to block all three phases. Still other relays may have settings that allow harmonics to desensitize the differential function (restrain) instead of blocking it completely.

A harmonic-restrained function that uses all the harmonics for restraint may be in danger of preventing a trip for an internal fault if the CTs should saturate. As was pointed out in Section 8.4, saturated CTs produce a predominant third harmonic in the current, but may also produce some even harmonics due to DC offset and/or remanent flux. Modern solid-state transformer differential protection systems include a high-set unrestrained differential overcurrent function which is intended to improve dependability for severe internal faults with CT saturation which could prevent operation of the harmonic-restrained function. This unrestrained differential overcurrent function is set so high that it will not operate undesirably (be secure) for inrush current or for faults on the other side of the transformer.

8.6 Three-phase transformer protection

8.6.1 Phase shift in Wye–Delta transformers

Protection of three-phase transformers requires that primary and secondary currents of the three phases be compared individually to achieve differential protection of the three-phase transformer. The major difference between three-phase transformer protection and that of three single-phase transformers is the necessity to deal with the effect of a wye–delta transformation. Under normal load conditions, the currents in the primary and secondary windings are in phase, but the line currents on the wye and delta sides of the three-phase transformer are out of phase by 30°. Since CTs are usually connected in the line—and not in the winding on the delta side—this phase shift causes a standing differential current, even when the turns ratio of the main transformer is correctly taken into account. Historically, the difficulty was resolved by connecting the CTs in such a manner that they undo the effect of the wye–delta phase shift produced by the main transformer. The CTs on the wye side of the power transformer were connected in delta, and the CTs on the delta side of the power transformer were connected in wye. Recall that a delta connection can be made in two ways: one whereby the delta currents lag the primary currents by 30° and the other whereby the delta currents lead the primary currents by 30°. It is of course necessary to use the connection which compensates the phase shift created by the power transformer. Modern digital relays include settings that can compensate for the phase shift internally. With such relays, it is desirable to connect the CTs on both sides in wye connection. This wye connection will allow the event recording feature of the relay to record the actual primary phase current for use during disturbance analysis.

The best method of achieving the correct connections and settings is by carefully checking the flow of the currents in the differential circuits when the power transformer is normally loaded. Under these conditions, when the CT connections are correct, there will be no (or little) current in the differential circuit. In addition to the phasing consideration discussed in the earlier text, it is also necessary to adjust the turns ratios of the CTs so that the delta connection on the wye side of the power transformer produces relay currents that are numerically matched with the relay currents produced by the wye-connected CTs. Thus, the delta CT winding currents must be $(1/\sqrt{3})$ times the wye CT currents. The entire procedure is best illustrated by an example.

Example 8.4 Consider the three-phase transformer bank shown in Figure 8.14. The transformer is rated 500 MVA at 34.5(delta)/500(wye) kV.

Assume that the transformer is carrying normal load, and that the current in phase a on the wye side is the reference phasor. Thus, the three-phase currents on the wye side are

$$I_a = \frac{500 \times 10^6}{\sqrt{3} \times 500 \times 10^3} = (577.35 + j0.0) \text{ A},$$
$$I_b = (-288.67 - j500.0)\text{A},$$
$$I_c = (-288.67 + j500.0)\text{A}.$$

The currents on the delta side of the power transformer are

$$I_{ad} = \frac{500 \times 10^6}{\sqrt{3} \times 34.5 \times 10^3} e^{j30°} = (7246.38 + j4183.69) \text{ A},$$
$$I_{bd} = (0.0 - j8367.39) \text{ A},$$
$$I_{cd} = (-7246.37 + j4183.69) \text{ A}.$$

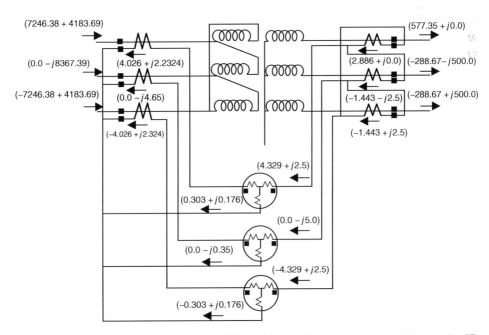

Figure 8.14 Connections for a three-phase differential relay. Note the polarity markings on the CTs. Relay taps will further reduce the differential currents.

Note that the power transformer is connected with a leading delta. The CTs on the delta side of the power transformer are to be connected in wye. We may therefore select the CT ratio on this side to be such that the CT secondary current will be less than 5 A when the primary current is 8367.39 A. Select the CT ratio of 9000:5. This produces CT secondary currents on this side of $(8367.39 \times 5/9000) = 4.65$ A. The three CT secondary current phasors are shown in Figure 8.14.

The CTs on the wye side of the power transformer are going to be connected in delta. Thus, the CT ratios must be such that the CT secondary winding currents will be close to $(4.65/\sqrt{3}) = 2.68$ A. This calls for a CT ratio of 577.35/2.68, or 1077:5. Selecting the nearest standard CT ratio (see Table 3.1 of Chapter 3) of 1000:5 produces CT secondary winding currents of magnitude $577.35 \times 5/1000 = 2.886$ A. This will produce a CT delta line current of magnitude $2.886 \times \sqrt{3} = 5$ A. Although this is not exactly equal to the line currents produced by CTs on the 34.5 kV side of the power transformer (4.65 A), this is the best that can be done with standard CT ratios. As in the case of single-phase transformers, the relay setting can be used to reduce this magnitude mismatch further.

The actual phasors of the CT secondary currents are shown in Figure 8.14. It should be noted that the CTs on the wye side of the power transformer are connected in such a manner that the currents in the relays are exactly in phase, and very small currents flow in the differential windings of the three relays during normal conditions. The currents are calculated with due attention given to the polarity markings on the CT windings.

8.6.2 Zero-sequence current contribution from Wye–Delta transformers

Another problem associated with differential protection of transformers with wye–delta configurations is dealing with the zero-sequence current through the wye-connected winding that does not flow into or out of the delta-connected winding.

Figure 8.15 shows the flow of sequence currents through this type of transformer during an external single-line-to-ground fault.

It can be seen that the line currents on the X (wye) side include all three sequence currents. However, on the H (delta) side, only positive- and negative-sequence currents flow in the line. The zero-sequence currents remain inside the delta winding and cannot be sensed by CTs on the line. In electromechanical relay implementations, the line CTs on the wye

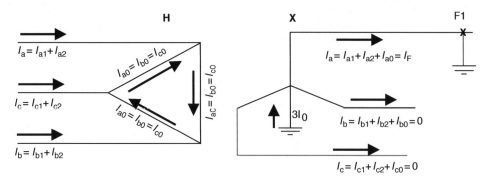

Figure 8.15 Sequence currents on delta and wye sides of a transformer.

winding side of the transformer will be connected in delta configuration (as in Figure 8.14) to manage the phase shift and will also cause zero-sequence currents to flow around the three CT secondary windings. In addition to managing the phase shift, the delta-connected CTs also act as a zero-sequence current trap, preventing the zero-sequence current on the wye winding side of the transformer from entering the differential relay. As noted in Section 8.6.1, in digital relay implementations, it is preferable for CTs on both sides of the transformer to be connected in wye configuration. In addition to the protection algorithm managing the phase shift inside the relay, the digital relay will also remove the zero-sequence current if necessary. Some relay settings include a setting "this winding is a source of zero-sequence current." If this setting is enabled, the zero-sequence current will not be included in the current differential calculation. Other relays may not include a setting with the same name, but the instruction manual will advise on how to manage zero-sequence current removal.

Note that the zero-sequence current removal can also be an issue for other types of transformer-winding configurations, including wye–delta–wye, wye–delta–delta, wye–zigzag, delta–zigzag, and auto-transformers with tertiary windings. The tertiary windings may be hidden or brought out to terminals. The differential protections can manage the zero-sequence currents with CT connections or relay settings in a similar fashion as for the delta–wye transformer described in the earlier text.

8.6.3 Multiwinding transformer protection

Although we consider multiwinding transformers in this section, which deals with three-phase transformers, similar considerations hold for single-phase transformers as well [1]. Consider the three-winding transformer shown in Figure 8.16. One winding is assumed to be delta-connected, while the other two are assumed to be wye-connected. For electromechanical relays, the CTs must of course be connected in wye on the delta side and in delta

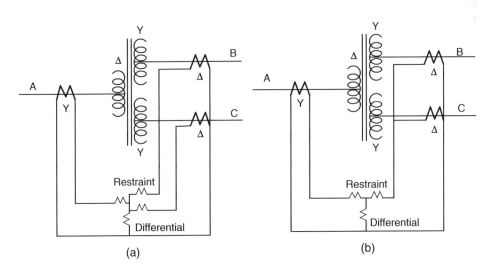

Figure 8.16 Protection of a three-winding transformer with (a) a three-winding relay and (b) a two-winding relay.

on the wye side of the power transformer. Digital relays must have the compensating settings properly adjusted. This will ensure that the phase shifts created in the currents of the power transformer and zero-sequence currents are compensated by the CT connections or relay settings, so that the secondary currents are once again in phase and the zero-sequence currents are removed from the differential calculation. The CT ratios are chosen, so that when any two windings are in service, equal secondary currents are produced. The third winding is assumed to be open-circuited under these conditions. The ratios of all three CTs are chosen in this fashion. The phasing of the CT and relay connections is checked as in Example 8.4.

It is interesting to note that under certain conditions a two-winding differential relay can be used to protect a three-winding transformer. If the transformer is connected to a source only on one side, the other two winding CTs could be paralleled to produce a net secondary current, which can then be used in a two-winding protection scheme. However, when more than one winding has a source behind it, paralleling two windings could be dangerous. Consider the case illustrated in Figure 8.16b. If the power transformer winding A is open, then an external fault on the side B would be supplied by the source on the side C. This should produce equal currents in the paralleled connection, and in the relay. However, because of unequal CT ratios, or unequal error in the two sets of CTs, there may be a residual current in the paralleled connection, which flows in the differential coil of the relay. It also flows in half of the restraint coil, but is clearly insufficient to prevent the relay from operating if the differential current is above the pickup setting of the relay. A much safer practice is to employ a specially designed three-winding differential relay, which utilizes three restraint coils, each being supplied by its own CTs. In the case discussed in the earlier text, this ensures that even if the differential coil of the relay has an unbalance current flowing in it, the restraint coils of the relay will carry the full fault currents flowing in windings B and C of the power transformer. The relay will thus be prevented from tripping for this external fault. The same considerations apply even to two-winding transformers if one of the windings is connected to CTs on two breakers on one side. Note that some modern relays are available with more than three restraint inputs, so that restraints can be connected to CTs on multiple breakers on one or more sides of a transformer.

Example 8.5 Consider the three-phase three-winding transformer shown in Figure 8.16. Let the three windings be rated at: 34.5 kV, delta, 500 MVA; 500 kV, wye, 300 MVA; and 138 kV, wye, 200 MVA. The rated line currents on the delta side have been calculated in Example 8.4 to be 8367.39 A. As before, the CT ratios for the wye-connected CTs on this side are 9000:5.

To determine the CT ratios on the 500 kV side, we must assume that the 138 kV side is carrying no load. In this case, the 500 kV side will carry 577.35 A, when the 34.5 kV side is carrying 8367.39 A. (This will be an overload for the 500 kV side, but we are using these figures only to calculate the CT ratios. It is not implied that with the 138 kV side unloaded, the transformer will continue to carry 500 MVA.) As the CTs on the 500 kV side are delta-connected, the CT ratios on this side are 1000:5, as in Example 8.4.

The 138 kV side, with the 500 kV side open, and the transformer loaded to 500 MVA (once again, only theoretically, to facilitate calculating the CT ratios), the line currents will be

$$I_{138} = \frac{500 \times 10^6}{\sqrt{3} \times 138 \times 10^3} = 2091.85 \text{ A.}$$

Since these CTs are going to be connected in delta, the secondary current should be $5/\sqrt{3} = 2.886$ A. This calls for a CT ratio of 2091.85:2.886. This gives the nearest available standard CT ratio of 3600:5.

When the transformer is normally loaded, that is, 500, 300, and 200 MVA in the three windings, the three winding currents will be 8367.39, 346.41 (3/5 of 577.35), and 836.74 (2/5 of 2091.85) A, respectively. The corresponding secondary currents produced by the CTs in the relay connection will be

$$\frac{8367.39 \times 5}{9000} = 4.65 \text{ A on the 34.5 kV side,}$$

$$\frac{346.41 \times 5}{1000} \times \sqrt{3} = 3.0 \text{ A on the 500 kV side,}$$

and

$$\frac{836.74 \times 5}{3600} \times \sqrt{3} = 2.01 \text{ A on the 138 kV side.}$$

The differential current will be $4(3.0 + 2.01 - 4.65) = 0.31$ A. With the ample restraint current provided by the three currents in a three-winding differential relay, this relay will not misoperate for this condition.

Now consider the possibility that a two-winding differential relay is used, with the CT secondaries on the 500 and 138 kV sides being paralleled, as in Figure 8.16b. Under normal loading conditions determined in the earlier text, there is no difference in the relay performance. However, consider the case of the 34.5 kV winding being open, and a fault on the 138 kV side being supplied by the system on the 500 kV side of the transformer. Let the fault current on the 500 kV side be 10000 A and that on the 138 kV side be $10\,000 \times 500/138$ or 36 231 A. The relay currents produced by the two sets of CTs are

$$\frac{10\,000 \times 5}{1000} \times \sqrt{3} = 86.6 \text{ from the 500 kV side}$$

and

$$\frac{36\,231 \times 5}{3600} \times \sqrt{3} = 87.16 \text{ A from the 138 kV side.}$$

Assume that the CT on the 500 kV side has a 10% error, while the CTs on the 138 kV side have negligible error under these conditions. Thus, the current in the differential winding and half-the-restraint winding of the relay will be

$$(87.16 - 0.9 \times 86.6) = 9.22 \text{ A.}$$

This is equivalent to 200% of the restraint because only half-the-restraint winding gets this current, and it being well above the normal pickup setting of 0.25 A for the differential relay, will trip the relay for this condition. It is thus not advisable to use a two-winding differential relay for this case.

8.6.4 Protection of regulating transformers

The regulating transformers may regulate the turns ratio or the phase shift between the primary and the secondary windings [2, 5, 6, 7]. The regulating transformers usually consist of two transformers: one to provide the magnetizing current for the transformers and the other to provide the variable turns ratio or the variable phase shift. In either case, the conventional percentage differential relay application is not suitable for the protection of these transformers. As the turns ratio or the phase shift changes, the balance between the primary and the secondary currents for through faults or load currents is no longer maintained. Full details of protection for phase-shifting transformer protection can be found in Reference [5]. In general, such units are sufficiently rare, and their protection requirements are too specialized, so that the user and manufacturer of the transformer should work together to provide protection suitable for regulating transformers. Some special CT locations and/or ratios may be required, so the protection requirements should be confirmed before the transformer design is finalized. Often, a sudden pressure relay (SPR; described in Section 8.8) provides the most sensitive protection, but even that cannot be used for that part of the regulating transformer which does the regulation, since during normal operation of the unit, the tap-changing mechanism produces sufficient arcing to cause pressure waves and gases which may trigger the SPR.

8.7 Volts-per-Hertz protection

Transformer cores are normally subjected to flux levels approaching the knee point in their magnetizing characteristic [8]. Typically, the rated voltage at rated frequency may be 10% below the saturation level. If the core flux should exceed the saturation level, the flux patterns in the core and the surrounding structure would change, and significant flux levels may be reached in the transformer tank and other structural members. As these are not laminated, very high eddy currents are likely to result, producing severe damage to the transformer. It is therefore desirable to provide a protection package which will respond to the flux level in the transformer. As the flux is proportional to the voltage, and inversely proportional to the operating frequency, the significant relaying quantity is the ratio of the per unit voltage to the per unit frequency. This is known as volts/hertz protection.

This protection is specially needed in the case of unit-connected generator transformers. If the turbine generator is shut down with the voltage regulator in service, the volts/hertz limit of the transformers (and indeed of generators as well) could be easily exceeded. Similar conditions could also be reached by load rejection with voltage regulators disconnected, or in manual position, or with faulty instrumentation in the regulator circuits. The volts/hertz capability of transformers is specified by manufacturers. A typical capability curve is shown in Figure 8.17. Many volts/hertz protections have two settings, a lower setting for alarm and a higher setting which may be used for tripping. Some volts/hertz protections include inverse time characteristics to more closely match the transformer capability than is apparent from Figure 8.17.

Figure 8.17 Volts/hertz capability of transformers, and relay settings.

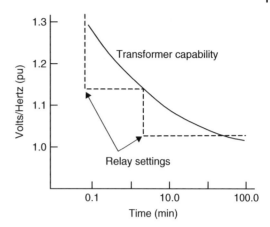

8.8 Nonelectrical protection

8.8.1 Pressure devices

A very sensitive form of transformer protection is provided by relays based upon a mechanical principle of operation. When a fault occurs inside an oil-filled transformer tank, the fault arc produces gases, which create pressure waves inside the oil. In the "conservator" type of tank construction [1, 2, 9], which is more common in Europe, the pressure wave created in the oil is detected by a pressure vane in the pipe which connects the transformer tank with the conservator. The movement of the vane is detected by a microswitch, which can be used to sound an alarm, or trip the transformer. This type of a relay is known as a Buchholz relay, named after its inventor.

In the United States, the more common transformer construction is of the tank type with a gas cushion at the top of the tank [7]. In such a transformer, the pressure wave is detected by a sudden pressure relay (SPR) mounted on the side of the transformer. The mechanical sensor consists of a bellows and a pressure equalizer, which together are insensitive to slow changes of pressure, for example those caused by thermal or loading changes in the transformer. However, a pressure wave created by a fault is detected by the relay, and can be used to trip or alarm. In fact, often these relays are found to be too sensitive, and may respond to the movement of transformer windings caused by severe external fault currents or movement of oil when pumps are switched on. Because of the aforementioned problems, not all utilities choose to connect sudden pressure relays to trip the transformer directly. Some may choose only to alarm, by the SPR, or block the SPR if an external fault is detected at the same time [9].

It should be noted that, aside from the incorrect operation just described, these pressure-type relays can only detect faults inside the transformer tank. Faults on the bushings, and connecting leads, do not create an arc in the insulating oil or gas, and must be protected by differential relays. In fact, a combination of pressure relays and differential relays provides an excellent protection system for a power transformer. In the presence of gas pressure relays, the differential relays may be made less sensitive.

8.8.2 Temperature devices

There are several temperature-detecting devices used for indication, recording, or control, and, on rare occasions, for tripping. Some devices simply measure the oil temperature, usually the top oil. Other devices use a combination of current, by placing a small search coil around a lead, and oil temperature to measure the total effect of load and ambient temperature. The critical temperature is referred to as the "hot-spot" temperature, and is the highest temperature that will occur somewhere in the winding. This hot-spot is not precisely known, since it varies with the physical structure of the transformer and the flow of cooling oil or gas. It is usually determined at the factory, and an arbitrary temperature, for example, $10\,^{\circ}\text{C}$, is specified to be added to the total temperature as indicated at any time. The temperature devices actuate alarms to a central dispatching office, to alert the operators, who can either remotely unload the transformer by opening the circuit breaker, or reconfiguring the network, or can dispatch an operator to the station. The hot-spot sensors are also commonly used to start and stop cooling fans and pumps. In extreme cases, when it is not possible to remotely remove the load, or send an operator to the station, an extreme high alarm will trip the bank or other breakers to reduce the load.

Some modern digital transformer protection systems can model the thermal characteristics of a transformer to calculate the hot-spot winding temperature. These devices accept inputs from ambient and transformer top oil temperature sensors to modify their calculations, depending on ambient and oil temperatures [1]. These devices may predict temperatures over a certain future time frame and alert operators to possible loss of life under overload conditions.

8.9 Protection systems for transformers

The protection of transformers is not simply a question of detecting faults within the winding, core, or tank. It is very often a question of how the transformer is integrated into the station or system. The detection and clearing of transformer faults are intimately connected to the primary switching that is available. There are too many variations of such switching arrangements to be all-inclusive, but a few typical examples will suffice to demonstrate the factors and logic involved.

8.9.1 Parallel transformer banks

It is not uncommon when two transformers are connected in parallel with no circuit breakers used to separate them that a single differential relay is used for the protection of both banks. On the face of it, this is reasonable, since the two banks are operated as a single unit. However, the problem of sympathetic inrush discussed in Section 8.4 presents difficulties to the common differential relay, when one of the transformers is energized with its switch, while the other transformer is in service. The sympathetic inrush is now likely to trip the two transformers. This is because the inrush current in the common lead becomes symmetrical very quickly, as shown in Figure 8.10b, and does not have a sufficient amount of the second harmonic in it to prevent the relay from tripping.

Another problem with two transformers sharing a single differential is that the protection sensitivity is reduced by a factor of two. Note that Figures 8.4 and 8.5 show a minimum pickup which is normally calculated in per unit of transformer rating. The sharing of the differential protection will effectively double the minimum pickup by a factor of two with respect to the individual transformer ratings. Both problems can be dealt with satisfactorily by the use of separate differential relays.

8.9.2 Tapped transformer banks

Connecting a transformer directly to the high-voltage transmission line is a practical and relatively inexpensive way of providing service to an isolated industrial or residential load. On low-voltage transmission systems, where circuit breakers are relatively inexpensive, the breakers are used both as locations for the CTs to detect a fault and define the zone of protection and as the way of clearing the fault. In Figure 8.18, a full complement of breakers is used. The transformer differential relay (87T) includes the high-side breaker and the low voltage leads, bus, and feeder breakers, as well as the transformer itself. Alternatively, a low-side breaker may be used and the transformer differential would be connected as shown by the dotted line in the figure. The high-voltage leads are protected with a bus differential (87B) and each line section has its own set of directional relays (67). In this configuration, a line fault will be cleared immediately by tripping the appropriate breakers, but the other line section continues to serve the transformer and its load.

A bus or transformer fault will, of course, de-energize the transformer and its low-voltage feeders. The remote breakers, however, will not trip, allowing those line sections and any connected loads to remain in service. The transformer is protected by overcurrent relay 51T from damage by heavy through current.

Figure 8.18 Full complement of circuit breakers for transformer protection.

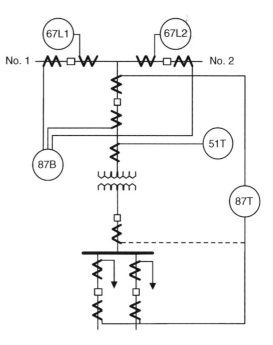

No. 1 ─────────────── No. 2

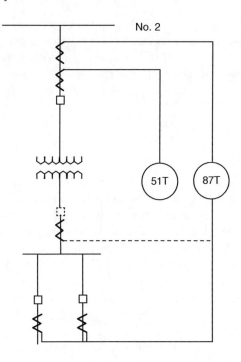

Figure 8.19 Direct connection of
transformer and transmission lines.

Figure 8.19 shows a variation in which the line circuit breakers are not used. In this con-figuration, a line fault will remove the transformer by opening the remote terminals. This is an acceptable condition only if the load being served can withstand the outage and there is no source on the low-voltage side. It is also necessary to set the remote relays to "see" into, but not beyond, the transformer to avoid tripping the high-side breakers for low-side faults.

The use of so many circuit breakers is often not justified, particularly at higher voltages. Figure 8.20 shows a very common configuration using motor-operated air break switches (MOABs) instead of circuit breakers. The transformer differential extends from the high-side bushings of the transformer to the low-side feeder circuit breakers or to a low-side transformer breaker. The use of the overcurrent relay 51T is the same as discussed in the earlier text. Since MOABs are not designed to interrupt fault current, in order to clear the fault in the transformer zone of protection, the remote line breakers must be tripped. As with the system shown in Figure 8.19, the relays at the remote terminal must be set to "see" into, but not beyond, the transformer. A common method of tripping the remote terminals is to provide a single-phase ground switch at the transformer, as shown in Figure 8.20. This is a spring-operated switch which is released by the transformer differential relay. The ground switch places a solid single-phase fault on the system, which the remote relays will see and trip their associated breakers. When the line is de-energized, the MOAB is opened automatically to remove both the ground and the faulted transformer from the system. The remote breakers then reclose, restoring the line to service.

There are many variations on this remote tripping scheme. Some utilities simply allow the transformer MOAB to open without the use of a ground switch. This presupposes that a heavy fault will be seen by the remote relays, and their breakers will trip before the MOAB

Figure 8.20 Use of motor-operated air break switches at a transformer tap.

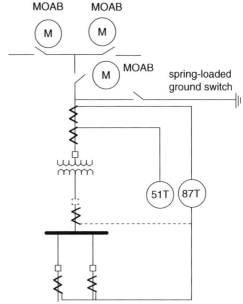

blades open. If the fault cannot be seen by the remote relays, the fault current is low and the MOAB will probably not be damaged. It is possible, of course, that the switch will be damaged anyway, but the cost savings by not providing the ground switch and its associated controls may justify the risk. If there are communication channels available, transfer trip schemes similar to those discussed in Chapter 6 can be used.

The use of MOABs in the line in place of circuit breakers results in a change in system operations that should be noted. When a line fault occurs, the remote breakers will trip from their respective relays. The line is de-energized; voltage relays at the tapped bank recognize this condition and open the switches. After a preset time delay, each remote circuit breaker will close and the one associated with the faulted line section will immediately retrip, the other remaining closed. Each MOAB then closes at slightly different times, and the one that reenergizes the faulted line results in the breaker tripping again. The potential device of the MOAB that allows the fault to be reenergized does not stay energized for more than the breaker tripping time (actually the timer is usually set for about 3 s) and that MOAB is locked out. The breaker closes one more time and stays closed, restoring the remaining line and transformer to service.

8.9.3 Substation design

Since a power transformer interconnects two or more voltage levels, its location requires special consideration in the design of a substation and the protection of all of the elements within it. Usually the various voltage levels are contained within separate areas, each with its own bus configuration and associated equipment and separated by considerable distances. Figure 8.21 shows a 345 kV/138 kV station with a full complement of circuit breakers. The protection is straightforward.

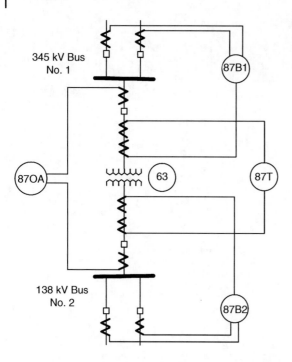

Figure 8.21 Substation transformer with a full complement of circuit breakers.

There are redundant transformer zones of protection, that is, an internal differential (87T) and an overall differential (870A). In addition to providing dependability, the two zones allow ready identification of the fault location, that is, within the transformer itself or on the high- or low-voltage leads. Variations of this configuration are very common. For instance, if the transformer and both circuit breakers are relatively close to each other, the internal differential may be omitted and a pressure relay used to indicate a fault within the transformer tank. If the low-voltage leads are very long, a lead differential may be added, as shown in Figure 8.22. This has the advantage of using a relay more suited to a bus or line differential than a transformer differential, that is, a relay without harmonic restraint.

Since extra-high-voltage (EHV) breakers are expensive, they are often omitted in station designs such as shown in Figure 8.22. As with the tapped transformer, the transformer is tied directly to the EHV bus through a MOAB. This is shown in Figure 8.22 with the associated zones of protection. The major disadvantage is the fact that a transformer fault must be cleared by opening all of the 345 kV bus breakers. This is usually acceptable, since transformer faults are relatively rare.

8.9.4 Station service

Most power transformers are built with a delta tertiary winding to provide a path for third harmonics and to help stabilize the neutral. Although the capacity of the tertiary need not be very large, it is a good source for the station auxiliary equipment such as circuit breaker air compressors, oil pumps, battery chargers, and so on. If the load taken off the tertiary is

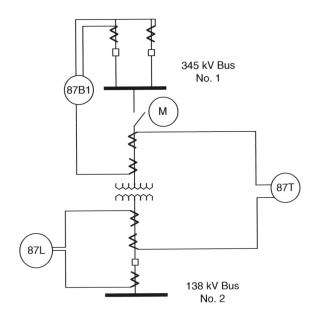

Figure 8.22 MOAB connection of a transformer to a 345 kV bus.

large enough to upset the transformer differential, it must be connected into the differential CTs, as shown in Figure 8.23. Protection of the 13.8 kV leads presents other problems that are discussed in Chapter 7. The difficulty arises first from the very large phase-to-phase fault currents that are present and then from the fact that this is an ungrounded circuit and must use a ground detector scheme as discussed in Section 7.2 (Chapter 7).

Figure 8.23 Protection of generating station service transformer.

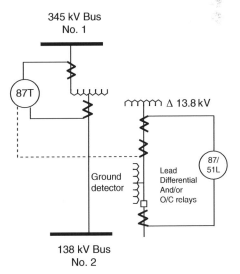

8.9.5 Generator station design

The most common configuration used presently for large fossil-fired generating plants is a unit-connected system. In this system, the generator, generator step-up (GSU) transformer, the unit auxiliary transformer(s) (UAT), and the reserve auxiliary transformer (RAT) or start-up transformer are all connected as a unit and their protection and controls are interrelated.

Figure 8.24 shows a typical design with the associated zones of protection. Since our concern here involves only the transformers of this design, we will only touch upon the protection and control aspects of the other equipment. They are discussed in more detail in the appropriate chapters. The GSU transformer is usually included in the unit and overall differentials (87U and 87OA). Since these see transformer inrush, they must be harmonic-restrained relays. It is possible, but not common, to use an internal transformer differential (87Tr) connected to the transformer bushing CTs. If this is done, then high- and low-voltage lead differentials should be added. If the leads are very long, that is, beyond 460–610 m, then

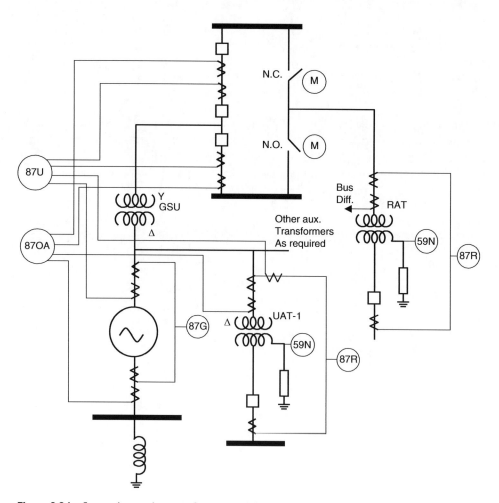

Figure 8.24 Generating station transformers and their protection systems.

line differential protection should be considered. The UAT has its own internal differential and is not included in the unit or overall differential. However, since there are no high-side breakers to clear a UAT fault, the generator must be tripped for such faults. For sensitivity in setting the two sets of differential relays, and to identify the fault location, the auxiliary transformer current is subtracted from the unit and overall transformer differential circuits.

Since the GSU transformer and UAT are connected to the generator as it is coming up to speed prior to synchronizing to the system, they will be energized at frequencies lower than the normal 60 or 50 Hz. Transformers are designed to operate at a given flux level and flux is related to both voltage and frequency. It is therefore necessary to provide volts/hertz relays to detect the condition either of excessive voltage, which is possible during start-up if the excitation is not monitored, or low frequency as the unit is starting.

Fossil-fired generators require auxiliary power to start pumps, fans, and pulverizers to fire up the boiler, lubricate the bearings, and so on. A start-up transformer (RAT) connected to the system is therefore required. This transformer is also available as a spare for the UAT and must be sized accordingly. There are many possible schemes to connect the RAT to the system and bus. One common scheme is as shown in Figure 8.22. The transformer is connected to the high-voltage bus (note that it is not the same bus to which the generator is connected) through MOABs, and all of our previous discussions concerning clearing faults when there is no circuit breaker apply.

8.10 Summary

In this chapter, we have developed the protection concepts for single- and three-phase transformers. We began with overcurrent protection, and after examining the effects of non-linearities in the magnetic circuits, arrived at the harmonic-restrained percentage differential relay as the protection system of choice for large power transformers. We considered the effect of the wye–delta connection of the power transformer on the protection system. We have also studied the protection problem of a transformer when it is a part of a complex power system, with different bus and breaker configurations.

Many of these concepts are further illustrated by the problems at the end of the chapter.

Problems

8.1 Consider the single-phase radial power system shown in Figure 8.25. The leakage impedance of the transformer is $(0.0 + j0.02)$ pu and the feeder impedance on the same base is $(0.01 + j0.06)$ pu. The transformer and the system base power is 100 kVA. The transformer voltage ratio is 11/3.3 kV. Time–overcurrent protection is to be provided by the relays R_{ab} and R_{bc}. Using the relays with characteristics shown in Figure 4.5 determine the CT ratios, pickup values, and the time dial settings for the two relays. You may assume that no other coordination considerations apply to this problem.

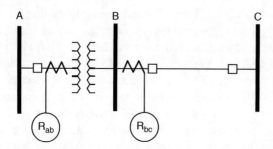

Figure 8.25 System for overcurrent protection of a transformer.

8.2 A single-phase transformer is rated at 110/33 kV, 50 MVA. It has an overload capability of 120%. The 110 kV side has a tap changer with (5/8)% step size, and with a range of 99–121 kV. Determine CT ratios for a percentage differential relay to protect this transformer. Assume a maximum CT error of 5% in either of the two CTs. What is the minimum slope of the percentage differential characteristic of the relay? Assume that both the relay windings are provided with taps for 3.0, 4.0, 4.8, 5.0, 5.2, and 6.0 A, and determine the appropriate taps which will minimize the required slope of the relay characteristic. What pickup current setting for the relay would you recommend?

8.3 If the transformer of Problem 8.2 had no tap changer, could you accept a lower slope? Explain your answer.

8.4 Assume a differential relay with a slope of 20%, and a pickup setting of 10.25 A (secondary) protects a 69/33 kV, 10 MVA transformer. Determine the CT ratios suitable for this protection system, assuming a 110% overload capability. What is the minimum primary fault current that the relay will detect when the transformer is operating at no load? What is the minimum fault current that the relay will detect when the transformer is carrying the maximum permissible overload?

8.5 Calculate the harmonics in the output of a CT which has symmetrical saturation as shown in Figure 8.12. Use $\alpha = 30°$, $90°$, and $120°$. If all three CTs in a delta-connected CT bank (on the wye-connected side of a power transformer) saturate in an identical fashion, what are the harmonic contents of the currents which flow in the differential windings of the three differential relays? You may assume that the line currents produced by the delta-connected CTs are obtained as the difference of two winding currents.

8.6 For the wye–delta transformer shown in Figure 8.14, assume that an internal three-phase fault develops at the bushings on the wye side (within the zone of protection of the differential relay) of the power transformer. On the transformer base, the system impedances on the wye and the delta sides are $(0.0 + j0.2)$ pu and $(0.0 + j0.3)$ pu, respectively. The leakage impedance of the transformer is $(0.0 + j0.08)$ pu on its own base. What are the currents in the restraining and operating windings of the differential relays? Will the relay operate?

8.7 For a phase-to-phase fault (outside the zone of the differential relay) on the wye side of the transformer of Problem 8.6, calculate the restraining and operating currents in each of the relays. Will the relay operate for this fault? How do the fault currents compare with the three-phase fault currents determined in Problem 8.6?

8.8 Consider a delta-zigzag transformer, as shown in Figure 8.26. Such transformers are often used as grounding transformers on power systems. Determine the appropriate connections of the differential relays to the CTs. Assume a normal load current to be flowing in the windings, and determine the differential currents resulting from the load currents. What should be the setting of the percentage slope to avoid tripping for this condition?

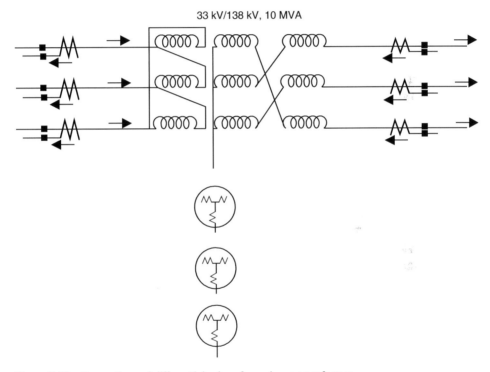

Figure 8.26 Connections of differential relays for a zigzag transformer.

References

1 IEEE Std C37.91-2021 (2021). *(Revision of IEEE Std C37.91-2008) IEEE Guide for Protecting Power Transformers*. New York, NY: IEEE.

2 Mason, C.R. (1956). *The Art and Science of Protective Relaying*. New York: Wiley.

3 Sonneman, W.K., Wagner, C.L., and Rockefeller, G.D. (1958). Magnetizing inrush phenomena in transformer banks. *AIEE Trans.: Part III: Power Apparatus and Systems* **77**: 884–892.

4 Hayward, G.D. (1941). Prolonged inrush currents with parallel transformers affect differential relaying. *AIEE Trans.* **60**: 1096–1101.

5 IEEE Std C37.245-2018 (2018). *IEEE Guide for the Application of Protective Relaying for Phase-Shifting Transformers*. New York, NY: IEEE.

6 Blackburn, J.L. (1987). *Protective Relaying*. New York: Marcel Dekker.

7 Westinghouse (1976). *Applied Protective Relaying*. Newark, NJ: Westinghouse Electric Corporation.

8 ANSI/IEEE standard C37. 106-2003 (2004). *(Revision of ANSI/IEEE C37.106-1987), Guide for Abnormal Frequency Protection for Power Generating Plants*. New York, NY: IEEE.

9 IEEE Power and Energy Society, Power System Relaying Committee Report (2014), *Sudden Pressure Protection for Transformers*, available online at http://www.pes-psrc.org/kb/published/reports/K6_SPR_Final.pdf

9

Bus, Reactor, and Capacitor Protection

9.1 Introduction to bus protection

Bus protection systems are more straightforward than transformer protection systems because the variables are reduced. There is no ratio or phase angle change or appreciable inrush. Paradoxically, bus protection has historically been the most difficult protection to implement because of the severity of an incorrect operation on the integrity of the system. A bus is one of the most critical system elements. It is the connecting point of a variety of elements and a number of transmission lines, and any incorrect operation would cause the loss of all of these elements. This would have the same disastrous effect as a large number of simultaneous faults. However, without bus protection, if a bus fault should occur, the remote terminals of lines must be tripped. In effect, this could create a worse situation than the loss of all of the elements at the bus itself for two reasons:

1) The loss of the remote ends will also result in the loss of intermediate loads.
2) As systems become stronger, it is increasingly difficult for the remote ends to see all faults owing to infeeds.

The major problem with bus protection has been unequal core saturation of the current transformers (CTs). This unequal core saturation is due to the possible large variation of current magnitude and residual flux in the individual transformers used in the system. In particular, for a close-in external fault, one CT will receive the total contribution from the bus, while the other CTs will only see the contribution of the individual lines. The basic requirement is that the total scheme will provide the degree of selectivity necessary to differentiate between an internal and an external fault.

Protection of high-voltage (HV) substation buses is almost universally accomplished by differential relaying [1]. This method makes use of Kirchhoff's law that all currents entering or leaving a point (the substation bus) must sum vectorially to zero. In practice, this type of protection is accomplished by balancing the CT secondary current of all of the branches connected to the bus and then bridging this balanced circuit with a relay operating circuit.

Power System Relaying, Fifth Edition. Stanley H. Horowitz, Arun G. Phadke and Charles F. Henville.
© 2023 John Wiley & Sons Ltd. Published 2023 by John Wiley & Sons Ltd.

9.2 Overcurrent relays

Differential relaying with overcurrent relays requires connecting CTs in each phase of each circuit in parallel with an overcurrent relay for that phase. Figure 9.1 shows the basic bus differential connection for one phase of a three-phase system. When conditions are normal, the bridge is balanced and no current flows through the relay operating circuit. For heavy loads, the CTs may not reproduce the primary current exactly and there will be an error current through the operating circuit. The relay must be set above this value.

When an external fault occurs, if all of the CTs reproduce the primary current accurately, the bridge is balanced as in the normal case and no current flows in the relay operating circuit. However, as discussed in the earlier text, if one of the CTs saturates (as is often the case), the bridge will not be balanced, the error current will flow in the operating circuit, and an incorrect trip will occur. When an internal fault occurs, this balance, as we would expect, is also disrupted and current flows through the operating circuit. This, however, is an appropriate trip situation despite the incorrect CT performance.

In general, this type of protection is rarely used in modern HV applications, because the additional installed cost of a more secure percentage-restrained differential protection is usually negligible. Also, the short time delay that is often required for security is usually not acceptable in HV applications. However, this type of protection may be found in medium-voltage or legacy HV applications. To minimize possible incorrect operations, the overcurrent relay may be set less sensitively and/or with time delay. The induction disk principle makes these relays less sensitive to DC and to the harmonic components of the differential current. Solid-state and digital relay designs must take these factors into consideration.

9.3 Percentage differential relays

To avoid the loss of protection that results from setting the overcurrent relay above any error current, as discussed in Chapters 7 and 8, it is common to use a percentage differential relay. These relays have restraint and operating circuits as shown in Figure 9.2. Only one

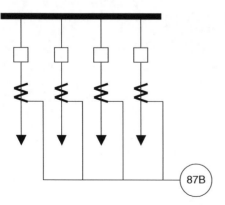

Figure 9.1 Differential with overcurrent relays.

87B

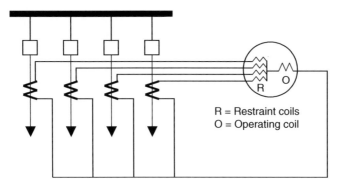

Figure 9.2 Percentage differential relay.

operating circuit per phase is required, but one restraint circuit for each phase of each circuit is necessary. This is conceptually similar to the transformer differential discussion in Chapter 8. Normally, one restraint input is connected to each branch that is a major source of fault current. Feeders and branches with low fault-current contribution may be paralleled on a single restraint input. The required current to operate the relay is proportional to the current flowing in the restraint circuits. In the case of legacy electromechanical relays, maximum security for external faults is obtained when all CTs have the same ratio, but satisfactory operation can be expected using high-quality auxiliary CTs. There are no relay taps in electromechanical bus differential relays. Digital bus differential relays have inputs electrically isolated from each other, which allows the use of CTs with different ratios on different branches. When different CT ratios are used with digital relays, the inputs are normalized by setting adjustment. For percentage-restrained differential protection, conventional CTs are used with the conventional CT ratios.

Similar to the case for percentage-restrained transformer differential protection systems (as discussed in Section 8.3 of Chapter 8), multi-slope bus differential protections are commonly used to provide higher sensitivity when the through current is small, and higher percentage restraint when the through current is large, and significant CT saturation is more likely.

In addition to the multi-slope characteristic of the percentage-restrained function, some modern digital relays have additional advanced security features. For instance, if one CT saturates due to high current with DC offset, the CT that saturates will not do so until some time after fault inception. Figure 9.3 shows the time relationship between operate and restrain currents when one of the two CTs saturate. CT1 does not saturate, but CT2 does (possibly it is less accurate or has higher burden). Figure 9.3a shows the case of an internal fault, where the operate and restrain currents start to rise simultaneously. Figure 9.3b shows the case of an external fault, where initially, before CT2 saturates, the differential current will be very small, but the restraining current will be very high. In the simulation shown in Figure 9.3b, the operate current does not start to rise until 7 ms after the fault starts, while the operate current rises immediately. This signature of initial low differential and high restraint identifies an external fault, and the relay may be automatically blocked, or switched to a higher (more secure) restraint characteristic.

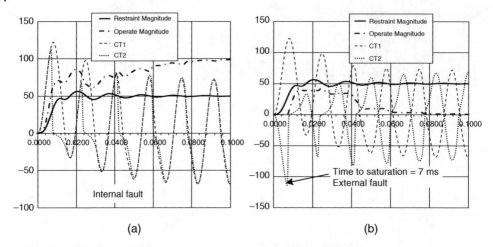

Figure 9.3 Restrain and operate currents: (a) internal fault, (b) external fault.

Another security enhancement that may be used by modern percentage-restrained differential protection is the use of phase comparators to check whether the phase angle of all currents entering a bus zone is the same. This would be the case for an internal fault. If one or more of the branch currents has opposite phase relationship(s) to one or more of the others, then the fault is external, and the protection may be blocked or switched to a high-security mode.

Modern digital bus differential protections also facilitate bus reconfiguration by electronic switching of current input connections into the protected zone(s) to match the status of primary switches. Thus, the protected zone can be automatically reconfigured to match a reconfigured primary zone. The flexibility, security, and dependability of modern digital percentage-restrained bus differential protection systems result in them being widely applied.

9.4 High-impedance voltage relays

Even with the use of percentage differential relays, the problem of the completely saturated CT for a close-in external fault still exists. To overcome this problem, a commonly used bus differential relay, particularly on extra-high-voltage (EHV) and HV buses, is the high-impedance voltage differential relay. This relay design circumvents the effects of CT saturation during external faults by assuming complete saturation for the worst external fault and calculating the error voltage across the operating coil. The relay discriminates between internal and external faults by the relative magnitudes of the voltage across the differential junction points [1, 2]. The connection for this relay is the same as shown in Figure 9.1 except that the relay includes a high-impedance voltage relay and associated equipment. The basic scheme is shown in Figure 9.4. The resistor R will be subjected to a high voltage during an

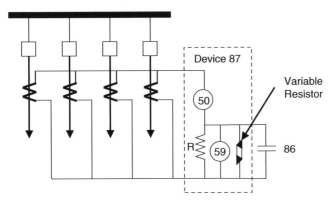

Figure 9.4 High-impedance relay.

internal fault, which will be detected by a voltage function (59). Function 59 may be tuned to 60 Hz to prevent the device from misoperating on DC offset or harmonics. Since this circuit would reduce the speed of operation of the relay, an instantaneous overcurrent relay (50) provides fast tripping for high-current faults. The variable resistor limits the voltage across the overvoltage relay and shunt resistor. Since the variable resistor may be subjected to a large amount of energy during the fault, a contact of an external lockout tripping relay (86) may be used to shunt the energy from the variable resistor after a fault has been declared.

The concept is to load the CTs with a high impedance to force the error differential current through the CTs instead of through the shunt resistor R. It is important that the CTs have fully distributed windings and are used at the full winding because secondary leakage reactance, if present, would be enough to increase the voltage across the operating circuit for external faults and cause misoperation. Because of the need to use the full ratio of the CT winding, this type of protection can limit the flexibility of expanding or modifying existing substations. For instance, if an existing station has all breakers at a particular bus supplied with 2000-5 CTs, and the owner wishes to extend the substation or replace an existing breaker with newly standardized bulk purchase breakers with 3000-5 CTs, it will be difficult to use the new breaker CTs with the existing high-impedance bus protection connected to 2000-5 CTs.

For the remainder of illustrations in this chapter, we will show differential relays connected to a common point of CT secondary wiring, as if the relays are high-impedance bus protections as in Figure 9.4, or differentially connected overcurrent as in Figure 9.1. However, it should be understood that percentage-restrained differential protection with discrete CT inputs as in Figure 9.2 may be used instead.

Example 9.1   Referring to the simplified circuit of Figure 9.5, a typical setting calculation is as follows:

Upon the initial occurrence of an internal fault, X_{ea}, X_{eb}, and Z_d can be assumed to be very large. Currents I_a/N and I_b/N will tend to be forced through the large impedances in such a

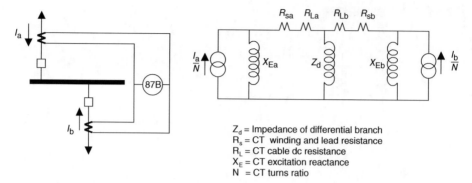

Z_d = Impedance of differential branch
R_s = CT winding and lead resistance
R_L = CT cable dc resistance
X_E = CT excitation reactance
N = CT turns ratio

Figure 9.5 Equivalent circuit for Example 9.1.

direction as to cause a high voltage to be developed across the differential branch, hence operating the relay. If an external fault occurs on line B, assuming no saturation, I_b/N will equal $-I_a/N$ and circulate around the outer loop, developing no net voltage across the relay. For the same fault, assuming complete CT saturation, X_{eb} is approximately zero, producing a net error voltage across the relay equal to $(R_{lb} + R_{sb})I_a/N$. This voltage, which is the basis for setting the relay, is generally much lower than that generated by the minimum internal fault. Note that smaller CT winding and lead resistances will allow lower and more sensitive settings. All the CTs should have the same ratio and should have the lowest effective secondary leakage. This is obtained with distributed windings on toroidal cores connected to the maximum tap. The use of auxiliary CTs is not recommended.

Example 9.2   Given the equivalent circuits shown in Figure 9.6 for external and internal faults, the voltage across 87B for the external fault is $I_{total} \times R_{lead}$ or $60 \times 2 = 120$ V. Set the relay at twice this value or 240 V. For an internal fault, the total current is 70 A times the relay resistance of 2600 Ω, or 182 kV. Obviously, such a high voltage cannot be developed without serious insulation damage. The variable resistor is used to limit the voltage by reducing the resistance as the voltage increases.

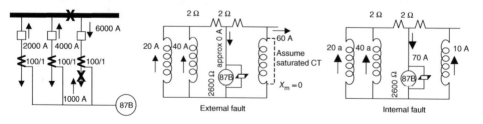

Figure 9.6   Equivalent circuits for Example 9.2.

9.5 Moderately high-impedance relay

This is a combination of the high-impedance voltage relay and the percentage differential relay. The percentage-restraint characteristic makes the relay application independent of the maximum external fault condition. The moderately high-impedance circuit makes the relay insensitive to the effects of CT saturation. The relay is usually connected into the system with a special auxiliary CT. When 5 A rated CTs are used, one special auxiliary transformer is required for each restraint circuit and one restraint circuit is required for each phase of each circuit connected to the bus. The special auxiliary CT permits the use of unmatched CT ratios to bring the overall ratios into agreement. This is a great help when adding breakers to an existing bus and the bushing CTs do not match [3].

9.6 Linear couplers and rogowski coils

This system uses linear couplers (air-core mutual reactors) in place of conventional iron-core CTs. They have linear characteristics and produce a secondary voltage that is proportional to the primary current. The secondaries of all of the linear couplers are connected in series as shown in Figure 9.7. This design solves the problem of saturation, since there is no iron in the CT. However, since the linear coupler is a special device, requiring low-energy relays, conventional CTs cannot be used. This may present problems with station or bus changes [4].

Example 9.3 The performance of a relay connected to a linear coupler is determined as follows:

$$V_{sec} = I_{pri} \times M,$$

where M is the mutual impedance of the coupler and is specified by the manufacturer.

Assume that it is equal to 0.005 at 60 Hz. $V_{sec} = 5\,V$ induced for 1000 primary amperes.

For an external fault, the sum of all of the induced voltages for all of the fault current flowing in is equal and opposite to the voltage for the current flowing out.

For an internal fault, all current is flowing in and the induced voltages add:

$$I_{rel} = \frac{V_{sec}}{Z_{rel} + Z_{coupler}}.$$

Figure 9.7 Differential relaying with linear couplers.

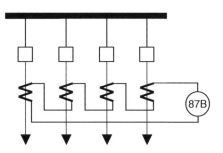

The load impedance is not significant compared to the 30–80 Ω relay impedance and 2–20 Ω coupler impedance. To set the relay, assume a three-phase external fault to establish the maximum nonoperating current and a phase-to-ground internal fault to establish the minimum operating current. A recommended factor of 25 is used between the maximum and minimum fault currents as shown in the following text. Assume the following values for the bus shown in Figure 9.7:

Maximum external fault = 20 000 A.

Minimum internal fault = 2000 A.

Total $Z_{coupler}$ = 30 + j45.

Z_{rel} = 55 + j20.

$I_{pri} = \dfrac{20\,000}{25} = 800\,A.$

$M \times I_{pri}$ = 800 × 0.005 = 4 V pickup setting.

Or, in terms of current, $I_{rel} = \dfrac{4}{107} = 0.037\,A\,(37\,mA)$ pickup setting.

Rogowski coils are like linear couplers in that they have no iron core that might saturate [5]. Like linear couplers, they are special devices that cannot be used in the same protection as conventional CTs. They are more immune to external magnetic fields than linear couplers and are used with modern low-burden digital relays. The relays may be connected measure the voltage across series-connected coils similar to the linear coupler connection in Figure 9.7, or with separate inputs to a current summing device.

9.7 Directional comparison

Historically, it was occasionally desirable to add bus protection to an existing substation where it is too costly to change out or add CTs. In such a case, it is possible to use the existing line and transformer grounding CTs to provide bus fault protection. With modern low-impedance percentage-restrained differential protections, it is usually possible to add bus protection sharing existing CTs if dedicated CTs are not available.

As we discussed in Sections 2.2.4, 4.5, and 5.5 from Chapters 2, 4, and 5, respectively, there are a number of directional relays that can compare the direction of current flow in each circuit connected to the bus. If the current flow in one or more circuits is away from the bus, an external fault exists. If the fault current flow in all of the circuits is into the bus, an internal bus fault exists. Such schemes were historically applied in this manner [6].

The scheme requires directional (tripping) relays, fault detectors (blocking), and a timer. Directional relays are used on each circuit connected to the bus and they are set to see beyond the immediate bus into the connected circuits. Since the relays must operate for all bus faults, they are set with the same philosophy as the zone 2 settings discussed in Chapter 5, that is, with all infeeds being present. Fault detectors are used on load circuits that do not contribute fault current to indicate that an external fault exists. Instantaneous overcurrent relays set above maximum load are commonly used for this purpose. The fault detector relay contacts are normally closed, and open to block tripping for any external fault on a load supply feeder. For phase faults, the tripping relays are connected to strong sources and to the bus tie breaker. For ground faults, the tripping relays are connected to power transformer neutrals. The timer is required to ensure the blocking relays have time to pick up for external faults before the tripping relays can trip. CT saturation is usually not a problem when comparing the direction of current rather than current magnitude as is done in other bus protection schemes. The CTs in each circuit do not have to have the same ratio and can be used for other purposes such as line relaying and metering.

The circuitry is complex and requires careful and periodic review and maintenance due to the number of relay contacts. The timer should be set for at least four or five cycles to ensure contact coordination. Contact bounce of electromechanical blocking relays must be checked, particularly at high-magnitude fault currents. The relay application and settings must be reviewed whenever system changes are made near the protected bus.

9.8 Partial-differential protection

Figures 9.8–9.11 show a variety of bus configurations that have significant impact on the connections and settings of the bus differential. A full bus differential requires that all circuits be connected to the relay, so the operating coil sees the vector sum of all of the currents as shown in Figures 9.1 and 9.2. Occasionally, however, the station design will result in the current in one or more of the primary circuits not being included in the summation of currents in the differential circuit.

In Figure 9.8, the connection to the load does not have a circuit breaker or CT. This load then presents a continuous error current to the bus differential relay and it must be set above this value and with time delay to coordinate with the load protection for faults on the load feeder. A close-in fault on this feeder is the same as a fault on the bus.

Figure 9.9 shows a transformer fed directly off the bus. This is a common configuration that saves a circuit breaker and is a practical condition, since bus and transformer faults are relatively rare. This results in a combined bus and transformer differential. In this case, the relay must be a transformer differential type to accommodate transformer inrush, as discussed in Chapter 8.

Figure 9.10 shows a more complicated configuration. It is common planning practice to provide duplicate feeds into one or more areas. The feeders are then split between the two buses. Bus or breaker maintenance, bus faults, or breaker failures will then only de-energize one bus, removing only one feeder into a given area. Normally, the two buses are operated as a single bus with the bus tie closed, but there are two bus differentials as shown in the figure. All sources of fault contribution must be included in the appropriate bus differential.

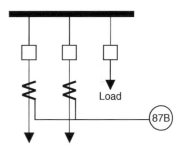

Figure 9.8 Partial bus differential.

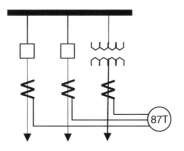

Figure 9.9 Transformer and bus differential.

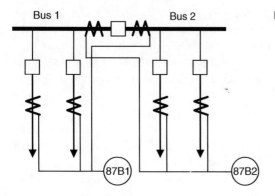

Figure 9.10 Two-bus relaying.

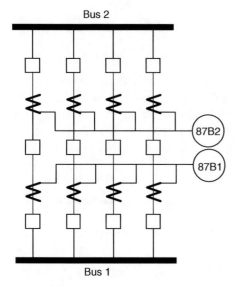

Figure 9.11 Breaker-and-a-half bus differential.

Figure 9.11 shows the bus configuration referred to as a breaker-and-a-half. This is a common EHV system bus configuration as described in Section 1.3 (Chapter 1). The primary advantage of this configuration is that a double-bus fault is unlikely due to the physical distance between the two buses, and any breaker or relay system can be removed for maintenance without removing any primary elements. From the protection point of view, each bus can be considered a single-bus, single-breaker configuration similar to Figures 9.1 and 9.2 and the same considerations apply.

9.9 Zone-interlocked blocking bus protection

This type of protection is not a full-differential protection, and is applied mostly for distribution bus protection [1]. It uses a bus overcurrent or partial-differential protection system as shown in Figure 9.8, but with additional logic to almost completely eliminate the need for

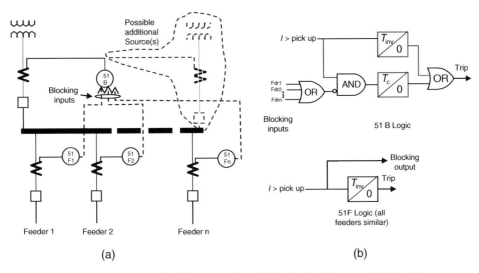

Figure 9.12 Zone-interlocked blocking protection: (a) single line diagram, (b) logic diagrams.

full coordination with the downstream feeder protection. Figure 9.12 shows an example implementation. Figure 9.12a shows the blocking signals from each of the distribution feeders being sent to the bus protection. The bus protection could be an overcurrent function (if there is only a single supply to the bus), or an overcurrent protection connected as a partial differential (if more than one source can supply the bus). Figure 9.12b shows how each of the feeder protections (51F) can output a blocking signal as soon as the measured current picks up to start the inverse time delay. All the blocking signals are connected in common through a single OR gate at the bus protection 51B. The output of the OR gate is then inverted and provided to one input of an AND gate. The second input to the AND gate is the pickup of the bus protection inverse time function. If the bus protection function picks up, and none of the feeder protections have picked up, then the fault must be on the bus, and the bus protection will trip the bus breaker(s) after a short time delay T_c. Time delay T_c is less than a few cycles long, and is intended to ensure that a blocking signal from any feeder (if present) has time to be recognized as an input to the bus protection.

This type of protection has an advantage over the full-differential protection because the bus protection may provide a backup function to the feeder protection, whereas a full differential would be blind to feeder faults. The zone-interlocked blocking scheme is only a few cycles slower than the full differential due to the timer T_c. This scheme is significantly faster than a simple partial-differential protection because there is no need for the full coordination delay of some hundreds of cycles that would be required for coordination of time overcurrent protections. This protection is not applicable to electromechanical relays, since they do not provide an output to indicate that the measured current is above the pickup (when the induction disk just starts to rotate).

The scheme as described so far with non-directional feeder protections could be applied only with conventional radial distribution systems with no sources of fault current on the feeders. If there are sources of fault current on the feeders, the feeder overcurrent protection blocking functions would have to be directional (looking out toward the feeder) to avoid undesirable blocking due to one or more of the feeder sources providing current to a bus fault. If there are fault current infeeds from any feeders, the bus protection must also trip any such feeder breakers.

9.10 Introduction to shunt reactor protection

Reactors are connected into a power system in either a series connection or a shunt connection. The series reactor is used to modify the system reactance, primarily to reduce the amount of short-circuit current available. The protection of series reactors is discussed in Section 5.10 (Chapter 5). The shunt reactor is used to modify the system voltage by compensating for the transmission line capacitance. In general, the protection of reactors is very similar to that of transformers. In considering the protection of shunt reactors, two configurations are involved [7]:

1) Dry-type, connected ungrounded-wye, and connected to the tertiary of a power transformer.
2) Oil-immersed, wye-connected, with a solidly grounded or impedance-grounded neutral, connected to the transmission system.

For oil-immersed types of reactor construction, there are two more considerations that have an effect on the protection:

1) Single-phase reactors, that is, each phase is in its own tank. These are usually applied on EHV transmission lines. There is no possibility of having a phase-to-phase fault within the reactor enclosure, although such a fault can occur in the bus and bushings.
2) Three-phase reactors where all three windings are in the same tank. These are primarily applied at lower voltages.

9.11 Dry-type reactors

The faults encountered in dry-type reactors are the following:

1) Phase-to-phase faults on the tertiary bus, resulting in high-magnitude phase current. These faults are rare, since the phases of the reactors are located physically at a considerable distance from each other.
2) Phase-to-ground faults on the tertiary bus resulting in low-magnitude fault current depending on the size of the grounding transformer and resistor. These faults are also rare, since the reactors are mounted on insulators or supports with standard clearances.
3) Turn-to-turn faults within the reactor bank, resulting in a very small change in phase current. Winding insulation failures may begin as tracking due to insulation deterioration which eventually will involve the entire winding. The result is a phase-to-neutral fault which increases the current in the unfaulted phases to a maximum of $\sqrt{3}$ times normal phase current.

Figure 9.13 Overcurrent relaying.

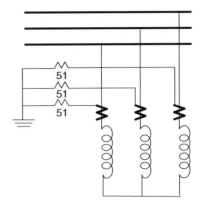

Figure 9.14 Percentage differential relay.

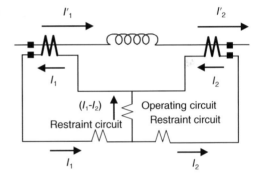

Fault protection for the dry-type reactor is achieved through overcurrent relays connected as shown in Figure 9.13 and differential relays as shown in Figure 9.14. This protection is the same as the overcurrent and differential relaying for generators and transformers. Overcurrent relays must be set above the normal reactor load current. They can detect phase-to-phase faults and phase-to-ground faults if the grounding impedance is low enough, and turn-to-turn faults if enough turns are involved to produce high enough fault currents. Since phase-to-phase and phase-to-ground faults also produce negative-sequence currents, a negative-sequence relay, connected the same as the overcurrent relay, can be used. Load is not a consideration with negative sequence, but there must be enough current to operate the relay. Non-directional negative-sequence overcurrent relays must be time-delayed to override unbalances on the external system.

Directional negative-sequence overcurrent protection can be applied to trip much faster than non-directional. For sensitive directional overcurrent protection, there might be too small an unbalance in the terminal voltage for reliable operation during a low-magnitude turn-to-turn fault. The directional principle must include reinforcement of the polarizing voltage by negative-sequence current flowing through an impedance. The negative-sequence directional relay will be subject to spurious negative-sequence currents caused by unequal saturation of CTs during energization. This saturation is caused by very long time constant decay of offset inrush currents. To avoid undesirable operation during energization, the sensitive directional negative-sequence overcurrent protection is normally disabled by an inrush tripping suppression logic for some seconds during reactor energization.

Fast and sensitive turn-to-turn protection for reactors on ungrounded or impedance grounded systems can also be provided by zero-sequence voltage unbalance protection. In this case, the zero-sequence voltage at the terminals is compared to the neutral voltage. This protection does not respond to unbalances on the weakly ground system which affect the neutral voltage and zero-sequence terminal voltage equally. However, it will respond to difference between these voltages caused by unbalance between the three phases of the reactor.

Differential relays can provide sensitive protection, but they do not see turn-to-turn faults, since the current entering a reactor with shorted turns is equal to the current leaving the reactor. Instantaneous relays are not usually applied, since the only fault location that will produce enough current to operate an instantaneous relay is at the phase end of the reactor or in the bus or bushings. These fault locations are usually protected by the bus differential relay.

9.12   Oil-immersed reactors

The failures encountered with oil-immersed reactors are the following:

1) Faults resulting in large changes in the magnitude of phase current such as bushing failures, insulation failures, and so on. Because of the proximity of the winding to the core and tank, phase-to-ground faults can occur, the magnitude being dependent upon the location of the fault with respect to the reactor bushing.
2) Turn-to-turn faults within the reactor winding, resulting in small changes in the magnitude of phase current.
3) Miscellaneous failures such as loss of oil or cooling.

Relay protection for faults producing large magnitudes of phase current is generally a combination of overcurrent, differential, and distance relaying. There are limitations to each of these protective schemes. Overcurrent relays must be set above the normal load current and a differential relay cannot detect a turn-to-turn fault.

Directional negative-sequence protection as described in Section 9.11 can also be used for turn-to-turn fault protection. Directional ground overcurrent protection similar to the directional negative-sequence protection can also be used, since the neutral of these reactors is usually grounded. Ground overcurrent is recommended to be measured through the neutral to ground connection, since the residual current from three-phase CTs often produces spurious zero-sequence current during inrush. An impedance relay can detect shorted turns, since there is a significant reduction in the 60 Hz impedance of a shunt reactor under such a condition. Figure 9.15 shows the connection for an impedance relay (device 21). Protection against low-level faults or mechanical failure involving the oil system is by pressure, temperature, or flow devices similar to those discussed in Chapter 8.

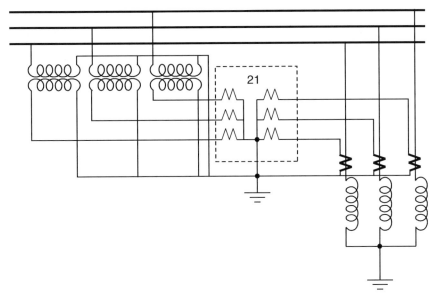

Figure 9.15 Connection for impedance relay.

One of the principal difficulties with shunt reactor protection is false relay operation during energizing or de-energizing the iron core. During these periods, DC offset with long time constants and low-frequency components of the reactor energization current cause most of the problems. High-impedance differential relays rather than low-impedance relays are recommended if this problem occurs. Also, sensitive turn-to-turn protection should be disabled during energization, and when a reactor is de-energized together with a line. The low-frequency "ring down" currents between the reactor and line shunt capacitance during de-energization can cause misoperation of turn-to-turn protection.

9.13 Introduction to shunt capacitor bank protection

Capacitors are also installed as either a series or a shunt element in power systems. The series capacitor is used primarily to modify the transmission line reactance for stability or load flow considerations. The protection of the series capacitor and its effect on the line protection are discussed in Chapter 5. The protection of fixed or switched shunt capacitor banks requires an understanding of the capabilities and limitations of both the capacitors and the associated switching devices. A variety of types of capacitor banks exist—externally fused, internally fused, fuseless, and unfused. Note that neither fuseless nor unfused banks have fuses, but the arrangement of the banks is different. Each bank consists of a number of capacitor cans, or units, and each unit has a number of capacitor elements inside it. Protection of these banks depends on the arrangement of the bank and whether or not the capacitor is fused [8]. Capacitor bank protective equipment must guard against a variety of conditions:

1) Overcurrents due to faults between the bus and the capacitor bank. The protection afforded is the conventional overcurrent relay applied at the breaker feeding the capacitor bank.
2) System surge voltages. The protection afforded is the conventional surge arresters and spark gaps.
3) Overcurrents due to individual capacitor unit failure. In the case of fused capacitors, the manufacturer provides the necessary fuses to blow for an internal unit failure. The fuse link should be capable of continuously carrying 125–135% of the rated capacitor current. In the case of internally fused capacitors, each of the dozens of elements inside a unit is fused, and the bank may operate continuously with several faulted elements removed by its own fuse. In the case of fuseless and unfused banks, when an element fails, the internal dielectric melts and the foils on each side of the element weld together to form a conductive path that allows the bank to continue in operation.
4) Continuous capacitor unit overvoltages.

Figure 9.16 shows the general arrangement for an externally fused high-voltage capacitor bank consisting of parallel units in layers to provide the required current capability and series connected layers to provide the desired capacitive rating.

In the case of externally fused banks, an overvoltage can be imposed across individual capacitor units as a result of the loss of one or more units, usually by the operation of a unit fuse. The overvoltage is the result of the increased impedance of the series section from which the faulty unit has been removed. As units are removed, the impedance of that section increases. However, since there are many layers in series, the effect of the increased

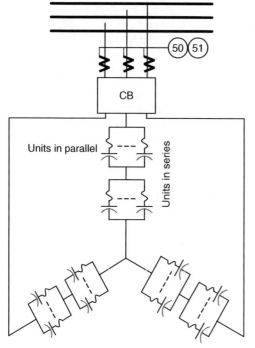

Figure 9.16 General arrangement for a high-voltage capacitor bank.

impedance in one layer does not decrease the phase current in the same relative proportion. The result of the slightly reduced current flowing through the more markedly increased impedance causes a higher voltage to appear across the remaining units in that layer [8].

In the case of other arrangements of banks, a number of elements inside one of more units may fail and be removed from service or shorted out with little impact on continued operation of the bank. Eventually, however, as the number of failures increases, voltages across remaining in-service units approach their maximum continuous rated voltage and the bank needs to be tripped off line. Reference [8] provides details on the different types of schemes that may be applied to different capacitor types and bank arrangements.

Example 9.4 Consider the general arrangement of an externally fused capacitor bank shown in Figure 9.16 with three parallel capacitors in each group and three groups in series in each phase. Assume that each capacitor is 1.0 per unit reactance and the bus voltage is 1.0 per unit.

With system normal, that is, no fuses blown, the effective parallel impedance is 0.333 pu and the sum of the three groups per phase is 0.999 pu. The total current is 1.0/0.999 = 1.01 puA and the voltage across each parallel group is 1.01 × 0.333 = 0.336.

Assume one capacitor develops a short and its fuse blows, removing it from the circuit. The effective parallel impedance of that group is 0.5, and the sum of the three groups is 0.5 + 0.333 + 0.333 or 1.16 pu. The total current is 1.0/1.16 = 0.86 puA.

In general, it is very difficult to protect against overvoltages by relays. If capital were unlimited and the need were great enough, a relay system could be designed using potential transformers across each series section. However, operating guides have been established that recognize that voltages in excess of 110% can be tolerated for emergency or infrequent short-time conditions. If there are two banks in parallel, there are several commercial schemes that detect a voltage unbalance, as shown in Figure 9.17.

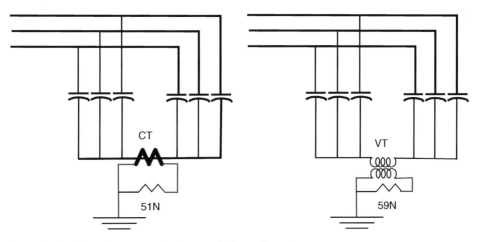

Figure 9.17 Neutral current and voltage unbalance detection.

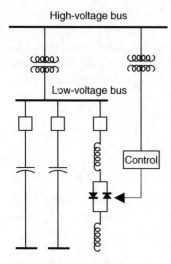

High-voltage bus

Low-voltage bus

Control

Figure 9.18 Typical SVC installation.

9.14 Static VAR compensator protection

Static VAR compensators (SVCs) are devices which control the voltage at their point of connection to the power system by adjusting their susceptance to compensate for reactive power deficiencies [9]. The basic reactive components of SVCs are shunt reactors and shunt capacitors (Figure 9.18). Reactor controls are either thyristor-controlled or thyristor-switched. Capacitors are either fixed or thyristor-switched.

As the load varies, a variable voltage drop will occur in the system impedance. This impedance is mainly reactive. Assuming the generator voltage remains constant, the voltage at the load bus will vary. The voltage is a function of the reactive component of the load current, system, and transformer reactance. An SVC can compensate for the voltage drop for load variations and maintain constant voltage by controlling the gating of thyristors in each cycle. With fixed capacitors and variable reactors, leading or lagging current can be provided to the bus and will correct the voltage drop or rise.

The protection scheme for SVCs is made up of a number of zones. In some cases, faults in a zone should shut down the entire SVC system. In other cases, the relaying can be coordinated so that only the faulted zone is cleared and the SVC kept in operation with limited capabilities.

Sometimes, these protective functions can be provided as part of the integrated protective system supplied by the manufacturer or they can be provided by the user. In either case, the settings must be coordinated between the manufacturer and the user.

9.14.1 Transformer protection

Differential relays, phase and ground overcurrent relays, gas pressure, low oil level, and temperature relays have all been effectively applied in accordance with standard transformer protection practice [10]. The transformer guide should be applied to transformers that are part of the SVC. The connection of the transformer windings determines the types of relay and their connection.

9.14.2 Bus protection

9.14.2.1 Phase Fault Protection

Conventional differential or time overcurrent relays are applicable [1]. Time overcurrent relays are usually used for backup protection. For some installations, the SVC bus is included in the protection zone of the transformer differential relays.

9.14.2.2 Ground Fault Protection

The voltage supplied to SVC buses may be grounded through a resistance or impedance in the main transformer or through a grounding transformer, or ungrounded. Grounding transformers are sized such that ground fault currents are limited to reduce damage, yet large enough to selectively operate ground relays. In addition, ground fault currents on the SVC low-voltage bus should be limited to 500–1500 A to prevent thyristor valve damage.

9.14.3 Typical protection schemes

Typical protection schemes are as follows:

1) Time overcurrent relays connected to CTs to measure the zero-sequence current in the main transformer or grounding transformer [10].
2) Time overvoltage relays connected across the broken-delta secondary winding of a voltage transformer as shown in Figure 7.11 (Chapter 7). This scheme does not provide fast location of a fault, since the ground fault could be anywhere in the low-voltage bus or any of its branches.

9.14.3.1 Overvoltage Protection

In most installations, relays connected to the bus voltage transformers are provided to protect the entire SVC system from excessive overvoltages. The capacitors in the SVC are vulnerable to overvoltage and therefore determine the relay settings.

9.14.3.2 Reactor Branch Protection

A thyristor-controlled reactor (TCR) or thyristor-switched reactor (TSR) is usually considered a separate zone of differential and coordinates with it. Conventional reactor differential and overcurrent relays are used [7]. In many designs, the reactor branch is connected to the SVC bus by a relatively slow motor-operated disconnect switch. The relaying would then trip the SVC main breaker, the faulted element would then be removed and the remainder of the SVC installation returned to service.

9.14.3.3 Capacitor Branch Protection

Conventional overvoltage and unbalance protection is applicable [8]. The details are the same as discussed in Section 9.12.

9.14.3.4 Filter Protection

Switching of SVC elements will introduce harmonics into the power system. Depending upon the particular design, the manufacturer will provide the necessary filters and will provide, or recommend, the protection required. The magnitude of the harmonic voltage generated depends on the type of SVC, the SVC configuration, the system impedance, and the amount of reactance switched.

In thyristor-swtiched capacitor (TSC) branches, air-core reactors connected in series with the capacitors limit the rate of change of inrush current generated from switching. This series combination also provides tuned frequency filtering to a specific harmonic order.

In TSR branches, the triple harmonic currents (3, 9, 15, etc.) are removed by a delta- or wye-ungrounded connection. The even harmonics are removed by symmetrical gating of the TCR thyristors.

9.14.3.5 Thyristor Protection

This protection is normally provided as a part of the thyristor control system. Typical protection is provided for overvoltage, overcurrent, temperature, and, where applicable, coolant flow and conductivity.

9.15 Static compensator

A static compensator (STATCOM) provides variable reactive power from lagging to leading, but with no inductors or capacitors for VAR generation [9]. This is achieved by regulating the terminal voltage of the converter. The STATCOM consists of a voltage source inverter using gate turnoff thyristors, which produces an alternating voltage source in phase with the transmission voltage, and is connected to the line through a series inductance. If the terminal voltage of the voltage source is higher than the bus voltage, STATCOM generates leading reactive power. If the voltage is lower than the bus voltage, STATCOM generates lagging reactive power.

SVCs have generally proven to have lower equipment costs and lower losses. STATCOMs have been used in transmission where land constraints, audible noise, or visual impact are of concern. STATCOM can provide both reactive power absorption and production capability, whereas an SVC requires individual branches for capacitors for VAR generation and reactors for VAR absorption.

9.16 Summary

In this chapter, we have discussed the protection practices related to buses, reactors, and capacitors. Bus protection is almost exclusively some form of differential relaying, although overcurrent, linear couplers Rogowski coils, zone-interlocked blocking, and directional comparison relays are covered. Dry-type and oil-immersed reactors use overcurrent, negative-sequence, percentage differential, or impedance relays depending on the specific application. The main protection for capacitors is a function of their type, arrangement, and installation within a station. In some applications, capacitor overvoltage protection is provided by fuses supplied by manufacturers. SVCs and STATCOMs are gaining increasing popularity as a means of controlling both inductive and capacitive reactive deficiencies at the point of their connection to the power system. The protection is a combination of protection required for both reactors and capacitors and the special requirements associated with the SVC and STATCOM design. In this chapter, we have described the possible protection schemes that are required. It is important to note that SVC and STATCOM protection is a complex combination involving station design, conventional equipment protection, and the specifics of the equipment design. Coordination with manufacturers is essential.

Problems

9.1 Design the bus protection for the system shown in Figure 9.19. Show the location of the primary CTs and draw the three-phase secondary wiring of the two bus differentials. Discuss the relay applications involved.

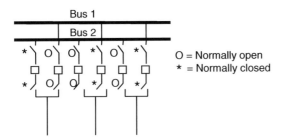

Figure 9.19 System for Problem 9.1.

9.2 Repeat Problem 9.1 for the bus configuration shown in Figure 9.20.

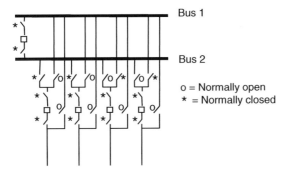

Figure 9.20 System for Problem 9.2.

9.3 Develop a procedure to remove one of the circuit breakers from the bus configuration shown in Figure 9.20 for maintenance, while continuously maintaining the bus protection specified in Problem 9.2.

9.4 Given three 765 kV, 50 M VAR (three-phase rating) shunt reactors, connected phase-to-ground, with the neutral solidly grounded, specify a CT ratio and the settings for time-delay overcurrent protection.

9.5 Design the differential protection for the reactor configuration of Problem 9.4.

References

1 IEEE Std. C37.234 (2021). *IEEE Guide for Protective Relay Applications to Power System Buses.* New York, NY: IEEE Power Engineering Society, IEEE.

2 Seeley, H.T. and von Roeschlaub, F. (1948). Instantaneous bus-differential protection using bushing current transformers. *AIEE Trans.* **67**: 1709–1718.

3 Forford, T. and Linders, J.R. (1974). A half cycle bus differential relay and its application. *IEEE Trans. PAS* **93**: 1110–1117.

4 Harder, E.L., Klemmer, E.H., Sonnemann, W.K., and Wentz, E.C. (1942). Linear couplers for bus protection. *AIEE Trans.* **61**: 241–248.

5 IEEE Std. C37.235-2007 (2008). *IEEE Guide for the Application of Rogowski Coils Used for Protective Relaying Purposes.* New York, NY: IEEE Power Engineering Society, IEEE.

6 Mason, C.R. (1956). *The Art and Science of Protective Relaying.* New York: Wiley.

7 ANSI/IEEE C37.109-2006 (Revision of IEEE Std. C37.109-1988) *Guide for The Protection of Shunt Reactors.* New York, NY: IEEE.

8 ANSI/IEEE C37.99-2012 (2008). *Guide for the protection of shunt capacitor banks.* New York, NY: IEEE Standards Association, IEEE Power Engineering Society, IEEE.

9 Grigsby, L.L. (ed.) (2001). *The Electric Power Engineering Handbook*, Chapter 4. CRC/IEEE Press.

10 IEEE Std C37.91-2021 (Revision of IEEE Std C37.91-2008) (2021). *IEEE Guide for Protective Relay Applications to Power Transformers.* New York, NY: IEEE.

10

Power System Phenomena and Relaying Considerations

10.1 Introduction

Historically, relays have been applied to protect specific equipment, that is, motors, generators, lines, and so on. In doing so, the system benefits by removing the targeted devices when they are faulted, thus eliminating the stress to the system itself and preventing further equipment damage and any associated expensive and lengthy repairs. In addition, the system benefits by not being forced into a temporary abnormal state. Systems, of course, must be robust enough to withstand removal of any element, but there is a limit to how far protection settings can anticipate the extent of the stress.

Systems can be forced into stressed, unplanned situations by operating beyond normally anticipated limits such as heavier than expected loads, abnormal weather, planned or unplanned equipment outages, or human error. These effects can result in wide-area blackouts with severe technical, economic, and social impact. This chapter examines the system phenomena that can result in such disturbances. Chapter 11 introduces the protective schemes and devices that address and seek to correct this problem. Stability—or its loss—is the overriding concern in power system operation. The system operates at close to its nominal frequency and voltage at all times, and all (synchronous) rotating machines connected to the power system operate at the same average speed and in step with each other. When any one of frequency, voltage, or synchronism deviates irrecoverably from normal, then stability is lost, and blackouts follow.

10.2 Power system rotor angle stability

Issues related to loss of synchronism between generators are in the class of rotor angle stability. For brevity, all references to "stability" in Sections 10.2, 10.2.1, and 10.2.2 refer to rotor angle stability. Other types of stability issues are discussed in Sections 10.3 and 10.4.

The generator speed governors maintain the average machine speed close to its nominal value while random changes in load and network configuration are constantly taking place, providing small disturbances to the network. When faults occur on major transmission lines or transformers, the resulting disturbances are no longer small, and severe oscillations in

Power System Relaying, Fifth Edition. Stanley H. Horowitz, Arun G. Phadke and Charles F. Henville.
© 2023 John Wiley & Sons Ltd. Published 2023 by John Wiley & Sons Ltd.

machine rotor angles and consequent severe swings in power flows can take place. The behavior of the machines in this dynamic regime is very complex, and a single concept of power system stability is not viable for all power system configurations. It has been found that three distinct classes of rotor angle stability can be identified for a power system [1]. The three definitions are as follows:

1) Steady-state stability: This is the property of a power system, by which it continues to operate in its present state, and small slow changes in system loading will produce small changes in the operating point. A steady-state unstable system will drift off to unsynchronized operation from its operating point when subjected to a slow small increase in load.
2) Dynamic stability: By this property, a small disturbance in the power system produces oscillations which decay in time, and return the system to its pre-disturbance operating state. A dynamically unstable system produces oscillations which grow in time, either indefinitely, or lead to a sustained limited oscillation.
3) Transient stability: By this property, the power system returns to synchronous operation following a large disturbance, such as that created by a sustained fault. A transient unstable system will lose synchronism as a result of the fault, with groups of machines accelerating or decelerating away from the synchronous speed.

Of these three stability concepts, steady-state stability and transient stability are of immediate interest in protection system design. Dynamic stability is largely determined by the gain and time constant settings of the various controllable devices (primarily the excitation systems of large generators) on the power system. In this chapter, we consider only the steady-state and transient instabilities and their influence on relaying.

10.2.1 Steady-state stability

Consider the simple system consisting of one machine connected to a power system, as shown in Figure 10.1. The internal voltage of the machine is E_s and the machine reactance is X_s.

The equivalent impedance (Thévenin) of the power system is X_t and power system equivalent voltage is E_2. The voltage E_s and the reactance X_s are determined by the context of the study: in steady-state analysis, one would use the field voltage E_f and the synchronous reactance X_d, while for transient studies, E'_q and X'_d would be appropriate. The total reactance between the machine internal bus and E_2 is $X = X_s + X_t$. The electric power output at the machine terminals (at bus S or at bus 2) is

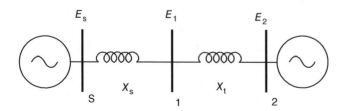

Figure 10.1 Simple system for steady-state stability analysis.

$$P_e = \frac{E_s E_2}{X} \sin \delta, \tag{10.1}$$

where δ is the angle by which the machine internal voltage E_s leads E_0. The mechanical power input to the machine is P_m, and in steady state, electrical power and mechanical power are in balance at a rotor angle δ_0, which is zero when the machine is operating in steady state at δ_0. The rate of change of the output power P_e with respect to δ is given by

$$\frac{\partial P_e}{\partial \delta} = \frac{E_s E_2}{X} \cos \delta, \tag{10.2}$$

which remains positive for $-\pi/2 \le \delta \le +\pi/2$. This is the range of steady-state stability for the system [2]. As a generator, our interest is in positive power output. Thus, the steady-state stability limit of interest is $\delta = \pi/2$.

The real and reactive power outputs of the machine (as measured at the machine terminals) are given by

$$P_1 + j Q_1 = E_1 I_1^*. \tag{10.3}$$

At the steady-state stability limit of the machine, the rotor angle is $\pi/2$, and it can be shown that, at the stability limit, P_1 and Q_1 satisfy the equation:

$$P_1^2 + \left[Q_1 - \frac{E_1^2}{2} \left(\frac{1}{X_t} - \frac{1}{X_s} \right) \right]^2 = \left[\frac{E_1^2}{2} \left(\frac{1}{X_s} + \frac{1}{X_t} \right) \right]^2. \tag{10.4}$$

This is an equation of a circle in the P_1–Q_1 plane, as shown in Figure 10.2a.

In the context of relaying to be discussed in Chapter 11, we are interested in determining the response of a distance relay when the machine is operating at its steady-state limit. We will next show that a circle in the P–Q plane maps into a circle in the apparent R–X plane. Whether or not a machine approaches a limit (such as a steady-state stability limit) defined by a circle in the P–Q plane can then be detected by the corresponding circle in the R–X plane, using a distance relay.

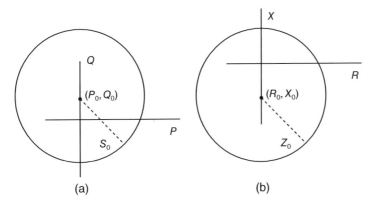

(a)   (b)

Figure 10.2   (a) A circle in the P–Q plane transforms into (b) a circle in the R–X plane.

Let us take a general circle in the P–Q plane, with its center at (P_0, Q_0), and a radius of S_0:

$$(P - P_0)^2 + (Q - Q_0)^2 = S_0^2.$$

For example, in the case of the steady-state stability limit of Equation (10.4), these parameters are given by

$$\left.\begin{array}{l} P_0 = 0 \\[4pt] Q_0 = \dfrac{E_1^2}{2}\left(\dfrac{1}{X_t} - \dfrac{1}{X_s}\right) \\[8pt] S_0 = \dfrac{E_1^2}{2}\left(\dfrac{1}{X_t} + \dfrac{1}{X_s}\right) \end{array}\right\}. \tag{10.5}$$

Consider the three circuits shown in Figure 10.3. Figure 10.3a shows a generator with output $P + jQ$ with a terminal voltage E. Figure 10.3b shows the generator with the same terminal conditions but now supplying an impedance load R' and X' connected in parallel at the generator terminal. The parallel impedances are next converted to series-connected R and X, which are related to the terminal conditions by

$$P = \frac{E^2 R}{R^2 + X^2} \text{ and } Q = \frac{E^2 X}{R^2 + X^2}. \tag{10.6}$$

Substituting Equation (10.6) into Equation (10.5), we obtain

$$(R - R_0)^2 + (X - X_0)^2 = Z_0^2. \tag{10.7}$$

This is an equation of a circle in the R–X plane with its center at (R_0, X_0) and a radius of Z_0. It can be shown that these parameters are given by

$$\left.\begin{array}{l} R_0 = \dfrac{P_0 E^2}{P_0^2 + Q_0^2 - S_0^2} \\[10pt] X_0 = \dfrac{Q_0 E^2}{P_0^2 + Q_0^2 - S_0^2} \\[10pt] Z_0 = \dfrac{S_0 E^2}{P_0^2 + Q_0^2 - S_0^2} \end{array}\right\}. \tag{10.8}$$

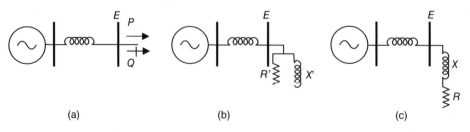

| (a) | (b) | (c) |

Figure 10.3 (a) Generator connected to a power system. (b) The generator supplying the same power to a parallel load. (c) The generator supplying the same power to a series-connected load. This is the apparent impedance seen by an impedance relay.

The circle in the impedance plane for the steady-state stability limit is shown in Figure 10.2b [2]. For this case, we may substitute the values for P_0, Q_0, and S_0 from Equation (10.5), using E for the machine terminal voltage, rather than E_1 as in Equation (10.5). We then obtain

$$\left.\begin{array}{l} R_0 = 0 \\[2mm] X_0 = -\dfrac{(X_t - X_s)}{2} \\[3mm] Z_0 = \dfrac{(X_t + X_s)}{2} \end{array}\right\}. \tag{10.9}$$

This has also been referred to in Section 7.8 and Figure 7.20 (Chapter 7) discussing loss of excitation. An explanation as to why the steady-state instability is referred to as a loss-of-field condition is necessary. Figure 10.4a shows the phasor diagram of the terminal voltage,

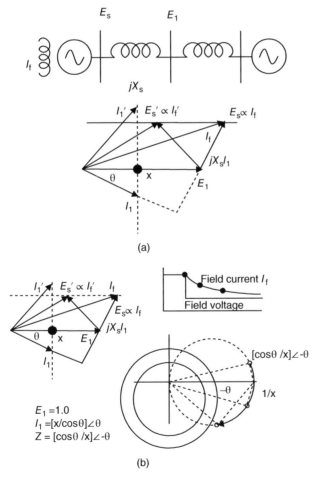

(a)

(b)

Figure 10.4 Relationship between reduction in field current and the phasor diagram of generator voltage and current: (a) changes in phasor diagram as field current decreases; (b) changes in apparent impedance seen by a distance relay at the generator terminal as the field current decreases.

stator current, and the internal voltage of the generator E_f. Also shown is the field circuit, with a field current I_f. Neglecting the effects of machine saturation, the voltage E_f and the field current I_f are proportional to each other. We may use the phasor E_f (magnitude) to represent the field current.

If the field current of the generator decreases for whatever reason, the output power P is not affected. Since the power P is equal to $EI\cos\theta$, where θ is the power factor angle, the projection of the stator current I_1 on the axis of E_1 must be constant as the field current changes. This is designated as "x" in Figure 10.4a. Now consider a reduction in the field current I_f. In order to maintain the phasor relationship between E_f, E_1, and I_1 under these conditions, as E_f changes (drops in magnitude) it must travel along the dashed horizontal line and the current I_1 must travel along the dashed vertical line as shown in Figure 10.4a. This keeps the output power constant at P, while the stator current goes from a lagging power factor angle to a leading position. The machine thus absorbs reactive power when the field current decreases.

The ratio of E_1 to I_1 is the apparent impedance $(R + jX)$ seen by a distance relay connected at the terminals of the generator. Keeping "x" constant, the apparent impedance travels along the circle, crossing over from the first quadrant to the fourth quadrant. The characteristics of the impedance relays which define the steady stability boundary are also shown in Figure 10.4b. It should be clear that as the field current of the generator drops, the generator goes from a lagging power factor to a leading power factor, and the apparent impedance seen by a distance relay approaches the steady-state stability boundary. Application of this principle to out-of-step relaying is considered in Chapter 11.

Example 10.1   Consider a 1000 MVA, 34.5 kV generator, with a synchronous reactance of 0.6 pu on its own base, connected to a power system through a 1000 MVA transformer with a leakage reactance of 0.1 pu on its own base. The system may be assumed to have a Thévenin impedance of 0.2 pu on a 100 MVA base. We are to determine the steady-state stability limit circles in the P–Q and R–X planes.

Converting all impedances to a common base of 100 MVA, the generator and transformer reactances are $0.6 \times 100/1000 = 0.06$ pu and $0.1-100/1000 = 0.01$ pu, respectively. Thus, in the notation of Figure 10.5, X_s is 0.06 pu and X_t is 0.21 pu. From Equation (10.5), assuming that the terminal voltage is 1.0 pu, it follows that the center of the stability limit circle in the P–Q plane is located at

$$P_0 = 0, Q_0 = \left(\frac{1.0}{2}\right)\left(\frac{1.0}{0.06} - \frac{1.0}{0.21}\right) = 5.95 \text{ pu}$$

and

$$S_0 = \left(\frac{1.0}{2}\right)\left(\frac{1.0}{0.06} + \frac{1.0}{0.21}\right) = 10.71 \text{ pu.}$$

On a 100 MVA system base, the center of the steady-state stability limit circle is located at $(0, 595)$ MVA and the radius of the circle is 1071 MVA. In the R–X plane, the center of the circle is located at

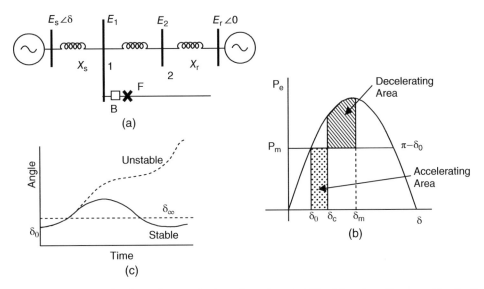

Figure 10.5 Principle of equal-area criterion of transient stability: (a) system diagram with a fault; (b) power angle curve during fault and after fault clearing; (c) stable and unstable swings.

$$R_0 = 0, \quad X_0 = -\frac{(0.21 - 0.06)}{2} = -0.075 \text{ pu}$$

and

$$Z_0 = \frac{(0.21 + 0.06)}{2} = 0.135 \text{ pu.}$$

The base impedance on the 100 MVA base is $(34.52/100) = 11.902\ \Omega$. Thus, in terms of primary ohms, the coordinates of the center of the R–X circle are $(0, -0.892)\ \Omega$ and its radius is $1.607\ \Omega$. If a distance relay is connected at the terminals of the generator, its settings must be made in terms of the secondary ohms. If we assume a current transformer (CT) ratio of 20 000:5 (i.e., 4000:1), we would have the full-load generator current of $1000 \times 103/(\sqrt{3} \times 34.5)$, or 16 735 A produce a secondary current of approximately 4.18 A. The voltage transformer (VT) ratio is $34.5 \times 103/(\sqrt{3} \times 69.3) = 287.4 : 1$. Hence, the impedance conversion factor is $n_i/n_e = 4000/287.4 = 13.92$. Thus, the distance relay zone for detecting the steady-state stability limit would be a circle, with its center at $(0, -12.41)\ \Omega$ secondary, and the radius of the circle should be $22.37\ \Omega$ secondary.

10.2.2 Transient stability

Equation (10.1) shows the electrical power output of the generator when its internal voltage angle is δ with respect to the angle of the equivalent voltage of the system to which it is connected. The accelerating power P_a operating on the generator rotor is the difference between the mechanical power input P_m and the electrical power output. The accelerating power is not zero when the network is disturbed from its steady state. A fault and its subsequent clearing are such a disturbance. The accelerating power sets the rotor of the machine in motion, and the law of motion in the rotational frame of reference is the familiar Newton's law:

$$\frac{2H}{\omega_s}\frac{d^2\delta}{dt^2} = P_a = P_m - \frac{E_sE_r}{X}\sin\delta, \tag{10.10}$$

where we have replaced E_0 with E_r, representing the receiving-end voltage. H is the inertia constant of the machine defined as the kinetic energy stored in the rotor at synchronous speed divided by the MVA rating of the machine, and ω_s is the synchronous speed of the machine. All other quantities in Equation (10.10) are in per unit. The units of the inertia constant are seconds, and for normal machines driven by steam turbines, the inertia constants vary between 4 and 10 s. For hydroelectric generators, the inertia constants range from 2 to 4 s. Equation (10.10) can also be put on a common system base, by converting all the voltages, reactances, and the inertia constant by the usual conversion formulas.

Example 10.2   Assume that the machine of Example 10.1 has an inertia constant of 7 s on its own base. On the system base of 100 MVA, the inertia constant becomes

$$H_{system} = 7 - \frac{1000}{100} = 70\,\text{s}.$$

Let the initial power transfer be 200 MW (2.0 pu on the system base), which results in an initial rotor angle of S_0 (assuming the voltages to be 1.0 pu), given by

$$2.0 = \frac{1}{0.27}\sin\delta_0; \quad \text{or } \delta_0 = 32.68°.$$

With the synchronous speed of 377 rad s^{-1} for the 60 Hz system, the equation of motion of the machine is given by

$$\frac{2 \times 70}{377}\frac{d^2\delta}{dt^2} = P_a = 2.0 - \frac{1}{0.27}\sin\delta$$

$$0.371\frac{d^2\delta}{dt^2} = 2.0 - 3.704\sin\delta$$

It is interesting to calculate the frequency of small oscillations of this system, as it is often quite close to the frequency of large oscillations produced by a transient stability swing. For small oscillations around δ_0, we may replace δ by $\Delta\delta_0 + \Delta\delta$, and using the usual assumptions for small $\Delta\delta_0$, and recognizing that 3.704 $\sin\delta_0$ is equal to 2.0, we get

$$0.371\frac{d^2\delta}{dt^2} - 3.704 \times \Delta\delta \times \cos\delta_0 = -3.117\Delta\delta.$$

This is the equation of motion of a pendulum, of the form

$$\frac{d^2x}{dt^2} + \omega^2x = 0,$$

with an oscillation frequency of $\omega = (3.117/\sqrt{0.371}) = 2.9$ rad s^{-1}. This corresponds to a small oscillation frequency of 0.461 Hz. Such low frequencies are typical of the electromechanical oscillations, or "swings," of large electric generators following faults and subsequent breaker operations.

As the steady-state equilibrium is disturbed by a fault, the effective reactance between the two machines changes, and it changes again when the fault is cleared. As the reactance changes, and the rotor angles cannot move instantaneously to compensate for this change, the rotor is subjected to an accelerating torque, which depends upon the angle δ. The resulting equation of motion can be solved by the familiar equal-area criterion [1, 2] in the case of a two-machine system.

Consider the simple two-machine power system shown in Figure 10.5. Before the occurrence of the fault F, the machine is operating at an initial rotor angle δ_0 and the accelerating power is zero as shown in the following equation:

$$P_a = P_m - \frac{E_s E_r}{X} \sin \delta_0 = 0. \tag{10.11}$$

If a three-phase fault occurs at the terminals of the machine as shown in Figure 10.5a, the electrical power output of the generator goes to zero and the accelerating power of the generator becomes equal to P_m. Under the influence of this power, the rotor angle increases until the fault is removed by the circuit breaker B. This is shown to occur at a rotor angle δ_c, as shown in Figure 10.5b. After the removal of the fault, the electrical power P_e is greater than P_m because the rotor angle δ_c is greater than δ_0. The rotor now begins to decelerate, and will begin to decrease after reaching a maximum value δ_m. The principle which governs this motion of the rotor is known as the "equal-area criterion," and refers to the fact that at the maximum angle reached by the rotor the area in the P–δ space while the rotor is accelerating is equal to the area when the rotor is decelerating.

It should be clear from Figure 10.5b that the maximum available decelerating area corresponds to a maximum angle $\delta_m = \pi - \delta_0$. If the total accelerating area is smaller than the maximum available decelerating area, the resulting rotor oscillation is stable as shown in Figure 10.5c by the solid line. Otherwise the oscillation is unstable, and rotor angle increases indefinitely, and the machine is said to have lost synchronization. An unstable rotor motion is shown by the dotted line in Figure 10.5c. The protection problem associated with transient instability is considered in Chapter 11.

It should be remembered that the principle of the equal-area criterion is applicable only to a system which can be represented by two machines connected by an impedance. For more complex systems, other techniques of analysis are needed—for example, transient stability simulation programs. However, even in those cases, there are two distinct classes of rotor behavior, stable and unstable, which must be detected and appropriate control action taken by out-of-step relays.

10.2.2.1 Apparent impedance during stability swings

As has been observed earlier, the stability swing is characterized by a movement of the rotor angle over wide ranges. During a stable swing the rotor angle oscillates around an equilibrium point, while during an unstable swing the rotor angle progressively increases (or decreases), and acquires values beyond the $\pm 2\pi r$ range. Consider the two-machine system shown in Figure 10.6. We will assume that the receiving-end voltage is the reference phasor, and the angle of the voltage E_s represents the position of its rotor. It is this angle which will

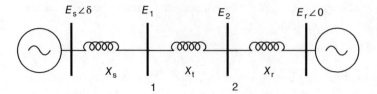

Figure 10.6 Two-machine system for analysis of phasors as δ is varied.

vary during a stability swing. In order to develop protection principles for instability detection, it is necessary to analyze the behavior of current and voltage phasors in the power system for stable and unstable cases.

The phasor diagram for this case is shown in Figure 10.7. The current in the line is given by

$$I = \frac{E_s E_r}{j(X_r + X_s + X_t)} = jB(E_s - E_r), \tag{10.12}$$

where B is the total susceptance. If we assume E_r to be the reference phasor, the current will describe a circle as the angle δ of E_s changes from 0 to 2π. If we further assume that the two voltages are of equal magnitude (1.0 pu), the circle described by the current will pass through the origin as shown in Figure 10.7. Now consider the voltages at the intermediate buses E_1 and E_2, shown in Figure 10.6. E_2 can be calculated from the voltage at the receiving bus:

$$E_2 = E_r + jX_r I = E_r + \frac{X_r}{(X_r + X_s + X_t)}(E_s - E_r), \tag{10.13}$$

or

$$(E_2 - E_r) + k E_s, \text{ where } k = \frac{X_r}{(X_r + X_s + X_t)}. \tag{10.14}$$

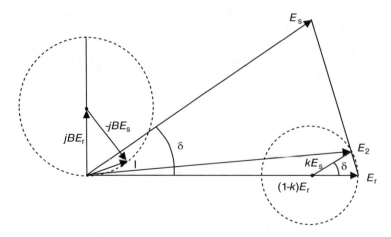

Figure 10.7 Phasor diagram for the system shown in Figure 10.6 for variable δ.

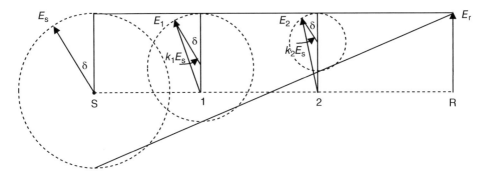

Figure 10.8 View of the phasor diagram for the system shown in Figure 10.6. Voltages at intermediate buses.

Again, as E_s describes a circle while δ varies between 0 and 2π, E_2 also describes a circle, as shown in Figure 10.7. Voltages at any other intermediate points—such as E_1—can be calculated similarly, using the appropriate k for each location. A somewhat different way of expressing the same relationships is shown in Figure 10.8.

The apparent impedance seen by a distance relay located at one of the intermediate points (say at bus 1 in Figure 10.6) is the ratio of the voltage and current at that location. The voltages and currents produced during a stability swing are balanced. As discussed in Chapter 5, all distance relays respond to a three-phase fault, or to a balanced condition, in an identical fashion. For example, the phase and ground distance relays located at bus 1 in Figure 10.6 will see an apparent impedance Z given by

$$Z = R + jX = \frac{E_1}{I}. \tag{10.15}$$

We note that if the current I is 1.0, the voltage and the apparent impedance are numerically identical. One could visualize the apparent impedance seen by the relay by plotting the phasor diagram with the current as a unit reference phasor. If we redraw the phasor diagram of Figure 10.7 with the current aligned with the real axis, and use the current magnitude as a unit, we get the R–X diagram of Figure 10.9. Note that, in this figure, the apparent impedance seen by a relay at a location is plotted from that point as the origin. This has the effect of creating an impedance diagram, which is obtained by a rotation of the phasor diagram.

The line R–S now represents the total reactance ($X_s + X_t + X_r$) of Figure 10.6. The apparent impedance seen by a distance relay located at bus R (if indeed a relay could be located there) is represented by R–O, while the apparent impedances seen by distance relays located at buses 1, 2, and S are 1–O, 2–O, and S–O, respectively (Figure 10.9). For a greater value of the angle δ, such as δ', the current is greater. However, setting the increased current equal to $(1.0 + j0)$, produces the new origin O', and the apparent impedances are the distances from R, 1, 2, and S to O', as shown in Figure 10.9. It should be clear that, as the rotor angle varies, the apparent impedances seen by distance relays would also change, their apex lying on the locus of the point O. For equal voltages at the sending and the receiving ends, the locus is a straight line, while for unequal voltages, the locus is a circle. For a stable swing, the

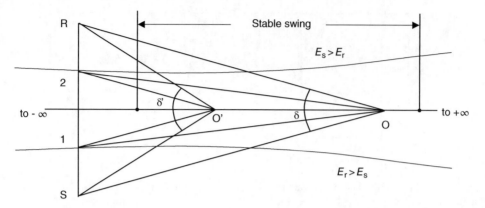

Figure 10.9 Locus of apparent impedances as seen during stable and unstable swings of a system.

excursion of the point O will be on a limited portion of the locus, while for an unstable swing, it will traverse from $+\infty$ to $-\infty$, as the rotor angle varies from $0°$ to $360°$.

It should be noted that the lengths of the segments S–1, 1–2, and 2–R represent the reactances of the three elements in the power system. This must be so, since these represent the voltage drops across each of these elements, and at a current of 1.0, the voltage drop equals the element impedance. Since the distance relay characteristic is determined by the impedance of the line, we are now in a very good position to follow the apparent impedance locus seen by distance relays that are set to protect a given line.

10.3 Voltage stability

Voltage stability is defined as "the ability of a system to maintain voltage so that when load admittance is increased, load power will increase, and so both power and voltage are controllable" [3]. Additionally, the associated important phenomenon of voltage collapse is defined as "the process by which voltage instability leads to loss of voltage in a significant part of the system" [4].

The primary causes of voltage instability are inadequate reactive support, heavy reactive power flows, and heavily loaded transmission lines. Figure 10.10 shows a simple radial system with a load admittance G supplied from a power source E_S. The power supplied to the load will be $S = V_R^2 G$. The power supplied to the load will increase as G is increased.

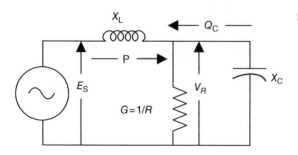

Figure 10.10 Radial load supply.

However, if we assume initially that no reactive support is available at the load ($e_C = 0$), the voltage V_R will decrease as G increases and the current supplying the load increases. At some point, the fact that the power is proportional to the square of the decreasing voltage will override the increase of power delivered due to the increasing of G.

The relationship of power delivered to voltage at the delivery point is readily visualized by a P–V curve (also known as a "nose curve") as shown in Figure 10.11. In Figure 10.11, the voltage at the load terminal (V_e in Figure 10.10) is plotted against the real power consumed by the admittance (e). The solid line graph of Figure 10.11 is plotted assuming the source voltage (E_S) and the reactance (X_L) between the source and the load are both 1.0 pu, and no reactive power (Q_C) is injected to boost the voltage V_R. It can be seen for this system that the maximum amount of power that can be transmitted is 0.5 pu, and this limit is reached when $V_R = 1/\sqrt{2} = 0.707$ pu. At this point, the magnitude of the load impedance (e) is equal to the magnitude of the source impedance (X) and the load is "matched" to the source. For this figure, the worst case allowed maximum and minimum operating voltages are assumed to be +5% and −10% respectively.

If reactive power is injected into the load bus, the voltage V_R will be boosted, and the power transfer limit is increased. In Figure 10.11, the broken line shows that injecting a varying degree of reactive power to maintain a combined load/capacitance power factor of 0.95 leading allows up to 35% more real power to be transmitted. Alternatively, if series compensation is added to the transmission system to decrease its total reactance by 17%, the dotted line shows the power transfer limit is again increased by 35%.

If the operating point of the system falls below the nose of the P–V curve, the power delivered decreases as the admittance is increased. In that case, voltage stability has been lost, and voltage collapse usually follows. The upper portion of the curve is the stable operating

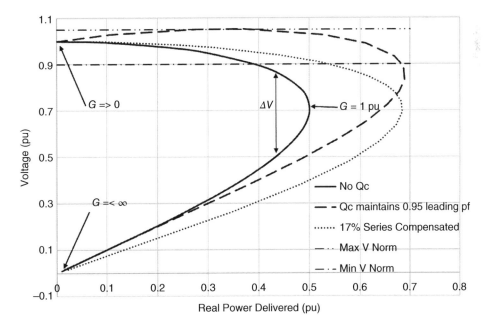

Figure 10.11 Voltage stability P–V curve.

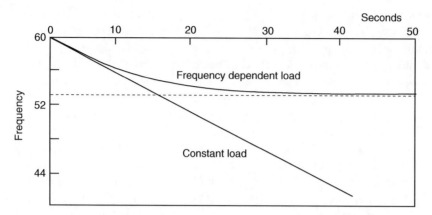

Figure 10.12 Frequency decay in a power system due to generation shortage: constant and frequency-dependent loads.

range calculated for the static system condition. More realistic studies use a dynamic analysis and show two different voltages for the same power delivered. The lower curve shows the performance for the unstable voltage condition, but for which a solution is possible. As ΔV tends to zero, the margin of power between the operating point and the point of maximum power approaches zero [3–5].

To provide sufficient reactive support, a variety of electronic devices are available (see Figures 9.17 and 9.18 in Chapter 9). In addition, conventional sources such as shunt capacitors, local generators, and synchronous condensers may also provide reactive power near the load. It must be noted that reactive power (volt-ampere reactives [VARs]), unlike real power (MW), cannot be transmitted over long distances. This is due to the fact that VARs are caused to flow by the difference in voltage, and voltages on a typical transmission system are only ±5% of nominal, and this small difference will not cause VARs to flow over long distances. Reactive power flowing over transmission lines can be measured by phasor measurement units placed at strategic locations. These measurements will provide a measure of power system proximity to voltage collapse. Real power can be transmitted over long distances through the coordinated operation of the interconnected grid, whereas reactive power must be generated at or near the load center.

Note from Figure 10.11 that injecting reactive power (Q_C) at the load center raises the load bus voltage, and the nose curve. Because of the rise in load voltage, it becomes more difficult to determine the approach to the critical point on the nose curve by simply monitoring voltage levels. For this reason, shunt reactive power sources for improving voltage stability must be applied with caution.

Figure 10.11 also shows that series capacitor compensation of transmission line reactance also extends the power transmission limit. Series compensation is often used to increase the transmission capability of long lines. However, as noted in Chapter 5 and Section 10.5, series compensation also introduces some challenges in transmission line protection.

Since VARs cannot be transmitted over long distances, the sudden loss of transmission lines results in the immediate need for local reactive power. If this is not available, the voltage will go down and may collapse. For these reasons, voltage rather than frequency has become an important indicator that the power system is under stress. Availability of reserve reactive power near the load center is also an important factor [5]. Many of today's loads are single-phase, small motors in contrast to years ago when air conditioning and appliance loads were not as prevalent and lighting and other resistive loads were. The slow tripping of stalled motors and the relatively slow re-acceleration of more robust motors result in low system voltage after a system fault has cleared, exacerbating the system stress. The voltage stability phenomenon in the transient region is often closely involved with angular stability discussed in Section 10.2.

Inadequate reactive power has been an exacerbating factor in several large blackouts as noted in Section 10.8 and the following additional incidents:

- 30 July 2012, India.
- 23 August 2003, South Sweden and Denmark.
- 10 August 1996, Western Interconnection US.
- 13 March 1989 Quebec

10.4 Dynamics of system frequency

When loss of generation, loss of load, or out-of-step relay actions as described in the earlier text cause a mismatch to develop between the load and generation in a power system or in a portion of a power system, the generators will either speed up (when generation is in excess) or slow down (when load is in excess). As the generators slow down, they must be protected from reaching dangerously low operating speeds, which may cause failures in turbine blades—especially those of the low-pressure section of a steam turbine. The mechanical resonance frequencies of these blades lie close to (but somewhat under) the system normal frequency. For a 60 Hz turbine generator system, a resonance may exist around 57 Hz, and should the generator frequency approach this value serious damage to the turbine blades may result. Station auxiliary system motors may also be damaged by continued operation at the reduced frequency. It is therefore imperative that the frequency decay of an isolated system be arrested before generator protective devices isolate the plant from the system, and further exacerbate a critical situation. It is the function of underfrequency load-shedding relays to detect the onset of the decay in system frequency, and shed appropriate amounts of system load, so that the generation and load are once again in balance, and the power system can return to normal operating frequency without disconnecting any generators from the system. Load-shedding schemes have become quite important in present-day systems, where there is a lack of adequate spinning reserve margins, and a shortage of tie-line capacity to make up for the lost generation by importing large blocks of power from the interconnection (preferred strategies for load shedding have been defined by a number of operating committees of the power engineering community, such as the North American Electric Reliability Corporation [NERC] in the United States.)

Consider an islanded system, which has an excess load, and which starts at $t = 0$ with a balance between load and generation at frequency f_0—usually close to the nominal operating frequency of 60 Hz. The system consists of many generators, and after the transient stability oscillations have died out, the frequency of all the generators may be assumed to become equal. We combine all the machine inertias into a total system inertia J_0 by the formula

$$J_0 = \frac{\sum J_i S_i}{\sum S_i}, \tag{10.16}$$

where J_i are the individual machine inertias in kg·m² and S_i are the corresponding machine power ratings. An aggregate rotor angle δ is defined as the center of angle (COA):

$$\delta = \frac{\sum \delta_i S_i}{\sum S_i}. \tag{10.17}$$

The equation of motion for the aggregate rotor is given by

$$J_0 \frac{d^2\delta}{dt^2} = T_a = T_m - T_e, \tag{10.18}$$

where T_a represents the accelerating torque, T_m the combined mechanical input torque to the system, and T_e the combined electrical load torque. Multiplying both sides by the aggregate rotor speed ω gives

$$\omega J_0 \frac{d^2\delta}{dt^2} = \omega T_a = \omega T_m - \omega T_e = P_m - P_e = \sum G_i - \sum L_i. \tag{10.19}$$

The inertia constant H introduced in Equation (10.10) is related to J. H is the kinetic energy stored in a machine at synchronous speed divided by its power rating. Let the total system power rating be S_0. Then,

$$H_0 = \frac{(1/2)J_0\,\omega_s^2}{S_0}. \tag{10.20}$$

Substituting for J_0 in Equation (10.18) gives

$$\omega \frac{2H_0 S_0}{\omega_s^2} \frac{d^2\delta}{dt^2} = P_m - P_e = \sum G_i - \sum L_i. \tag{10.21}$$

Note that if we approximate ω by ω_s, and convert the right-hand side of Equation (10.12) to per unit by dividing by S_0, we get Equation (10.12). Assuming ω equal to ω_s is a valid approximation in transient stability studies, where the frequency does not change by a large amount. However, this approximation is not acceptable here, as we are primarily interested in determining the change in system frequency. We identify the system frequency as the rate of change of the aggregate rotor angle:

$$\frac{d^2\delta}{dt^2} = \frac{d\omega}{dt} = 2\pi \frac{df}{dt}. \tag{10.22}$$

And, thus, Equation (10.15) becomes

$$f\left(\frac{2H_0}{f_s^2}\right) S_0 \frac{df}{dt} = \sum G_i - \sum L_i. \tag{10.23}$$

If the generators have an average power factor rating of p, then

$$\sum G_i = pS_0, \tag{10.24}$$

and defining a relative load excess by a factor L:

$$L \equiv \frac{\sum L_i - \sum G_i}{\sum G_i}. \tag{10.25}$$

Equation (10.17) becomes

$$f\frac{df}{dt} = \frac{pL}{2H_0}f_s^2. \tag{10.26}$$

The abovementioned equation can be integrated directly, and using the boundary condition that at $t = 0, f = f_0$, we get

$$f = f_0\sqrt{1 - \frac{pLf_s^2}{2H_0f_0^2}t}. \tag{10.27}$$

As t goes to infinity, the frequency will decrease at an ever increasing rate. In reality, the load is not independent of frequency, and decreases with a reduction in frequency. Thus, as the frequency decreases, at some reduced frequency it balances the generation, and the frequency stabilizes at that point. Consider a load which depends upon the frequency, with a constant decrement factor d:

$$d = \frac{\% \text{ load reduction}}{\% \text{ frequency reduction}}. \tag{10.28}$$

Let $\sum L_i$, the total system load, depend upon frequency in this fashion. Then, if $\sum L_i$ is equal to P_{10} at the initial frequency f_0, at some other frequency it will be given by P_1, where

$$d = \frac{(1 - P_1/P_{10})}{(1 - f/f_0)}. \tag{10.29}$$

In terms of the factor d, the load excess factor L at some frequency f is related to the factor L_0 at frequency f_0 by

$$L = L_0 - (1 + L_0)d\left(1 - \frac{f}{f_0}\right). \tag{10.30}$$

Substituting this value of L in Equation (10.19), the differential equation for the frequency decay in this case is given by

$$f\frac{df}{dt} = -\frac{p}{2H_0}\left[L_0 - (1 + L_0)d\left(1 - \frac{f}{f_0}\right)\right]f_s^2. \tag{10.31}$$

After some manipulations, the solution to this equation, with the boundary condition of $f = f_0$ at $t = 0$, is given by

$$(f_0 - f) - f_0 \left[\frac{L_0}{(1 + L_0)d} - 1 \right] \ln \left[\frac{1}{1 - (f_0 - f)(1 + L_0)d/L_0 f_0} \right]$$
$$= \frac{p}{2H_0} (1 + L_0) d \frac{f_s^2}{f_0} t. \tag{10.32}$$

This equation provides a limit f_∞, as $t \to \infty$:

$$f_\infty = f_0 \left[1 - \frac{L_0}{(1 + L_0)d} \right], \tag{10.33}$$

which is precisely the value of the frequency at which the frequency-dependent $\sum L_i$ balances $\sum G_i$. Equations (10.27) and (10.32) are plotted in Figure 10.12 for assumed values of $p = 0.85$, $H_0 = 10.0$, $L_0 = 0.3$, and $d = 2.0$. It can be seen that with a constant load, and a load excess factor of 0.3, the frequency continues to decay, whereas with a load decrement factor of 2.0, the frequency decline is arrested at 53.1 Hz. Similar curves for other cases can be found in Reference [2].

It is sometimes necessary to determine the average rate of change of frequency during a range of frequency variation, so that load-shedding relay settings could be determined. Recall that Equations (10.26) and (10.31) give us the rate of change of frequency at a single value of frequency. We may calculate an approximate average rate of change over an interval by taking an average of the rates at the two ends of the interval, or we may calculate the average rate by taking a ratio of the frequency difference and the time interval. For example, if the average rate of frequency change over a frequency interval (f_1, f_2) is needed, using Equation (10.27), the average rate of frequency change is given by

$$\frac{f_2 - f_1}{t_2 - t_1} = \frac{f_2^2 - f_1^2}{(t_2 - t_1)(f_2 + f_1)} = \frac{f_s^2}{f_1^2} \frac{pL}{H} \frac{(f_2 - f_1)}{(1 - f_2^2/ - f_1^2)}. \tag{10.34}$$

If f_s and f_1 are approximately equal, we may make their ratio equal to 1, and the average rate of frequency change R becomes

$$R = \frac{pL}{H} \frac{(f_2 - f_1)}{(1 - f_2^2/f_1^2)}. \tag{10.35}$$

For f_2 smaller than f_1, R is negative, indicating a decreasing frequency. The rate of decay R is used in setting the load-shedding relays, as is seen in Section 11.3 (Chapter 11).

10.5 Series compensation for increased power transfer

Series capacitors are applied to improve stability, increase power transfer capability, reduce losses and voltage drop, and provide better load division on parallel transmission lines. They may be installed at one end of the line, at both ends, or in the middle. The impedance value of a series capacitor is typically 25–75% of the line impedance. Capacitor overvoltage protection is part of the manufacturer-supplied protection and usually consists of parallel power gaps or metal-oxide varistors (MOVs), the purpose of which is to limit the voltage across the capacitor if fault or load current produces voltages high enough to damage the capacitor. A bypass breaker is also applied for protection (to limit the required energy absorption capability of an MOV) and operating flexibility.

The effect of series capacitors on other relays must be considered. In particular, distance relays are adversely impacted by the discontinuity a capacitor introduces in the line imped-ance (see Section 5.10 in Chapter 5). Other relaying schemes must take into account the possibility of a power gap or MOV failure, unsymmetrical gap flashing, or MOV conduction. Current differential or phase comparison schemes will work properly with series capacitors, since the comparison in a series circuit will not change. Transmission line charging current and the phase relationship of current at the two ends of the line must be considered.

Series reactors are typically applied for better load division on parallel paths or to limit fault current. They should have their own protection because certain internal faults may not be detected by the line relays. Such protection usually includes differential and distance relays. Series reactors can be bypassed with a circuit switcher or other switching device. Relay setting changes will usually be required when the reactor is bypassed. This can be accomplished with an adaptive relay system.

Transmission capacity in a corridor can be increased with the use of phase shifting trans-formers. It is recognized that power flows on AC networks in proportion to phase angle difference between the sending and receiving ends. Phase-shifting transformers inserted in the transmission corridor help provide an increased phase angle spread, thereby increas-ing power transfer. Power electronic devices can also provide effective power transfer con-trol by incorporating phase angle shifts.

10.6 Independent power producers

The advent of deregulation or re-regulation in the late 1990s [6] allowed entrepreneurs to own power-producing facilities and connect to existing utility transmission or distribution lines. Termed independent power producers (IPPs) or nonutility generators (NUGs), these entities were intended to provide additional capacity and offered a potential significant sav-ings to consumers. The result was never realized fully, as residential consumers were hes-itant to connect to a nonutility service. However, the IPPs did attract industrial and commercial load, often contracting with facilities remote from the local utilities' service area. It should be noted that the number of transmission connections (often tapped onto transmission lines) is increasing. IPPs are encouraged to connect into the transmission grid to provide greater access to "green" power sources. Control of these power producers is in the hands of the owners of the IPP, and there remains a degree of uncertainty about their availability to meet the load demand. Also, some IPPs have drop-out policies in order to protect the IPP equipment. The dropping out of these sources at critical loading conditions adds another dimension to the responsibility of the grid to meet the load demand at all times.

Despite the uncertainty regarding the eventual advantages of this concept, the immediate impact on system operation was significant. In general, nonutility-owned generation on the distribution system introduces several problems. The generation is not directly dispatched by the utility control center and therefore is not easily included in the generation/load schedule. The primary aim of the IPP is of course profit, and quality of service becomes a secondary issue. As a result, stringent contractual obligations are imposed to allow the utility dispatcher to effect a reliable operating agenda. A promising solution to this problem is being implemented by many utilities. This is the creation of microgrids [7] in the distribution system. Such grids are basically self-sufficient islands where the generation and load are matched. In the event of trouble in the distribution system, the microgrids are automatically invoked, maintaining service to a majority of customers.

In addition to the dispatching problem, there are important relaying problems introduced. With generators dispersed throughout a distribution system, the direction of fault current becomes unpredictable. Replacing time-delay or instantaneous overcurrent relays with directional relays wherever a problem of directionality exists is of course possible at a significant cost. In addition, it is essential that the IPP and the utility cooperate fully to ensure that the magnitude of the fault contribution is known and the timing of the installation is known. Another solution, but one that must be covered by contract, is the permission for distributed generators to be tripped at the first indication of a fault, before the distribution relays can operate. This can be done by local relays or a transfer trip scheme involving a communication system.

10.7   Islanding

When groups of machines in a power system lose synchronism, they separate from the power system, and system islands are formed. Each resulting island should have a balance between generation and load in the island, so that islands can continue to operate at nominal frequencies, and in time can be reconnected into an integrated network. In practice, such a balance between loads and generation within the islands may not exist, in which case some additional corrective actions (such as load or generation shedding) must be employed to achieve some sort of normalcy within the islands.

As is seen in Chapter 11, island formation and restoration are accomplished by automatic load shedding and restoration schemes. In addition, operating procedures are determined from offline planning-type simulations for resynchronizing the islands. As with many other relaying functions, the power system is often in a state which has not been anticipated during the planning studies, and consequently some of these schemes may not operate as expected in practice.

In recent years much emphasis has been placed on system integrity protection schemes (SIPS), also known as remedial action schemes or system protection schemes. These schemes take into account the prevailing power system conditions by obtaining measurements of key parameters in real time. The most promising technique is to use wide-area measurements (such as synchronized phasor measurements) and determine appropriate islanding and restoration strategies based upon these measurements. However, this whole field is in its infancy, and much theoretical work needs to be done (and is being done) before practical solutions based on real-time measurements can be found.

10.8 Blackouts and restoration

Wide-area electric power outages, commonly referred to as blackouts, have been more prevalent in recent years and are receiving worldwide attention, both in the popular news media and in engineering organizations. Analysis of the events leading to and resulting in a power outage over an extensive geographic area concludes the following sequence [8]:

1) The system is stressed beyond normal conditions. This can be the result of unusual weather resulting in unanticipated power demand, unplanned or abnormal equipment outages, or heavier than normal transmission line loads.
2) Generator or transmission line outages as a result of a fault. Normally the system is designed to withstand such outages, but coupled with point (1) the entire system is further stressed.
3) Inappropriate control actions. The control operators are trained to respond to abnormal system stress and performance, but if their information is inaccurate or incomplete, they can respond incorrectly exacerbating the situation.
4) Cascading. As a result of points (1), (2), and (3), additional system elements are lost.
5) Loss of synchronism.
6) Blackout.

Analysis of blackouts has revealed that they are never the result of a single event. It is the outcome of a stressed system, correct or incorrect equipment or relay operations, and possible personnel error. The fundamental system design criterion of being able to withstand one or more outages and the relay philosophy of high dependability at the cost of somewhat reduced security may, under abnormal system conditions, be the contributors to further stress and eventual outage.

It should be further noted that the effect of open access and deregulation or re-regulation may be a factor in the higher incidence of blackouts. In recent years, there has been a shift in the load–generation relationship with competition among power producers and an accompanying focus on profitability that is replacing concerns for high reliability and good service. This has resulted in a decrease in infrastructure improvements and restrictions on information exchange.

Among the many blackouts that have occurred, several offer interesting and instructive information that should be studied to avoid or minimize future occurrences. The detailed sequence of operation is available in the reports cited in the references, but the description in the following text presents the initiating cause, overall impact, and lessons learned.

10.8.1 Northeast blackout, 9 November 1965

Popularly referred to as the "granddaddy" of them all, this power failure affected 50 million people from New York City up through New England and parts of Canada [9]. The primary cause was the setting of remote breaker-failure admittance relays set 9 years earlier to provide remote breaker-failure protection. On 9 November 1965, a phase angle regulator was adjusted in upper New York to correct voltage, resulting in an increase in load from the Adam Beck plant in Niagara feeding five lines to Ontario, Canada. The increase in load caused the remote breaker-failure relays to trip one line, increasing the load on the remaining four lines which then also tripped. The resulting load distribution back into the United States caused transmission lines to overload and generators to trip by loss-of-field or inadequate governor response, eventually blacking out the entire Northeastern United States. The lessons learned resulted in a complete review of utility planning practices, establishing the NERC, and establishing underfrequency load-shedding standards throughout the United States and Canada, reevaluating load ability of admittance relays and relay performance during system stress.

10.8.2 New York City power failure, 13/14 July 1977

This was a classic case of multiple events, each of which was rare and improbable, but all combined to black out New York City for 2 days resulting in riots, looting, and millions of dollars lost in business and social impact [10]. Two lightning storms, 10 min apart resulted in losing four 345 kV lines feeding the city from Westchester. Two of the lines tripped due to carrier problems, a line from New England tripped due to a damaged directional relay contact, one Indian Point nuclear unit tripped due to an incorrect breaker-failure timer setting, reclosing was prevented on all lines due to restrictive check-synchronizing relay settings, and the control room operators in separate control rooms were not fully aware of the situation and delayed manually shedding load. As a result of this blackout, tower grounding counterpoises were improved, the two control rooms were reconfigured into one, reclosing settings were revised, load-shedding procedures were reviewed and are now tested periodically, and barge-loaded generators on the East River surrounding New York City are started as soon as weather reports indicate impeding weather problems.

10.8.3 French blackout, 19 December 1978

At 8:00 a.m., there was one operator on shift waiting for a shift change at 8:30 a.m. Load in excess of forecast accompanied by deficiencies in generation capacity led to excessive reactive load and low voltage on generator auxiliary buses tripping several units. Underfrequency load shedding and planned system islanding did not operate as expected and planned. As a result, better planning of reactive supply and a revision in automatic distribution transformer load tap changing have been instituted. Islanding is now automatic and load tap changing is manually directed from the control center when the system is stressed.

10.8.4 Western US system breakup, 2 July 1996

Temperature and load were very high in the Western United States, as were power exports from the Pacific Northwest to California and the AC and DC ties [11]. A flashover to a tree on a 345 kV line tripped the line, and a ground unit on the adjacent line misoperated. A special stability control scheme tripped two units when two of the three lines were out, which should have stopped the cascade. However, a voltage depression followed by a zone 3 operation due to load led to controlled and uncontrolled load shedding, which led to a breakup into five small islands and blackout in five western states. Recognizing that a power system can never be 100% reliable, and trying too hard may be too costly, the best approach was considered to be defense in depth. This includes a vigorous tree-trimming program, a thorough review of system restoration and protective relaying settings including undervoltage load shedding, automatic reactive power compensation and improved online security assessments, and data exchange among utilities and control centers.

10.8.5 US/Canada blackout, 14 August 2003

This event started in the early afternoon of a summer day with no unusual weather or line loadings. A generator in Ohio was shut down according to schedule [12]. The change in power flows and voltage resulted in a 345 kV line sagging into a tree despite operating within its nominal rating. This resulted in other line overloads and voltage reduction. Unfortunately, the control operators responsible for this area did not recognize the urgency of the situation. A supervisory control and data acquisition (SCADA) system was not operating, and a zone 3 relay on a key interconnection operated on load as well as other lines relaying out, creating islands within the Eastern Interconnection. The largest island was undergenerated and collapsed despite extensive underfrequency load shedding. Voltage collapse occurred throughout the area. A smaller, overgenerated island survived and was used as a starting point to enable restoration. Over 50 million people and the entire Northeastern United States and part of Canada made this the largest blackout ever experienced, surpassing the 1965 event. Analysis of all of the events was initiated by NERC and joint US/Canadian committees resulting in a series of recommendation concerning relaying settings, SCADA performance, and operator training.

10.8.6 Italian power failure, 28 September 2003

The blackout was initially triggered by a tree contact on a 380 kV line. Several attempts to reclose automatically failed as did a manual attempt 7 min later. The Swiss control center tried to relieve overloads in Switzerland by reducing Italian imports, but it was insufficient. This further overloaded other Swiss lines causing interconnections between Switzerland and Italy to trip, resulting in low voltage in Northern Italy and tripping several Italian power plants. Two-and-a-half minutes after being disconnected from Europe, Italy suffered the severest blackout in its history.

10.8.7 Restoration

The societal cost of a massive blackout is estimated to be in the order of $10 billion per event as described in the "Final Report of the August 14, 2003 Blackout in the United States and Canada." The modern grid will be far less vulnerable to such occurrences due to use of improved multiple technologies to identify and immediately response to threats to reliable service.

In addition, blackouts can be made less likely if the system design is such that extreme system stress is avoided. Designing a ductile system instead of a brittle system would be very helpful [13]. Although random events are beyond control, loss of reactive power and its effect on voltage must be considered. Recent advances in digital devices make more intelligent control possible as has the introduction of SIPS discussed in Section 11.2 (Chapter 11).

Planned islanding, that is, separating segments into areas with balanced load and generation, would allow much of the system to maintain service and reduce the blacked out area to a minimum and ease the resynchronizing effort [13].

10.9 Summary

In this chapter, we have discussed system concerns which impact protection practices. Chapter 11 presents the specific relaying that is required in response to these concerns. The chief concerns in most power system protection applications are the following issues (individually or in combination):

- Loss of synchronism from rotor angle instability.
- Voltage collapse due to voltage instability.
- Blackouts due to insufficient balance of load and generation causing unrecoverable frequency decline.

A summary of some of the world's most catastrophic blackouts has been presented, primarily to examine the relaying involved, and describe the lessons learned.

Problems

10.1 Prove Equation (10.5). To do this, first determine the terminal voltage of the generator, given that its internal voltage phase angle is $\pi/2$ with respect to the infinite bus. Then find the line current and, using Equation (10.4), prove the required result.

10.2 Show that Equation (10.9) follows from Equation (10.8), with the definitions given in Equation (10.10). What is the result in the R–X plane, if the circle in the P–Q plane passes through the origin?

10.3 Show that, as t tends to ω, the frequency of the power system reaches a limit given by Equation (10.33).

10.4 For a system with a capacity of 10 000 MW, an average power factor of 0.8, and an average H constant of 15 s, calculate the rate of change of frequency decay upon a loss of a 1000 MW generating unit. Use the exact formula, Equation (10.26), and compare your result with that obtained with the approximate formula, Equation (10.35).

10.5 If the load decrement factor d for the system of Problem (10.4) is 2.5, what is the final frequency of the system, assuming no load shedding takes place?

References

1 Stevenson, W.D. Jr. (1982). *Elements of Power System Analysis*, 4e. New York: McGraw-Hill.

2 Elmore, W. (ed.) (2004). *Protective Relaying Theory and Applications*, 2e. New York, NY: ABB, Marcel Dekker Inc.

3 IEEE Committee Report (1990). *Voltage Stability of Power Systems; Concepts, Analytical Tools, and Industry Experience*, IEEE publication 90 0358-2-PWR, New York. Piscataway, NJ: IEEE.

4 Taylor, C.W. (1994, Technology & Engineering). *Power System Voltage Stability*, 273. McGraw-Hill.

5 Begovic, M.M. and Phadke, A.G. (1990). Dynamic simulation of voltage collapse. *IEEE Trans. Power Syst.* **5** (4): 1529–1534.

6 FERC (1996). *FERC Order 888, Final Rule, open access docket RM95-8-000*

7 Hatziargyriou, N., Asano, H., Iravani, R., and Marnay, C. (2007). Microgrids. *IEEE Power Energy* **5** (4): 78–94.

8 Horowitz, S.H. and Phadke, A.G. (2006). Blackouts and relaying considerations. *IEEE Power Energy* **4** (5): 60–67.

9 US Federal Power Commission (1965). *North East Power Failure*. Washington, DC: US Government Printing Office.

10 Federal Energy Regulatory Commission (1978). *The Con Edison power failure of July 13 and 14, 1977*. Final staff report. US Department of Energy.

11 Taylor, C.W. and Ericson, D.C. (1997). Recording and analyzing the July 2 cascading outage. *IEEE Comput. Appl. Power* **10** (1): 26–30.

12 US-Canada Power System Outage Task Force (2003). *Final Report*, www.doe.gov.

13 Horowitz, S.H. and Phadke, A.G. (2003). Boosting immunity to blackouts. *IEEE Power Energy* **1** (5): 47–53.

11

Relaying for System Performance

11.1 Introduction

The traditional goal of protective devices is to protect power system equipment. This is achieved by detecting a fault or undesirable performance and taking corrective action which in most cases involves tripping appropriate circuit breakers. Thus, one could say that the principal task of relays is to disconnect equipment or subsystems from the overall power system. Normally, this is the appropriate action. Systems are designed to be robust, that is, to withstand the removal of one or several elements without unduly stressing the overall system. This was discussed in Chapter 1 and is the basis for preferring dependability over security. It is as well to recognize that the built-in, inherent strength of the power system is the best defense against catastrophic failures. However, if the system is already stressed for whatever reason, such as equipment outages, heavier than normal loads, extreme weather, and so on, this corrective action may exacerbate the situation and result in wide-area outages. The condition that will result in a stressed system that eventually could become a blackout is covered in detail in Chapter 10. In this chapter, we shall discuss the specific relay actions that are designed explicitly to avoid or minimize such outages.

11.2 System integrity protection schemes

As discussed in the earlier text, a conventional protection scheme is dedicated to a specific piece of equipment (line, transformer, generator, bus bar, etc.). However, a different concept must be developed which would apply to the overall power system or a strategic part of it in order to preserve system stability, maintain overall system connectivity, and/or avoid serious equipment damage during major events. For several years, the concept was called special protection system (SPS) or remedial action system (RAS). In 1988, a survey by Conférence Internationale des Grands Réseaux Électriques à Haute Tension (CIGRE) reported a total of nine schemes in service in 18 utilities throughout the world. This survey was updated in 1996 [1] by a joint working group of CIGRE and Institute of Electrical and Electronics Engineers (IEEE), and it was recognized from the utilities' response that the schemes were no longer "special" but were in fact included in virtually every utility's overall protection practice.

Power System Relaying, Fifth Edition. Stanley H. Horowitz, Arun G. Phadke and Charles F. Henville.
© 2023 John Wiley & Sons Ltd. Published 2023 by John Wiley & Sons Ltd.

In 2002, a joint working group of the Power System Relaying and Control Committee and CIGRE updated this survey [2] and renamed the concept system integrity protection scheme (SIPS) with the following definition: "To protect the integrity of the power system or strategic portions thereof, as opposed to conventional protection systems that are dedicated to specific power system elements." SIPS was intended to be broader than just SPS and RAS, since other system protection schemes such as underfrequency load shedding and out-of-step protection were not included in some definitions of SPS and RAS.

Typical of such schemes are:

1) Underfrequency load shedding.
2) Undervoltage load shedding.
3) Out-of-step tripping and blocking.
4) Congestion mitigation.
5) Static var compensator (SVC)/static synchronous compensator (STATCOM) control.
6) Dynamic braking.
7) Generator runback.
8) Black start of gas turbines.
9) System separation.
10) Load shedding to prevent equipment damage due to overload.

In large synchronously interconnected regions such as those that exist in North America, regional reliability organizations such as the Western Electricity Coordinating Council take a special interest in SIPS. The reliability authorities may also issue regulations governing the application and operation of these schemes. They may also review the application, design, and operation of proposed SIPS to make sure that they are reliable, and that failure of a SIPS to operate when required in one operating authority does not adversely impact any of the neighboring authorities. The IEEE has issued a standard guide for engineering, implementation, and management of SIPS [3].

It is beyond the scope of this book to describe the details of all types of SIPS, but the reader should be aware that they are active, effective, and should be studied. However, some of the more popular schemes are covered in the following text.

11.3 Underfrequency load shedding

Section 10.4 (Chapter 10) derived the equations governing the frequency decay following a loss of generation or increase of load. For gradual changes, the governor response is sufficient to maintain synchronism, but with sudden loss of generation the governors are not fast enough, and if all generators are operating at maximum capacity, spinning reserve is not available and it is necessary to automatically drop load. The load-shedding relays may be electromechanical, solid-state, or computer-based. The measuring element senses a frequency equal to its setting, and will operate after a certain amount of time has elapsed after the frequency passes through its setting on its way down. The relay time, the delay time, and the breaker tripping time all add up to a delay of the order of 10 cycles or more. If the rate of change of frequency can be estimated, as with Equation (10.35) (Chapter 10), then the frequency at which the load is actually tripped can be determined. The setting of the next step of the load-shedding relay can then be made with a certain safety margin. This procedure is illustrated by the following example.

Example 11.1 For a system with 10 000 MW connected load, a generating station delivering 1500 MW is lost due to a contingency. If the aggregate inertia constant is 5 s, determine the settings of the underfrequency relays which will accomplish a load-shedding plan to drop 1500 MW of load in two steps of 750 MW each. The total relay plus interrupting time of 15 cycles may be assumed at each step.

We will set the first step of the load shedding at 59.5 Hz. The load excess factor immediately after the generating station is lost is

$$L = \frac{10\,000 - 8500}{8500} = 0.1765.$$

The average rate of frequency decline in the range 60–59.5 Hz is given by Equation (10.34) (Chapter 10), assuming a power factor of 0.85, as

$$R = \frac{0.85 \times 0.1765}{5.0} \times \frac{59.5 - 60.0}{1 - 59.5^2/60.0^2} = -0.9039 \text{ Hz s}^{-1}.$$

At this rate, 59.5 Hz will be reached in (0.5/0.9039) or 0.553 s. With a tripping delay of 15 cycles (0.25 s), the frequency will decline further by 0.25/0.9039 or 0.226 Hz. Thus, the load of 750 MW will be shed at an actual system frequency of (59.5/0.226) or 59.274 Hz. The actual load shedding of 750 MW will thus occur (0.726/0.9039) = 0.8032 s after the generating plant is lost.

Allowing for a safety margin of 0.2 Hz, the next step of load shedding can be set at 59.3 − 0.2 = 59.1 Hz. Returning to the original disturbance, the excess load ratio now becomes

$$L = \frac{9250 - 8500}{8500} = 0.0882.$$

The average rate of frequency decline in the range 59.27–59.0 Hz is

$$R = \frac{0.85 \times 0.0882}{5.0} \times \frac{59.0 - 59.27}{1 - 59.0^2/59.27^2} = -0.4454 \text{ Hz s}^{-1}.$$

At this rate, 59.1 Hz will be reached in (0.174/0.4454) = 0.39 s. The load will be shed in an additional time of 15 cycles; hence, the frequency at which the load will be shed will be 59.1 − 0.4458 × 0.25 = 59 Hz. The two-step load shedding will be accomplished in a total time of (60.0 − 59.274)/0.9039 + (59.274 − 59.0)/0.4458 = 1.409 s. At this point, a complete balance between the load and generation will be established, and the frequency will begin to return to its normal value of 60 Hz.

The actual relay characteristic furnished by the manufacturer must be consulted for the detailed design of the load-shedding scheme. Additional details of the procedure to be followed in practical cases can be found elsewhere [4].

The size and steps of the load-shedding scheme must be established by studying the possible system scenarios which may require load shedding. Consider a system with an installed plant capacity of 10 000 MW, with the largest single-unit capacity being 1000 MW, there being up to two such units in one generating station. Let us further assume that the system imports 600 MW from its tie lines with the neighbors. A credible set of contingencies could be the loss of one or two generators, and the tie line. Thus, the steps in which the generation could be lost would consist of:

1) 600 MW (loss of the tie line).
2) 1000 MW (loss of one generator).
3) 1600 MW (loss of one generator and the tie line).
4) 2000 MW (loss of two generators).
5) 2600 MW (loss of two generators and the tie line).

In order to meet these contingencies, one could design a load-shedding scheme of five steps:

1) 600 MW.
2) 400 MW.
3) 600 MW.
4) 400 MW.
5) 600 MW.

Appropriate steps should be selected so that 2600 MW can be dropped before a dangerously low frequency of 57 Hz is reached. The general principles of determining the relay settings are illustrated in Example 11.1.

The load-shedding relays are installed in distribution or subtransmission stations, where feeder loads can be controlled. Loads throughout the system are classified as being critical or noncritical. Only noncritical loads are subject to shedding. The "firmness" of the load must also be considered. Commercial and residential loads are daily and seasonally variable, ranging from 25 to 100% of full load. Industrial loads are firmer, with smaller variations. Shedding more than the minimum desired in each step may be applied to compensate for variability of some loads. Load shedding schedules are coordinated between synchronously interconnected regions to avoid overstressing interconnections.

Feeders or stations which meet the power rating needed for load-shedding steps are identified, and are assigned to appropriate steps of the load-shedding relays. In some instances, load shedding may be controlled from a central location under the control of the supervisory control system or by centralized SIPS for faster action in the case of severe generation/load imbalance. This is particularly useful when the load shedding is called for before the system is islanded, and the frequency begins to decline.

Load that has been shed must be restored when the system frequency returns to normal. Automatic load-restoration systems are in service, which accomplish this function. Load restoration must also be done in steps, with sufficient time delays, so that hunting between the load-shedding and restoration relays does not occur. There is divided opinion as to the order in which the load is restored. Either the first-in-first-out or the first-in-last-out strategy may be used.

11.4   Undervoltage load shedding

The problem of voltage stability has been discussed in Section 10.3 (Chapter 10) and some mitigation techniques are described in Reference [5]. This problem is relatively new to system protection due to the changes in the characteristic of the total system load and increased loading of transmission lines. The initial application of electric energy involved heating and lighting, that is, primarily resistive load. When the balance between generation and load was upset, it was the system frequency that changed, and either an increase in generation or underfrequency load shedding would restore the frequency. However, as air conditioning and other small motor appliances proliferated, it was the system voltage that was affected. A motor is essentially a constant kilovolt-ampere device so the current would increase when the voltage went down, decreasing the voltage even further. Power system deregulation has also resulted in increased power transmission from distant sources, thereby increasing reactive power consumption of transmission lines and increased system stresses, especially during contingencies.

As an alternative to building more transmission lines, the solution, of course, is to either add additional reactive support or remove the load. The SVC and STATCOM described in Sections 9.13 and 9.14 (Chapter 9) automatically provide the additional reactive support to the level of their capability. If it does not correct the situation sufficiently, an undervoltage load-shedding (UVLS) scheme must be employed. Traditionally, UVLS schemes may be applied in a distributed flat architecture, that is, the system functions locally and independently, much the same way as an underfrequency load-shedding scheme operates. The measurement and disconnecting devices are located within a given substation without any recourse to communication and/or a control center.

For localized UVLS schemes, it is desirable to use different bus voltages within the same high-voltage level, as the true system voltage condition will appear on all buses of the station. Connecting the relays to sense the voltage on all three phases ensures that a fault on one phase is not misinterpreted as a system undervoltage event. Even with three-phase sensing, time delays are normally required to avoid undesired load shedding due to voltage drops during three-phase faults. In some systems, shunt capacitors, SVCs, and STATCOMs bring the voltage level of the nose of the curve shown in Figure 10.11 (Chapter 10) to a level very close to normal operating voltage. In such cases, localized UVLS schemes may not be fast enough to avoid voltage collapse. In many cases, a centralized hierarchical SIPS may be required.

For schemes with hierarchical characteristics involving multiple stations, additional factors such as reliability of the communication system, overall throughput and added monitoring, and alarming systems must be considered. Whether the UVLS is local or hierarchical, the design must maintain operating flexibility for measuring, arming, performing, and maintaining security.

The phenomenon of voltage stability is very often illustrated by the so-called nose curve, as shown in Figure 10.10 (Chapter 10), where the receiving-end voltage in a transmission system is plotted against the receiving-end real power. The nose curve is a static description of the transmission system capability. If dynamic aspects are taken into consideration, which is necessary for a complete analysis, the nose curve has to be modified.

11.5 Out-of-step relaying

In Section 10.2.1 (Chapter 10), we discussed the system phenomenon associated with steady-state stability. We now are interested in determining the response of a distance relay during power swing conditions. Referring to Figure 10.6 of Section 10.2.2 (Chapter 10), consider the distance relay used to protect line 1–2, at the bus 1 end. We will assume that the receiving-end system consists of additional lines in series: 2–3 and 3–4. We will also reintroduce the resistances of the transmission lines. The impedances are now as shown in Figure 11.1. Each of the lines is assumed to be protected by a three-zone stepped distance mho relay. For the sake of clarity, characteristics of only two sets of relays, R_{12} and R_{23}, are shown in Figure 11.1. Also shown is the excursion due to a stable and an unstable swing. It can be seen that both the swings may penetrate one or more zones of one or more distance relays. If no other measures are taken, and the impedance stays in a zone long enough, one or more relays in this system will trip. It should be noted that when the power system is more complex, the stable and unstable swings are no longer along the arcs of circles. More often, they have complex shapes, an example of which is shown in Figure 11.1 as "other swings."

There are two aspects to the problem of out-of-step relaying. First, a trip under stable swings is unwarranted, and should be avoided at all costs. Second, during an unstable swing, a trip is usually required, but it should be permitted at locations of choice. Thus, an uncontrolled trip during unstable swings should also be avoided. Both of these requirements are satisfied by the out-of-step relays. Consider the power system shown in Figure 11.2. The generator output powers are approximately equal to the total load on the system. Let us assume that G_1 is approximately equal to the sum of two loads L_2 and L_4, while G_3 is approximately equal to L_5. If this system becomes unstable due to some transient event, it would be desirable to break the system along the dotted line, so that the two halves of the system will have an approximate balance between generation and load. It is thus desirable to detect the out-of-step (unstable) condition at all locations, and block tripping for all lines, except for lines 2–3, 2–5, and 4–5, where tripping should be initiated. The out-of-step relaying tasks can be broken into two stages: one of detection and the other of

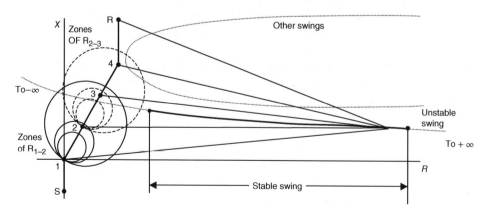

Figure 11.1 Impedance loci for stable and unstable swings and their impact on step-distance relays.

tripping or blocking. The detection function is served by out-of-step relays, which consist of multiple characteristics. Remember that a fault causes the apparent impedance to come into the zones of a step-distance relay, and so does a stability swing. We must be able to distinguish between a fault and a swing. The primary point of distinction is the speed with which the two phenomena occur: a fault trajectory moves into a zone almost instantaneously (within a few milliseconds), while a stability swing moves more gradually. If we

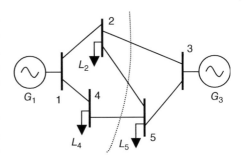

Figure 11.2 Out-of-step blocking and tripping functions for relays.

assume an oscillation frequency for the swing of the order of 0.5 Hz as suggested by Example 11.1, the excursion from the farthest to the nearest point on the impedance trajectory would take one half the period of the 0.5 Hz oscillation, or about 1 s. We may thus time the passage of the impedance locus in the R–X plane, and if it travels relatively quickly, declare it to be a fault, and take the corresponding action. By contrast, if the transition is gradual, we would declare it to be a stability swing. We then must determine if it is a stable or an unstable swing. This is usually done by simulating numerous stability swings of the system in question with a transient stability program. The impedance trajectory seen by the distance relay for each of the stable cases is examined. It will be found that all stable swings come no closer than a certain minimum distance from the origin of the R–X plane. This suggests that a zone with a setting smaller than this minimum distance should be chosen to detect an unstable swing. If the trajectory is detected to be that due to a stability swing, and further if it encroaches upon this small impedance zone, an unstable swing can be concluded. If a stability swing does not encroach into this zone, the swing is determined to be a stable swing. Upon detecting an unstable swing, an out-of-step condition is declared, and appropriate tripping or blocking action may be initiated depending upon the requirements of the system-splitting strategy established according to the considerations discussed with reference to Figure 11.3.

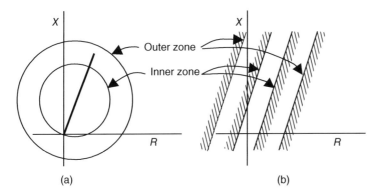

Figure 11.3 Relays for detecting out-of-step condition using distance relays and timers.

Typical out-of-step relay characteristics are shown in Figure 11.3. Figure 11.3a shows the use of two concentric impedance circles, while Figure 11.3b shows the use of straight lines, known as "blinders" (the blinder characteristics are obtained using a rotated reactance characteristic discussed in Chapter 5). The inner zone is used to detect an unstable swing as discussed in the earlier text. A timer is set slightly greater than the longest dwell time of an unstable swing between the inner and the outer zones. The outer zone is used to start a timer, and if the impedance locus crosses the inner zone before the timer runs out, an unstable swing is declared. If the inner zone is never entered after the outer zone is entered, a stable swing is concluded. No tripping is permitted in the case of a stable swing, while tripping or blocking is initiated after an unstable swing is detected. If the outer and inner zones are entered in very quick succession, a fault is declared.

It should be noted that the stability swing is a balanced phenomenon, and hence any zero- or negative-sequence currents accompanying the transient should disable the relaying out-of-step logic, and revert to fault-detecting logic.

Example 11.2   For the system of Example 10.1 (Chapter 10), determine appropriate zone settings for an out-of-step relay to be located at bus 1. Assume that the "system" Thévenin impedance of 0.2 pu is actually made up of three line sections 1–2, 2–3, and 3–4, having positive-sequence impedances of 0.1, 0.05, and 0.05 pu (Figure 11.4 shows the system and R–X diagram). Assume that the line resistances can be neglected. All stable swings are found to produce rotor angle excursions ranging between 15° and 110°, while unstable swings are expected to produce pole-slipping conditions.

We will assume equal voltages at the sending and receiving ends of the system. Noting that the impedance line S–R is vertical (i.e., there is no resistance in the lines), we can calculate the distance between 1 and O, as the subtended angle changes between 15° and 110°,

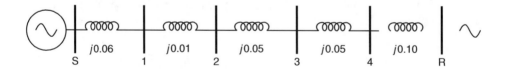

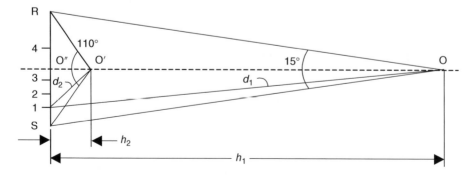

Figure 11.4   System and R–X diagram for Example 11.2.

by trigonometry. The height of the triangle with a base of 0.27 pu (total system reactance) for these two angles is

$$h_1 = \frac{0.27/2}{\tan 15°/2} = 1.025 \text{ pu and } h_2 = \frac{0.27/2}{\tan 10°/2} = 0.094 \text{ pu.}$$

The distances from point 1 for the two conditions are now determined as

$$d_1 = \sqrt{1.025^2 + (0.135 - 0.06)^2} = 1.028 \text{ pu}$$

and

$$d_2 = \sqrt{0.094^2 + (0.135 - 0.06)^2} = 0.12 \text{ pu.}$$

Parameter d_2 determines the nearest excursion of a stable swing for the relay located at bus 1. Allowing for a safety margin, a setting for the inner zone of about 0.1 pu may be adopted. The outer zone may be set at 0.15 pu. Although oscillation periods of about 2 s are expected for this case, the timer settings to determine whether a fault or an unstable swing exists may be set at a short value, say about 20 ms. If the transition between the outer and the inner zones is faster than 20 ms, a fault would be indicated. Another timer setting— say about 100 ms—would help distinguish between a stable and an unstable swing. If the outer zone is entered, but the inner zone is not entered even after 100 ms have elapsed, a stable swing would be detected. If the outer zone is entered, and the inner zone is entered within 20–100 ms after that, an unstable swing would be detected and appropriate blocking or tripping commands issued.

It should be noted that the zone settings determined in the earlier text are on the system base. The conversion to primary ohms, and then to secondary ohms, must be made to determine the actual relay settings. Also, a supervision scheme must be provided, which would bypass the out-of-step condition detection, if an unbalance in the currents is detected.

It is conceivable that for some high-impedance faults, a fault may enter the outer zone, and stay there long enough, before evolving into a solid fault and entering the inner zone. If the timing of these events is accidentally matched to the out-of-step relay settings, such an event may be interpreted as an unstable swing. Of course, if the fault is an unbalanced fault, the out-of-step relay would not be operative. In any case, such a combination of events is extremely unlikely.

11.6 Loss-of-field relaying

The steady-state stability limit and its equivalence to reduction in field current of a generator have been discussed extensively in Section 10.2.1 (Chapter 10). It should be remembered that the critical conditions for the generator are the characteristics in the fourth quadrant of the $P–Q$ plane. There are two possible limiting conditions: the end-iron heating limit or the steady-state stability limit. The former is a characteristic of the generator, and is fixed for a given machine. The latter is dependent on machine reactance and the equivalent system impedance as seen at the machine terminal. Whichever of these limits is "smaller" in the *P–Q* plane determines the limit which must be honored in operating the machine (see Equation [10.5] where the machine impedance and the system impedance are seen to affect the radius of the limiting circle in the *P–Q* plane). A "smaller" limiting circle in the *P–Q* plane corresponds to a "larger" circle in the *R–X* plane, and that should be the setting used for the offset impedance relay used to detect the loss-of-field condition.

A two-zone offset impedance relay connected at the machine terminal as shown in Figure 10.4 (Chapter 10) is used to detect the loss-of-field condition. The inner zone is the actual limit (determined either by the end-iron heating condition or steady-state stability condition), and a somewhat greater outer zone is set to create an alarm that the limit (i.e., inner zone) is in danger of being breached. The alarm usually alerts the generator operator that some corrective action needs to be taken before the actual limit is reached.

It should also be remembered that if the operative limit is due to steady-state instability, it may change as the power system changes, because a weaker power system means increased equivalent impedance. Thus, as a power system begins to lose transmission facilities during a cascading event, the loss-of-field relay setting determined when the system was normal may not be appropriate for the prevailing system conditions, and may lead to an unstable condition without a warning from the loss-of-field relay. The only solution to this problem is to use an adaptive adjustment (Section 11.7) to the loss-of-field relay depending upon the system equivalent impedance.

11.7 Adaptive relaying

Adaptive relaying is a concept that recognizes that relays that protect a power network may need to change their characteristics to suit the prevailing power system conditions [6, 7]. Normally, a protective system responds to faults or abnormal events in a fixed, predetermined manner that is embodied in the characteristics of the relays. It is based upon certain assumptions made about the power system. Thus, one may have assumed a level of load current which normally would not be exceeded at the relay location. However, there may be occasions when the actual load exceeds this assumed value and the setting of the relay(s) is inappropriate. It is not possible to ensure that all relay settings will be appropriate for the power system, as it operates at any given instant or evolves in time.

Even if it were possible to foresee all contingencies and define boundaries of required performance for a relay, the setting to meet this condition may be less than desirable for the normal system parameters. It may, therefore, be desirable to make relays adapt to changing conditions. Of course, not all features of all relays must be adaptive. Adaptive elements exist in many modern electromechanical or solid-state relays. For instance, by automatically adjusting its speed of operation to the actual operating current, a time overcurrent relay is adapting to changing fault locations (see Section 4.3 and Figure 4.7 in Chapter 4).

A harmonic restraint transformer differential relay can recognize the difference between energizing the transformer and a fault within the transformer as discussed in Section 8.4 (Chapter 8).

Perhaps, in a less obvious fashion, adaptive relaying and control can be seen in the complex circuitry associated with many circuit breaker bypass schemes as discussed in Section 1.3 (Chapter 1). Current transformer and tripping circuits are reconfigured using auxiliary switches to make the measuring and tripping circuits compatible at each stage of the switching procedure. Such examples support the view that we have, indeed, been enjoying adaptive protection and control, but they do not encompass the present excitement about the full range of adaptive relaying made possible by computer relays. Existing practices represent permanent provisions for well-considered scenarios or hard-wired solutions to a variety of operating states of the power system.

To completely describe the concept of adaptive relaying, it is necessary to include the possibility that the protection systems will permit making automatic changes within the protection system itself as the power system undergoes normal or abnormal changes during its course of operation. With this concept of adaptive relaying, therefore, it is not necessary to anticipate all contingencies nor is it necessary to make compromises in setting relays. Adaptive relaying is defined as follows: "it permits and seeks to make adjustments automatically in various protections in order to make them attuned to prevailing power system conditions."

Several other adaptive relaying ideas can be found in the literature. Among these is the idea of adaptive out-of-step relaying. This falls under the general problem in trying to predict the instability of a power swing as it is developing. This is discussed in detail in Chapter 10. With accurate synchronized phasor measurements from several buses, the goal of real-time instability predictors seems achievable (as discussed in Chapter 13). One could then foresee the use of adaptive out-of-step relays which would change their function from tripping to blocking, or vice versa, as a power swing develops.

The adaptive relaying philosophy can be made fully effective only with digital, computer-based relays. These relays, which are now the protective devices of choice, contain two important features that are vital to the concept:

- Their functions are determined through software.
- They have a communication capability which can be used to alter the software in response to higher-level supervisory software or under command from a remote control center.

It is, of course, necessary to recognize that any relay or communication system may fail at some time and appropriate fallback positions and safety checks must be built in to adaptive relays.

11.7.1 Examples of adaptive relaying

All relay settings are a compromise. In every phase of the development of a system's protection, a balance must be struck between economy and performance, dependability and security, complexity and simplicity, speed and accuracy, and credible versus conceivable. The objective of providing adaptive relay settings or characteristics is to minimize compromises and allow relays to respond to actual system conditions.

11.7.1.1 Load effect

The ability to detect low-grade faults is limited by the necessity of not tripping during emergency load conditions. On distribution and subtransmission systems, overcurrent relays must be set sufficiently above load and below fault currents to allow for both dependability

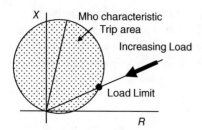

Figure 11.5 Load encroachment on the tripping characteristic of a mho relay.

and security. On higher-voltage systems, distance relays must not encompass the load impedance (Figure 11.5).

11.7.1.2 End-of-line protection

For instantaneous overcurrent relays, when the remote breaker opens, the current the relay sees will change. In the nonadaptive case, the relay must be set for the worst case currents, that is, for the system configuration that provides the maximum fault current. For any other configuration, the current may be less and the instantaneous relay may not pick up (see Example 4.6 in Chapter 4). With adaptive relaying, the system configuration can be communicated to all terminals and the resulting current change can be determined and the relay setting changed accordingly.

In the case of impedance relays, the opening of the remote breaker can be communicated to the remote relays and the zone 1 setting can be changed to overreach, thus including the 10–20% end of the line that was heretofore protected in zone 2 time. This is used in Europe and referred to as zone acceleration.

11.7.1.3 Multiterminal transmission lines

Probably one of the best known relay problems is the multiterminal transmission line (Figure 11.6) and the compromise settings that must be made, as discussed in Section 5.8 and Example 5.6 (Chapter 5). Zone 1 relays must be set without infeed. This reduces the reach when there is infeed, but it is in the safe direction. Zone 2 relays are set with all infeeds present. Again, this results in a greater overreach when there is no infeed, which is safe, but the overreach must be coordinated with all other zones to ensure correct coordination. With adaptive relaying, communicating the conditions at all of the

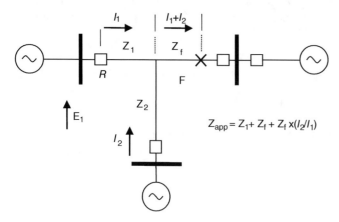

Figure 11.6 Three-terminal line. Apparent impedance seen by the relay is not the true impedance to the fault.

terminals allows the relays to adjust their settings to either ignore any infeeds when a related breaker is open or send the exact current infeed quantity and adjust the remote settings accordingly.

11.7.2 Concluding remarks

The field of adaptive protection is a growing field. A number of other subjects, such adaptive out-of-step relays, adaptive reclosing, and adaptive transformer protection, can be found in the technical literature. The interested reader should continue to watch for newer developments in this field.

11.8 Hidden failures

A study of major outages by the North American Electric Reliability Corporation (NERC) and an analysis of other events revealed that a defect (termed a hidden failure) has a significant impact on the possibility of false trips and extending a "normal" disturbance into a major wide-area outage. Hidden failures are defined as a "permanent defect that will cause a relay or relay system to incorrectly and inappropriately remove a circuit element(s) as a direct consequence of another switching event."

- Relay malfunction that would immediately cause a trip (Z_1 in Figure 11.7) is not a hidden failure.
- Relay contact that is incorrectly always open or closed (as opposed to changing its state as a result of some logic action caused by a system event which would initiate a trip). Timer contact that is incorrectly closed (T_2 or T_3); then closing Z_2 or Z_3 will result in an immediate trip, losing coordination time (Figure 11.7).
- Receiver relay (contact 85-1) in a carrier blocking scheme that is always closed, resulting in a false trip for an external fault.

None of these hidden failures becomes known until some other event occurs. This unwanted additional interruption is particularly troublesome when the system is already stressed by severe transmission overloads, insecure system topology, voltage difficulties, or decreased generation margins.

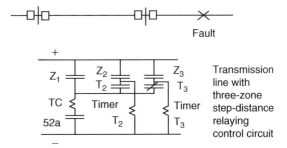

Figure 11.7 Hidden failure in a three-zone step-distance relay. The hidden failure is a permanently closed third zone timer contact.

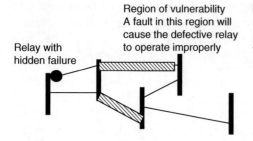

Region of vulnerability
A fault in this region will
cause the defective relay
to operate improperly

Relay with
hidden failure

Figure 11.8 Regions of vulnerability for a stepped-distance relay. A hidden failure such as a permanently closed timer contact would produce the region of vulnerability shown.

11.8.1 Region of vulnerability

The existence of a hidden failure does not, by itself, always result in an incorrect and unde-sirable operation. There is also a component that consists of a physical region in the network such that a given relay failure would not operate incorrectly for faults in that region. It is the combination of the hidden failure and a fault in its region of vulnerability that results in extending the area outage. Figure 11.8 shows a step-distance scheme with zones 1, 2, and 3 reach settings. As discussed in Section 5.2 (Chapter 5), zone 1 operates instantane-ously for faults within its setting and zones 2 and 3 operate after a time delay. Assuming the timer contact is incorrectly closed all of the time, a fault within the zone 3 setting would trip incorrectly without a time delay. In other words, there is a zone of vulnerability for which hidden failures may cause incorrect and undesirable trips, but beyond which they would have no effect. Relaying schemes operating on a differential principle such as phase comparison or transmission line current differential do not have zones beyond their trip-ping area.

11.9 Distance relay polarizing

Section 4.6 (Chapter 4) discusses the need to polarize a relay to differentiate between a fault in one direction and another. Distance relays can be self-polarized, that is, using the voltage from the same phase as the faulted phase, or cross-polarized, that is, using unfaulted phase voltages. Self-polarized relays may also use memory action to allow for the reduced voltage during a fault.

Consider the operation of a mho relay discussed in Chapter 5. One method of obtaining a mho characteristic is to generate two voltages from available voltage and current signals: an operating voltage and a polarizing voltage. For a distance relay connected at a bus, the avail-able signals are line terminal voltages and line currents. For this discussion, we will disre-gard the exact identity of the voltages and currents, assuming that signals appropriate for the fault being protected are chosen for this purpose (Section 5.4, Chapter 5). Let the voltage at the relay terminal be E and the current in the line be I. In a self-polarized mho relay, the operating and polarizing quantities are

$$E_{\text{op}} = E - I Z_{\text{c}},$$
$$E_{\text{pol}} = E,$$

$$(11.1)$$

where Z_c is the relay setting. The condition for the relay to operate is that the phase angle between the polarizing and operating quantities must be greater than 90°. The operating conditions can also be expressed in terms of operating and polarizing impedances obtained by dividing Equation (11.1) by the relay current:

$$Z_{op} = \frac{E}{I} - Z_c,$$
$$Z_{pol} = \frac{E}{I}. \tag{11.2}$$

The ratio E/I is known as the apparent impedance Z_{app} seen by the relay. Thus, Equation (11.2) can be rewritten as

$$Z_{op} = Z_{app} - Z_c,$$
$$Z_{pol} = Z_{app}. \tag{11.3}$$

Consider the phasors E and I shown in Figure 11.9a. The R–X diagram for these phasors is shown in Figure 11.9b. For the relay to operate, the angle between the impedances Z_{op} and Z_{pol} must exceed 90°, which defines the mho characteristic as a circle passing through the origin, and with Z_c as its diameter. This is the familiar mho characteristic passing through the origin.

A self-polarized mho relay does not operate reliably when the voltage at the relay location is low because of the fault being close to the relay location. In order to make the relay operate reliably, voltages from the unfaulted phases can be used to recreate the phasor relationship required to produce the mho characteristic. This is achieved by constructing a voltage at the relay location which reproduces the prefault voltage from voltages of the unfaulted phases. In effect, the source voltage of the faulted circuit is to be recreated in this fashion. Using this recreated voltage for the polarizing signal is equivalent to using the source voltage E_s for the system shown in Figure 11.9a. This would be correct if the prefault load is quite small, so that the prefault voltage at the relay location can be approximated by the source voltage.

Using the source voltage for the polarizing quantity, one gets

$$E_{pol} = E_s = E + I Z_s. \tag{11.4}$$

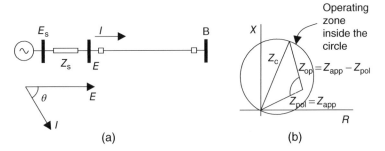

(a) (b)

Figure 11.9 (a) and (b) Phasor diagram and Mho characteristic of a self-polarized distance relay. Voltage at the relay location is used to polarize the relay.

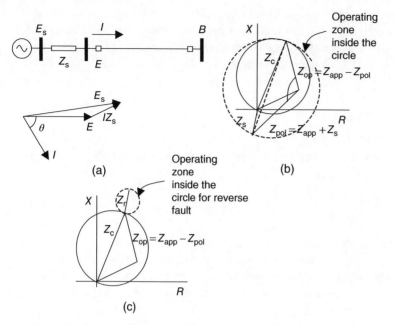

Figure 11.10 (a)–(c) Phasor diagram and R–X diagram for polarizing known as 100% cross-polarization.

Or, converting to the polarizing impedance:

$$Z_{pol} = Z_{app} + Z_s. \tag{11.5}$$

The corresponding phasor diagram and R–X diagram are shown in Figure 11.10. This type of polarizing is known as 100% cross-polarization, since the entire polarizing signal is derived from unfaulted phases. It is of course possible to have other combinations of phase voltage to achieve different types of polarization.

Note that the characteristic of Figure 11.9b is expanded, and thus has a reduced loadability. This is the principal effect of which one should be aware. If the power system becomes weaker, thereby increasing the system source impedance Z_s, the characteristic expansion (ballooning) would be greater. It should also be noted that there is an apparent loss of directionality in the expanded characteristic. This is of course not a correct interpretation of the case. One must redraw the phasor diagram resulting from a current flow in the reverse direction. When this is done, it can be shown that for reverse faults the operating zone is shifted to a circle at the remote end of the relay setting, having a diameter equal to the source impedance of the remote system. So, although the cross-polarized mho characteristic may balloon, it does not cause a failure of directionality.

Finally, it should be remembered that, in the case of a three-phase fault near the relay location, there are no unfaulted phase voltages available for cross-polarization. In this case, polarized (self or cross) distance relaying cannot be used reliably, and one must use overcurrent protection, or use a memory voltage to reconstruct the prefault voltage phasor.

11.10   Summary

In this chapter, we have responded to the system problems addressed in Chapter 10. The introduction of a new term, "system integrity protection," replaces the expression "special protection schemes," since these schemes have become fairly common in response to system stresses and there is little "special" about them. Underfrequency and undervoltage load shedding have become standard system responses to potential, or actual, system stresses and are presented in some detail.

The theory and definition of dynamic, steady-state, and transient stability were presented in Chapter 10, and in this chapter, we have explored the relays that can respond to these system events. System stress is often accompanied by unusual relay response and this effect on distance relays is covered. Two new concepts, adaptive relaying and hidden failures, have been presented in detail.

Problems

11.1   Prove that the impedance locus for a transient swing is an arc of a circle, when the sending-end voltage and the receiving-end voltage have unequal magnitudes. This is illustrated in Figure 11.1.

11.2   Assume that the transmission lines in Example 11.2 have an impedance angle of 80°. The sending-end and the receiving-end systems may be assumed to be purely reactive. Determine the zone settings for the out-of-step relay for this system.

References

1 Anderson, P.M. and LeReverend, B.K. (1996). Industry experience with special protective systems. *IEEE Trans. Power Syst.* **11** (3): 1166–1179.

2 IEEE Power System Relaying Committee (2002). *Industry Experience with System Integrity Protection Schemes*, Industrial survey on SIPS, IEEE PSRC System Protection Subcommittee.

3 IEEE Std C37.250 (2020). *IEEE Guide for Engineering, Implementation, and Management of System Integrity Protection Schemes*. New York NY: IEEE.

4 Stevenson, W.D. Jr. (1982). *Elements of Power System Analysis*, 4e. New York: McGraw-Hill.

5 IEEE Power System Relaying Committee report (1996). *Voltage Collapse Mitigation*. On line at http://www.pes-psrc.org/kb/published/reports/Voltage%20Collapse%20Mitigation.pdf

6 Phadke, A.G., Thorp, J.S., and Horowitz, S.H. (1988). *Study of adaptive transmission system protection and control*, US Department of Energy, ORNL/Sub/85-2205C.

7 Rockefeller, G.D., Wagner, C.L., Linders, J.R., and Rizy, K.L. (1996). *Adaptive power system protection*, US Department of Energy, ORNL/Sub/85-2201C.

12

Switching Schemes and Procedures

12.1 Introduction

In this book, so far, we have covered the synergy involved with protective relays on individual equipment and on entire systems. To understand the entire panoply of protective devices requires that we also discuss peripheral disciplines associated with power system relaying. In this chapter, we cover aspects of protection that must be included when designing or applying a protective scheme. These are subjects that are not relays or relay applications by themselves, but are areas that must be considered such as testing, computer programs for relay setting, and associated schemes such as breaker failure, reclosing, and single-phase breaker operation.

12.2 Relay testing

Relay reliability in its dual aspects of dependability and security is the most important quality of a protective relay and relay systems. Yet, despite this requirement, a relay may have the unique distinction of being the only power system equipment that is designed and installed with the hope that it is never called upon to operate. In fact, over the lifetime of a given relay, which may extend to 40 or 50 years, it may never be asked to respond to the conditions for which it was applied, or if it does operate, it may fully complete its function in a few milliseconds. It is, therefore, not surprising that there is a continuing concern over the possibility that a relay may not be ready to do its job at the desired instant, and with the speed that is required. As a result, it is generally accepted that protective relays must be periodically tested and maintained. It is ironic, though, that for almost all of the relays presently in service, such testing, calibrating, and other maintenance only prove that the relay would have done its job up to the time it was tested. There is, of course, no guarantee that this condition will continue when the relay is returned to service. In fact, it is not uncommon that the testing itself may leave the relay in an undetected, degraded condition. It cannot be stressed enough that too much testing is as bad as, or worse than, too little testing, and whatever tests are performed, they should clearly be limited to what is essential to be verified.

Power System Relaying, Fifth Edition. Stanley H. Horowitz, Arun G. Phadke and Charles F. Henville.
© 2023 John Wiley & Sons Ltd. Published 2023 by John Wiley & Sons Ltd.

12.2.1 New relay designs or applications

Our concern here is not with testing a new design, or a prototype of a new relaying concept. Such testing requires exhaustive simulations and studies, usually including actual short circuits on an operating power system. It is not too difficult to design a relay to operate for specified conditions, that is, to prove its dependability. It is far more difficult to design a relay to be secure, that is, one that will not operate for all of the other conditions that it will be exposed to in service. Laboratory facilities can usually reproduce high voltages or high currents, but not both simultaneously. In most cases, the laboratory equipment does not have the necessary power rating. In addition, the transient voltages and currents on the primary system that accompany normal switching, fault initiation, and clearing are so variable and site-specific that it is almost impossible to reproduce them in a laboratory environment, or to simulate them with a high degree of accuracy. However, new designs are normally tested by the manufacturer using digital simulators to reproduce the secondary voltages and currents that a relay might see when applied to a wide variety of power system configurations.

In addition, users with critical and/or unusual applications may choose to test a relay for specific applications. These applications could include extra-high-voltage (EHV) lines with or without series and shut compensation, single-phase tripping and reclosing, and unusually high resistance faults. Some users have their own digital simulation test facilities, while others may use consultant or manufacturer test facilities. It should be noted, however, that such tests are costly and only justified under critical applications.

The high penetration of digital disturbance-recording devices in modern power systems has increased the availability of records that may be used to test a relay's response to a wide variety of disturbances. Since these records are from real-life events, they provide an excellent source of test facilities for new designs and new applications. They are also very useful for investigating relay performance. The records need not necessarily be played back to the physical relay, but may also be played back to a detailed transient model of the relay. Unfortunately, sufficiently accurate transient relay models are not usually available to users, though, of course, they will be available to manufacturers. Many users faced with questionable relay performance may send the record(s) of the disturbance back to the manufacturer for further investigation.

12.2.2 Production relays

For a proven relay design, there are two general types of tests that are periodically performed: acceptance or commissioning tests and periodic tests.

12.2.2.1 Acceptance or commissioning tests during initial installation

Initial commissioning tests should include a thorough visual inspection to be sure that the relay was not physically damaged in shipping. When the relay is connected to its current and/or voltage transformers, it is necessary to make primary circuit checks, that is, to inject current in the transducers from a low-voltage source, say 230 V and 30 A. This should be a variable source, and with sufficient current capacity to check polarities of connections but not to simulate fault current. Complete secondary wiring checks are then performed to ensure that the relay(s) are connected according to the drawings. Initial calibration and settings should then be completed. Of course, manufacturers' instruction manuals provide valuable information concerning the specific relays being tested and should be rigorously followed. Many utilities perform rigorous checks when the relay is received, and then the relay is brought to the station. Regardless of the checks completed at a laboratory, final checks including breaker trip and other input/output validation tests must be made with the relay in place.

In addition to pre-energization tests, protection systems should also be tested when the protected equipment is first energized and loaded. Modern digital relays will allow users to inspect the currents, voltages, and status of various inputs. The value can be read directly from the user interface and compared with measurements from other instrument transformers and status indicators. For other relays without such interface, the measured current and voltages can be checked by temporary instrumentation at the relay inputs. These tests will confirm correct connections to instrument transformers and equipment status switches.

12.2.2.2 Periodic tests

Periodic tests are made to check calibration, overall mechanical and electrical conditions, and circuit continuity of the trip circuit. Historically, periodic relay testing has been the result of experience with electromechanical relays. These test procedures and practices are gradually being modified as a result of increased experience with solid-state and digital relays. Electromechanical relays are prone to failures due to environmental factors that may not affect solid-state or digital relays. Factors, such as aging, that change a component's characteristic or value, wear and binding of bearings due to temperature variations, dirt on disks and cups that affects their operating time, and oxides that form on contacts to increase their resistance and prevent circuit continuity are all potential defects that will alter a relay's performance. As a result of these failure possibilities, and since most of these defects cannot be detected with the relay in service, many test procedures call for removing the relay from its case for inspection, and possibly dismantling part of the device for cleaning and adjustment. After the preliminary examinations and bench tests are completed, the relay must be restored to its case on the switchboard panel for final calibration and trip tests. This is essential, particularly with electromechanical relays, where alignment, balance, and shielding affect the relay performance. To assist in testing the relay in place, all electromechanical relay manufacturers provide internal test facilities such as a removable test block, or switches. These separate the internal parts of the relay from the external wiring. The tester must then reconnect those elements that are to be tested through pre-connected jumpers or switches. For example, current and voltage coils can be reconnected to check settings or directionality, leaving the contacts disconnected, so that the relay operation will not trip the circuit breaker. This test facility also allows the tester to test one phase of a three-phase relay, while keeping the other two phases in service. The increased use of solid-state relays significantly changes the testing requirements. The physical factors associated with dirt and alignment no longer apply, although temperature effects, aging, and electrical deterioration remain a problem. However, solid-state relays allow much more continuous monitoring of internal voltages and logic status than their electromechanical counterparts. Indicating lights actuated by internal relay elements can trace the flow of logic during a fault or test, and provide a good indication of possible problems. In addition, functional testing facilities are provided with trip cutout switches, so the relay can be tested in service without tripping the circuit breaker. As a result, the intervals between periodic tests are being extended, and the time and effort required to do the testing greatly reduced. The digital relay offers a completely new dimension in periodic testing and calibration. Inasmuch as the digital relay can completely and continuously self-check itself, periodic tests can be virtually eliminated. Since the digital relay is always performing its calculations and making a decision, and usually there is no fault and no action is required, it is able to report that it is (or is not) operating correctly. The relay could then be maintained only when there is an indication of a problem.

Note, however, that digital relays cannot continuously check 100% of a protection system. There are components of the complete protection system that are external to the digital relay. Such components include instrument transformers, circuit breakers, teleprotection systems, and interconnecting wiring. There are, however, many features of external equipment that can be checked by examination of disturbance records that digital relays can provide. In Chapter 13, we identify many power system issues that can be identified and investigated through these records. Much of the protection system can also be periodically checked through these records. In addition to the relay itself, such periodic checks include performances of circuit breakers, instrument transformers, and peer-to-peer communications.

12.2.3 Monitoring

A common practice in evaluating the performance of protection systems is with station digital fault recorders (DFRs), particularly in EHV systems and with electromechanical relays where the internal relay parts cannot be monitored. Interestingly, the increasing use of solid-state and digital relays has not resulted in a decrease in such monitoring. In fact, the opposite has occurred. In addition to individual digital relay recordings, DFRs allow much more of the power system and protective system to be monitored, especially the coordination between various protection systems, communication links, and circuit breakers.

12.3 Computer programs for relay setting

Relay settings are the ultimate operating variable of any protective system. As has been discussed throughout this book, each relay has a setting philosophy that is unique to its design and to its function within a given protective scheme. The settings of some relays, such as a zone 1 of a mho-type relay, involve relatively simple calculations that depend only upon the impedance of the line being protected and the values of the potential and current transformers (CTs). By contrast, single-input time-delay overcurrent (TDOC) relays actually involve far more complicated calculations that depend upon varying fault currents, system configurations, and the operating performance of other relays with which they must coordinate. This is not a trivial task and requires a great many separate studies that must be combined, as well as attention to system and load changes.

As computers developed from mainframe to minicomputers and microcomputers, they became more readily available to individual engineers. Their ability to manage a wealth of information and perform calculations and follow logical instructions and constraints made them logical candidates to do relay settings. The first computer programs addressed the problem of coordinating TDOC relays at lower voltage levels [1], but gradually the entire protective system at a given voltage level or in a given area became the subject of computer programs [2, 3]. One factor in accepting the computer as a major calculating and decision-making tool is its ability to access several separate programs and combine their results in a relay-setting program. For instance, load flows are periodically generated for planning and operational purposes. By combining the results of those studies with a file of relay settings, it is now practical to avoid relay misoperations when the relay's loadability is exceeded. Similarly, short-circuit studies that are updated as a result of system changes can automatically verify relay settings or suggest revisions. Digital relays have the ability of being remotely accessed, and it is conceivable, but not very likely in the near term, that their relay settings could be automatically revised as load-flow and short-circuit studies are completed. Part of the reluctance to totally accept computer-generated settings without some control by the relay engineer is the amount of engineering judgment that goes into each relay-setting decision. Another issue with remote accessibility of computer relays, which is now addressed by reliability regulatory authorities and reviewed by relay engineers and IT specialists, is the fear of malicious hackers getting access to relay programs and doing damage to the power system. This is an important topic for regulations and design of countermeasures against hacker attacks.

12.4 Breaker failure relaying

Section 1.4.4 (Chapter 1) discusses the application of primary and backup relays. Primary relays operate for a fault in their zone of protection in the shortest time, and remove the fewest system elements to clear the fault. Backup relays operate in the event that the primary relays fail. As discussed in Section 1.4.4 (Chapter 1), backup relays can be local or remote. Our interest here is in a subset of local backup relays, referred to as breaker failure relays. We will now consider this important topic in greater detail. Figure 12.1 shows a simple representation of lines and transformers around a bus. For the moment, we have shown one circuit breaker per line or transformer. Later on, we will consider the variations in breaker failure needed to accommodate different types of buses. A fault at F on line B–C should be cleared by primary relays R_{bc} and R_{cb} and their respective circuit breakers. Now consider the possibility that circuit breaker B_1 fails to clear the fault. This failure may be caused by the failure of the primary relays, by the failure of CTs or potential transformers (PTs) providing input to the primary relays, by the failure of the station battery, or by the failure of the circuit breaker. As discussed in earlier chapters, remote backup function is provided by relays at buses A, D, and E to clear the fault F, if it is not cleared by circuit breaker B_1. However, remote backup protection is often unsatisfactory in modern power systems [4]. First, it must be slow enough to coordinate with all the associated primary relays. Thus, the remote backup function at bus A must coordinate with the zone 2 relays of lines B–C, B–D, and the transformer B–E. Second, because of the possible infeeds at the remote stations, it may be difficult to set the remote backup relays to see fault F from

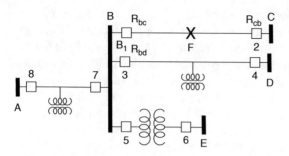

Figure 12.1 Breaker failure relaying for circuit breaker B_1.

stations A, D, and E. Finally, the power supplied to the tapped loads on lines A–B and B–D is unnecessarily lost due to remote backup operation.

A preferred method of protection against the failure of the primary relays at stations B and C is to provide a second set of relays at these locations, represented by R'_{bc} at station B in Figure 12.2. These relays operate more slowly than R_{bc} and R_{cb}, but trip the same circuit breakers. They cover the failure of the primary relays and their associated CTs, the secondary windings of the associated PTs, and, as shown in Figure 12.3, the direct current (DC) distribution circuit of the primary relays. However, relays R'_{bc} do not cover the failure of the circuit breakers themselves. To guard against this contingency, breaker failure relays are provided. When this scheme was first introduced [5], a separate protective relay was provided using an independent set of CTs, PT secondary windings, and DC circuit to trip the appropriate circuit breakers. In Figure 12.1, for example, breakers 3, 5, and 7 would be tripped by the breaker failure relays. Subsequent developments [6] replaced the independent relay with the basic control circuitry shown in Figure 12.3. In this scheme, any relay or switch which initiates a trip starts a timer known as the breaker failure timer. The timer is supervised (i.e., controlled) by an overcurrent relay (50-1), which drops out when the current through the breaker goes to zero. If this does not happen for any reason, the timer times out and energizes the lockout relay 86-1, which trips and locks out the circuit breakers 3, 5, and 7. The basic control circuitry shown in Figure 12.3 has since been updated with a number of optional supplementary features which are presented in Reference [7].

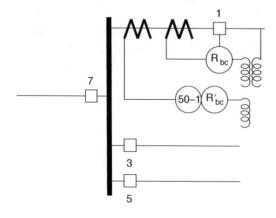

Figure 12.2 Primary, backup, and breaker failure relays in a single-bus, single-breaker scheme.

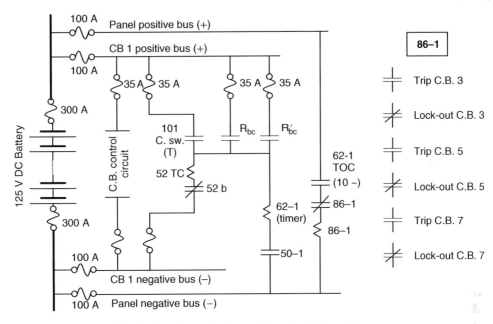

Figure 12.3 DC distribution for primary, backup, and breaker failure relays.

The earliest designs of the circuit breaker failure logic used a breaker auxiliary switch in place of the overcurrent relay. This switch is operated by mechanical or hydraulic linkages, and is designed to mimic the main contacts of the circuit breaker. However, the auxiliary switch has been found to be an unreliable device, especially when the circuit breaker itself is experiencing difficulties in clearing the fault, a condition for which the breaker failure relaying is supposed to provide a remedy. The mechanical linkages may break the auxiliary switches or the main contacts may be frozen, or—for some reason—become inoperative, while the main set of breaker contacts continues to function normally. There are some instances, however, when the breaker may be required to trip, even when there is no fault current to be interrupted. In such cases, a current detector cannot be used. The most common example of this condition is the trip initiated by turbine or boiler controls. Abnormalities in the pressure or temperature of the boiler, or some other mechanical critical element of the boiler–turbine system, may require that the station breaker be tripped, and in such cases it would be improper to use only an overcurrent detector to supervise the breaker failure protection. The setting of the breaker failure timer is determined by two constraints. The minimum time must be longer than the breaker clearing time, plus a security margin. Since the breaker failure timer does not start until a tripping relay (or switch) has initiated the trip, the operating time of the tripping relay is not a factor in determining the timer setting. This is a decided advantage of the present scheme. The maximum setting for the timer must be less than the critical clearing time for faults at that substation, as determined by transient stability studies [8]. Breaker failure timer settings are generally between 6 and 10 cycles for a 60 Hz power system. This discussion is centered upon the single-bus, single-bus arrangement shown in Figure 10.1 (Chapter 10). In reality, for other bus arrangements, a more involved procedure is required. This is explained in the following example.

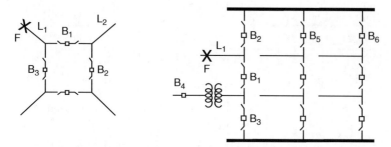

Figure 12.4 Breaker failure relaying for a ring bus and a breaker-and-a-half bus.

Example 12.1   Consider the ring bus and the breaker-and-a-half bus shown in Figure 12.4. Assume that breaker B_1 has failed in either system, following a fault at F. In the ring-bus scheme, the breaker failure relay must trip and lock out breaker B_2. Breaker B_3 will be tripped normally by the relaying responsible for the fault F. The breaker failure relay must also send a trip and lockout signal to the remote breakers of L_2, otherwise the line will continue to feed the original fault. The remote end of line L_1 is of course cleared by its relaying in response to the fault at F.

In the case of the breaker-and-a-half arrangement, for a failure of breaker B_1, breaker B_2 will be cleared by the line relaying of line L_1. Breakers B_3 and B_4 must be cleared and locked out by the breaker failure relay. The remote end of L_1 will clear normally as before. If breaker B_2 should fail in this case, the breaker failure relay of that breaker must clear all the breakers on that bus (i.e., breakers B_5 and B_6). Breaker B_1 and the remote end of line L_1 should open normally.

12.5   Reclosing

Section 1.4.1 (Chapter 1) discusses the dependability and security of relay systems. It should be recognized that these definitions are different when referring to the system itself. Reliability of a system refers to the design of the system, that is, the number of transmission lines and number and location of the generation and load. The security of the system refers to the actual operating configuration, that is, the actual state of the system with reference to outages. Maintenance of lines, breakers, generators, and so on is a planned event and must be considered in the normal operating procedures. Manual reclosing is therefore used.

Outages due to faults, however, are unplanned, and remedial measures must be incorporated. Automatic reclosing is then used to restore the system to its normal operating configuration. Automatic reclosing encompasses several schemes.

1) High-speed reclosing (HSR) refers to the operation of the circuit breakers without any intentional delay, the actual reclose time depending on the voltage class and type of circuit breaker. At 800 kV, HSR is about 15–20 cycles; at 138 kV, it would be 20–30 cycles.
2) Delayed reclosing introduces an intentional delay to allow fault arcs and ionized air to dissipate or to accommodate line switching. The time can be in cycles or seconds.

3) Synchronized reclosing refers to the application of a synchronizing device which monitors the voltage angle across the reclosing circuit breaker. Generators require full synchronizing, that is, comparing the voltage magnitude and angle between the system and the generator. The synchronizing relay monitors the generator voltage through 360° and the slip frequency, giving a closing signal allowing for the circuit breaker closing time. Circuit breakers connecting two elements of a transmission system use only check-synchronizing relays to confirm that the two elements are within a given angle, usually 20°–60°. Section 10.8 (Chapter 10) discusses the impact that a small angle can have on restoration following a blackout.

4) Interlocks are used when it is required that certain predetermined system conditions or elements are present before reclosing. It is common to reclose a transmission line from a preferred terminal, such as a remote bus rather than a generator, so a line voltage check may be used. Current or power through a parallel line may be used to ensure that both terminals of the line are synchronized. Reclosing a transformer only from the high-voltage side may use a breaker or voltage interlock. Also, during maintenance, a "manual-only" interlock is used to prevent the breaker from reclosing automatically during the maintenance period.

12.6   Single-phase operation

If one phase of a three-phase transmission line is open in response to a single-line-to-ground fault, the two healthy phases maintain synchronizing torque, allowing the system to maintain synchronism and continue to serve load. As mentioned in Sections 1.4.4 and 1.6 (Chapter 1), this is a common practice in Europe and Asia. The disadvantage, however, is the fact that inherent capacitance between the separate phases of the transmission line and the line to ground tend to maintain the fault arc resulting in a secondary arc that will maintain the original fault and result in an unsuccessful reclose. The remedy can be to delay the reclose until the secondary arc extinguishes itself, which may be undesirable from a system integrity viewpoint. It is also not uncommon to add shunt reactors between the separate phases and phase-to-ground to compensate for the inherent shunt capacitances. Shunt reactors are also beneficial on EHV transmission lines to compensate for the high voltage produced by the shunt capacitors on an open-ended line. Another disadvantage of single-phase tripping is the negative sequence that is produced by the unbalanced system. If allowed to persist, this will affect adversely generators on the system and must be removed by reclosing the open phase or tripping the remaining two phases.

12.7   Summary

In this chapter, we have presented some peripheral topics closely associated with protective reclosing. Testing production relays, as well as new designs, is essential to provide the necessary trust and history that the relays will perform as required. The increased use of computer relays is considered in this chapter in a preliminary way, but this subject is covered more exhaustively by specific manufacturers' instruction manuals. Breaker failure relaying and reclosing are essential elements of a protective relaying scheme and should be included in any discussion of relaying. Single-phase tripping and reclosing is not a generally applied concept in the United States, but is relatively common in Europe and Asia.

References

1 Albrecht, R.E., Nisja, M.J., Feero, W.E. et al. (1964). Digital computer protective device coordination program: I General program description. *IEEE Trans. PAS* **83** (4): 402–410.

2 IEEE Power System Relaying Committee (1991). Computer aided coordination of line protective schemes. *IEEE Trans. Power Deliv.* **6** (2): 575–583.

3 Ramaswami, R., Damborg, M.J., and Venkata, S.S. (1990). Coordination of directional overcurrent relays in transmission systems: a subsystem approach. *IEEE Trans. Power Deliv.* **5**: 64–71.

4 Kennedy, L.F. and McConnell, A.J. (1957). An appraisal of remote and local backup relaying. *AIEE Trans.* **78**: 735–741.

5 Barnes, H.C., Hauspurg, A., and Kinghorn, J.H. (1956). Relay protection for the Ohio valley electric corporation 330kV system. *AIEE Trans.* **75**: 613–625.

6 IEEE Power System Relaying Committee (1982). Summary update of practices on breaker failure protection. *Trans. PAS* **101** (3): 555–563.

7 IEEE Std C37.119-2016 (Revision of IEEE Std C37.119-2005) *IEEE Guide for Breaker Failure Protection of Power Circuit Breakers*. New York, NY: IEEE Standards Association, IEEE.

8 Stevenson, W.D. Jr. (1982). *Elements of Power System Analysis*, 4e. New York: McGraw-Hill.

13

Monitoring the Performance of Power Systems

13.1 Introduction

The importance of monitoring the performance of power system and equipment has steadily increased over the years. In the beginning, transmission lines had more capacity than was utilized and, in general, transmitted power from point to point, with few parallel paths. In addition, relays were provided with targets to indicate which relays operated and which phases were faulted. The evaluation of system faults, therefore, was relatively straightforward. As systems matured, transmission lines were connected in networks and more heavily loaded, so the analysis of faults became more complex and monitoring of equipment performance became essential for reliability and maintenance. As early as 1914, Charles Steinmetz [1] presented a paper in which he stated that the analysis of the performance of power systems was not possible without oscillographs. This attitude was common, since it was recognized that the operation of relays and circuit breakers (CBs) was too fast for the steady-state meters and recorders then in use and certainly beyond an operator's ability to track an event. The analog signals associated with fault currents and voltages were captured on galvanometers reflecting a light beam across a photographic film. These devices were variously called "oscillographs," "oscilloperturbographs," or "analog recorders." They were costly and introduced a significant time delay between the event and the ability to view the record; the record requiring transporting from the device to a photographic laboratory and then to the engineers who could analyze the record. As a result, oscillographs were difficult to justify and, except at power plants, were used sparingly. At power plants, the need to quickly analyze electrical failures to direct repair and to restore service justified the cost. In addition, the ability to develop and analyze the oscillogram was available on site.

At substations, a portable oscillograph that could be located wherever and whenever a particular problem had to be investigated was the most common application. This situation eventually changed. Photographic technology improved so that the record could be developed under ordinary light. This allowed the relay engineer to immediately examine a record at the station. As higher transmission line voltages were introduced and relays and stations became more complex, oscillographs became standard equipment at key locations throughout the system. The timing of CB and relay operations was displayed by devoting some of the traces to recording digital signals such as CB auxiliary or pilot relay contacts. Eventually, separate devices called sequence of events recorders (SERs) were used to monitor this aspect of system and equipment performance.

Power System Relaying, Fifth Edition. Stanley H. Horowitz, Arun G. Phadke and Charles F. Henville.
© 2023 John Wiley & Sons Ltd. Published 2023 by John Wiley & Sons Ltd.

With the advent of digital relays, the situation changed dramatically. Not only could the relays record the fault current and voltage and calculate the fault location, they could also report this information to a central location for analysis. Some digital devices are used exclusively as fault recorders. They have the ability to calculate parameters of interest and adjust individual traces for closer examination. Some of the algorithms and methods that are used will be discussed in the following text.

The recordings are used to determine what phenomena the signals represent, and the underlying events which have produced the actual signals. The IEEE Power and Energy Society (PES) Power System Relaying and Control Committee has produced several publications that provide helpful information on the application and uses of these recordings [2–4]. The analyses of the waveforms would help determine and analyze events such as:

- Failure of relay systems to operate as intended.
- Fault location and possible cause of fault.
- Incorrect tripping of terminals for external zone faults.
- Determination of the optimum line-reclose delay.
- Determination of the magnitude of station ground-mat potential rise (GPR) that influences the design of communication circuit protection and station grids or assists in the quantification of GPR and its direct current (DC) offset.
- Determination of protection communications systems performance.
- Determination of optimum preventive maintenance schedules for fault-interrupting devices.
- Deviation of actual system fault currents significantly from calculated values.
- Impending failure of fault-interrupting devices and insulation systems.
- Current transformer (CT) saturation and capacitor voltage transformer (CVT) response.

Fundamental frequency voltage and current signals of a power system contain a wealth of information about the system state. There are transient components superimposed on the fundamental frequency waveforms in the form of spikes and higher and lower frequencies that accompany faults and other switching events. These are revealed in the oscillographic recordings and are an essential element in analyzing power system and equipment performance.

Oscillographs have been used by power system engineers for many decades, until digital sample and event recording systems have replaced the oscillographs.

13.2 Early oscillograph analysis

It is well to review the oscillographic analysis techniques as a step in the evolution of system measurements and subsequent event analysis. In what follows in this section, we will give examples of analysis of some system events and resulting waveforms—although admittedly we will use a simplified oscillographic record to illustrate the method.

An example of a real oscillographic recording is shown in Figure 13.1. Oscillographs consist of light-sensitive paper which is driven at a constant speed. Galvanometers with mirrors provide light beams which are deflected by the voltage applied to the galvanometers in

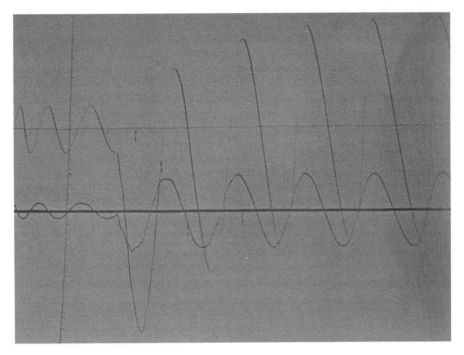

Figure 13.1 An example of optically recorded oscillograph showing the onset of a transient event.

proportion to the magnitude and polarity of the voltage. Both voltage and current signals are recorded in this fashion; the currents being converted to a voltage signal by passing through a precision resistor. The channels are calibrated in (for example) volts per centimeter. The vertical displacement of the actual trace is then measured to determine the magnitude of each relevant signal. The timescale is determined by the speed of travel of the paper. The relative phase angles between traces are determined directly by measuring the distance between waveforms of interest.

During any switching operation and during system faults, transients are generally present and the 60 Hz signal must be inferred. In Sections 13.2.1 and 13.2.2, we will discuss (simulated) oscillographic waveforms resulting from two events. It should be noted that these examples (and Problems 13.1 and 13.2 at the end of this chapter) use fundamental frequency signals in order to illustrate the methods of oscillographic analysis.

13.2.1 Voltage reduction during faults

Figure 13.2 shows how an oscillogram could be used to determine the normal operating voltage and the voltage reduction during a fault. In this figure, there are two cycles of pre-fault voltage, two-and-a-half cycles of reduced fault voltage, and then voltage recovery to normal voltage after the fault is removed. In the actual oscillogram, the trace may start earlier, and would continue to the end of the record, which is normally set for 60–70 cycles.

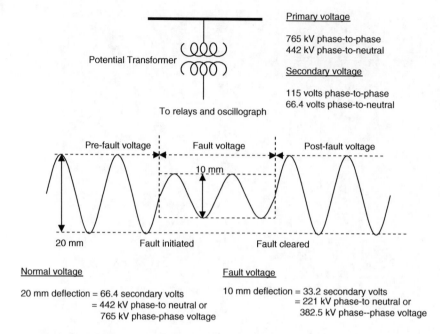

Figure 13.2 Voltage reduction during a fault.

13.2.2 Phase-to-ground fault

Figure 13.3 shows the traces associated with a phase-to-ground fault. There are two cycles of pre-fault normal load flow and two-and-a-half cycles of increased current during the fault. When the remote CB opens, there is a redistribution of fault current. Although the total fault current will be reduced, the contribution from the remaining end may increase due to infeed (see Example 4.6). Figure 13.3 shows this increase from the local end after the remote end opens, until the fault is finally cleared by opening of the local end. Prior to the fault, there is no ground current shown. Actually, as discussed in Chapter 4, there may be some slight pre-fault ground current due to unbalance. There are two cycles of ground fault current with both CBs closed, an increase when the remote end CB opens and then zero when the local CB opens. Since the two ends of the line will not open or close simultaneously, by comparing the oscillograms from the two ends, the exact CB sequence can be determined. The first CB to open will remove the fault current from that end of the line, while the second oscillogram will continue to show fault current. The first CB to close will result in an indication of charging current until the other end closes and load current is restored.

Figure 13.3 shows only two cycles of fault current, which is the time it takes for the relays and the CB to operate and clear the fault. Figure 13.3 also shows how to calculate the secondary and primary load and fault current from the oscillogram. These values can then be compared with planning and short-circuit studies to verify the system values that have been used.

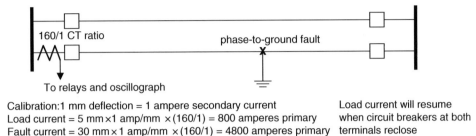

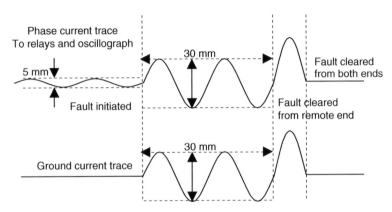

Figure 13.3 Phase-to-ground fault.

13.3 Disturbance analysis

As noted in Section 13.1, recordings from digital relays and dedicated digital recorders are now widely available, and are routinely used for disturbance analysis. Note that some recordings from digital relays may not be actual recordings of sampled data, but recordings of the filtered data after processing by the relay to remove non-fundamental frequency information. A recording of filtered data may look like a traditional oscillographic recording, but is in fact a plot of a series of points extracted from the raw data and processed to remove all information that is not fundamental frequency. For instance, Figure 13.4 shows recordings of the three phase currents observed during energization of a power transformer. The left-side recordings are the raw sampled data points and the right-side recordings are the processed fundamental frequency signals.

It can be seen from Figure 13.4 that the recording of the filtered data is of little help in understanding the specific incident being recorded, but the raw data readily show typical transformer inrush currents.

In spite of the inability of the plotting of the filtered data to show all the information available from a specific disturbance, this type of recording is very helpful in analyzing relay performance. Digital relays usually use fundamental frequency measurements to determine the presence (or not) of a fault within its protection zone. This is because normal short-circuit studies model the faulted power system in the steady state. The short-circuit studies derive the fundamental frequency currents and voltages before, during, and after the fault to

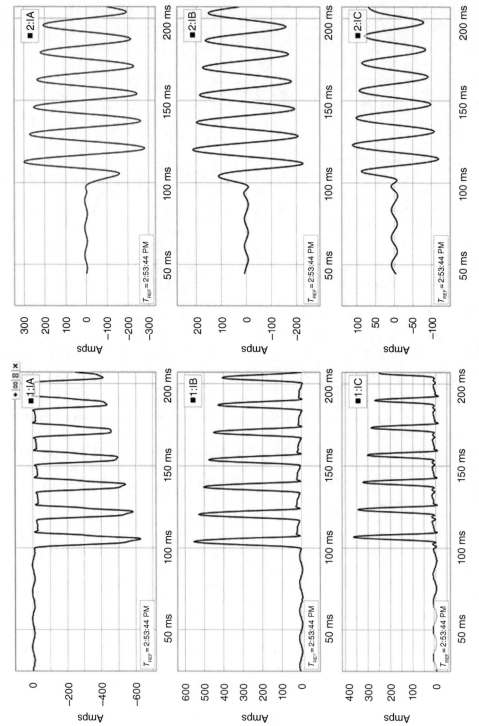

Figure 13.4 Transformer inrush currents raw sampled points (left side) and filtered data (right side).

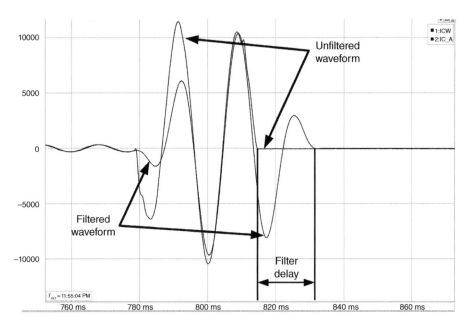

Figure 13.5 Comparison of timing of filtered and unfiltered waveforms.

facilitate the application of appropriate protection settings. Therefore, the recording of filtered data is helpful in validating the power system model as well as the protection performance.

When using the recordings of filtered data, care must be applied in two respects.

First, one must pay attention to the scaling of the recorded quantity. The peak values shown on the right-hand side of Figure 13.4 are not in fact the magnitude of the phasor. The magnitudes of all data points have been multiplied by $(1/\sqrt{2})$ so that the peak value is actually the rms value of the current. This has been done to facilitate comparison with fault studies which normally produce rms values of currents and voltages. The raw data points on the left side of Figure 13.5 are the actual sample data points; so the indicated peak values are the actual peak values. The manufacturer of the relay and/or the display program will provide information as to what scaling is used for plotting phasor recordings.

Second, the filtering delay makes it difficult to determine actual time points when the disturbance starts or ends. Figure 13.5 compares the time of two different recordings of a current for the same fault. This figure shows the current during a single line-to-ground fault on a 500 kV transmission line. It can be seen that at the start of the short circuit, the filtered waveform rises slowly compared to the unfiltered one. It is difficult to see when the filter stabilizes after the start of the fault. At the end, when the CB opens, it is easier to see the filter delay as the unfiltered waveform stops sharply at current interruption, but the filtered recording takes approximately one cycle to decay to a zero value.

Note that the recordings in Figure 13.5 are from two different instruments, and the scaling is the same in both recordings.

Figure 13.5 shows only two cycles of fault current, which is the time it takes for the relays and the CB to operate and clear the fault on this 500 kV line. This is often the case with extra-high-voltage (EHV) systems with high-speed protection and one- or two-cycle CBs. In high-voltage (HV) systems, with three-cycle CBs and nominal one-cycle relays, a fault should be cleared in about four or five cycles. Any longer time than this would alert the relay engineers to inspect either the relays or the CB. Of course, slower CBs or relays would take longer. Every system should have an expected clearing time to which the recording is compared.

13.3.1 Recording triggers

Digital recorders are, by nature, automatic devices. The time frame involved in recognizing and recording system parameters during a fault precludes any operator intervention. The most common initiating values are the currents and voltages associated with the fault itself. The phase currents will increase and the phase voltages decrease during a fault, so sensitive overcurrent and undervoltage detectors are used. In normal operation, there is very little ground current flowing, so a ground overcurrent sensor can be set very sensitively and initiate the recording. It is also common to use specific contacts such as CB auxiliary switches or relays or other recorders to trigger a record upon the operation of a particular device or other recorder. The majority of recordings are captured by digital relays. The relays will usually automatically trigger a recoding when a trip signal is issued. It is also useful to trigger a recording when a time-delayed element starts. Since time-delayed tripping may take longer than a few seconds, a single recording may not be long enough to capture the beginning of the disturbance, and the preliminary recording from the start of the timed protection can also be helpful. Recordings can also be initiated manually to capture normal operating values. During system faults, depending upon the recorder design, some cycles of pre-fault data can be retained and displayed. This is, of course, very useful in establishing the instant that a fault occurs and in correlating several records, particularly from separate stations. Another, and better, way to correlate separate devices is to use synchronized sampling clocks, either within a station or across the entire system. This is becoming more popular and will be discussed in more detail in Section 13.4.

13.3.2 Current increase and voltage reduction during faults

Figure 13.6 shows a recording of the three phase currents and voltages during a three-phase fault on a 230 kV line. The voltages are measured on the source side of the line CB. This recording shows the pre-fault load current and voltage at the start of the record. Then it shows the increased current and reduced voltage during a fault. After the CB opens to clear the fault, the currents go to zero as the fault is cleared from this terminal, and the bus voltage returns to normal.

Envelopes have been manually drawn around the peaks of the current to show the different degrees of transient offset of each of the three phase currents. This recording can also be used to determine the normal operating voltage and the voltage reduction during a fault. In this figure, there is one cycle of pre-fault voltage, three cycles of reduced fault voltage, and then voltage recovery to normal voltage after the fault is removed. In the actual recording, the trace would start earlier, and continue to the end of the record, which has been

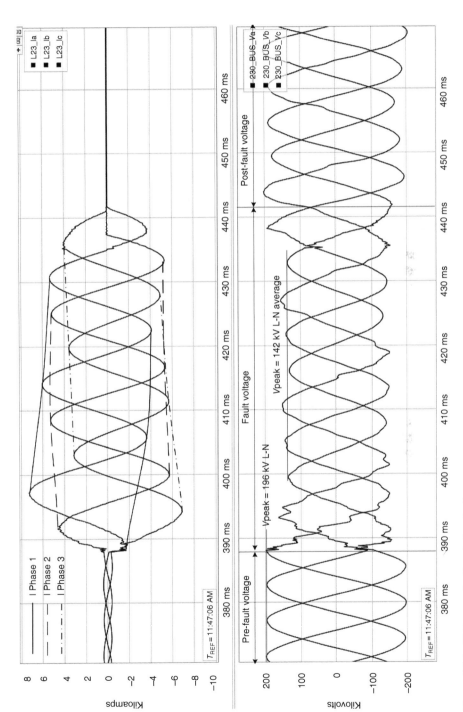

Figure 13.6 Voltage reduction during fault.

truncated for this figure. The figure shows the peaks of the pre-fault phase to neutral peak voltages are 196 kV. This converts to $196/\sqrt{2} = 141.6$ kV rms phase to neutral or 244 kV rms phase to phase. During the fault, the voltages on the three phases are not all exactly equal. This is commonly observed and may be due to small unbalances in arcing fault resistance, or small unbalances in the three phase impedances due to imperfect line transpositions between the substation and the fault location. However, the average peak phase to neutral voltage is 142 kV, which converts to 174 kV rms phase to phase, or 71% of the pre-fault voltage.

The recording of voltages in Figure 13.6 may be compared to the simulated oscillographic records of Figures 13.2 and 13.3 to observe the differences between real and simulated recordings. The digital recording of Figure 13.6 shows non-fundamental frequency transients that are not present in Figures 13.2 or 13.3. Further, digital recording of measured values eliminates the historical need to use a scale to measure distances on oscillographic traces as discussed in Sections 13.2.1 and 13.2.2.

13.3.3 Phase-to-ground fault

It will have been noticed in Figure 13.6 that it was non-trivial to measure the fundamental frequency values of the currents during the fault because of the transient offset. Notwithstanding the limitations of plots of filtered values, such recordings make it easier to measure fault voltages and currents in the presence of real-life transient waveform distortion. Figure 13.7 shows the traces of currents associated with the filtered values during a phase-to-ground fault. There are three cycles of pre-fault normal load flow and three-and-a-half cycles of increased current during the fault. When a remote CB opens, there is a redistribution to increase fault current at this local terminal as discussed in Section 13.2.2. Figure 13.7 shows this increase from the local end when the remote terminal opens when the fault is cleared there. Prior to the fault, there is very little ground current (only about 10% of load current). There are two cycles of ground fault current with both CBs closed, an increase when the remote CB opens and then zero when this local terminal CB opens. Figure 13.7 may be compared with the simulated Figure 13.3 which doesn't show filter delays when transitioning between different states of pre-fault, fault, and fault with remote terminal open.

Recorded values can be compared with planning and short-circuit studies to verify the system values that have been used.

13.3.4 Phase-to-ground fault and successful high-speed reclose

Figure 13.8 shows a simulated successful high-speed reclose (HSR) of CB A following a phase-to-ground fault on line A–B. Note that both bus and line potentials are recorded. This may be very useful for legacy light beam oscillographic recorders. Since most faults are line faults, the bus potential would be a continuous trace, although the magnitude would be reduced during the fault. This provides a convenient timing trace. Referring to the figure, note that the simulated line potential trace goes to zero when CB A opens, but the bus potential trace continues and allows us to determine the HSR time (HSR is a practice established by each utility as discussed in Section 1.4.5 of Chapter 1). Note that modern digital recordings have the recorded samples time stamped. This removes the need to count cycles on a healthy voltage signal for timing purposes.

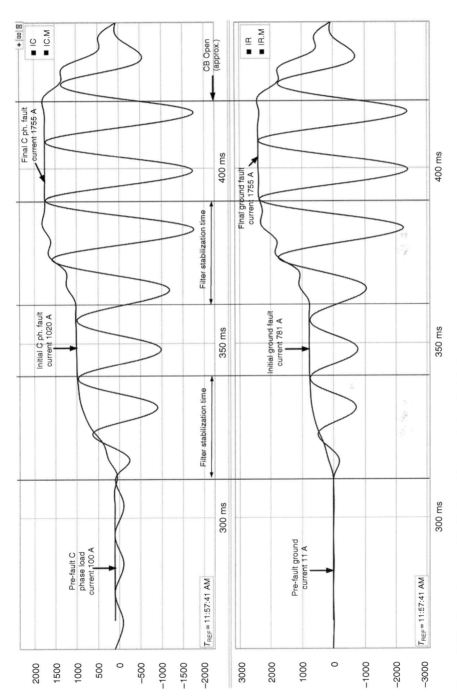

Figure 13.7 Phase-to-ground fault cleared by remote terminal first.

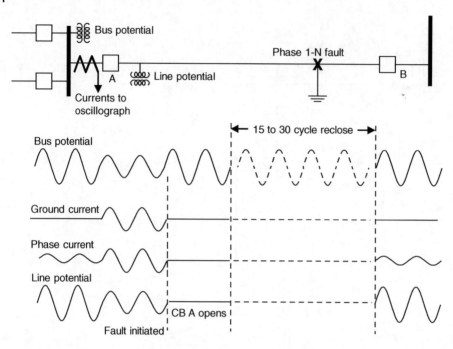

Figure 13.8 Single-phase-to-ground fault with successful HSR (CB A, circuit breaker A).

Before the fault, there is no ground current (or a very small amount depending upon the system unbalance), load current, and normal line and bus potential. After the fault is initiated, the phase and ground currents increase and the line and bus potentials decrease. When CB A opens, all line parameters go to zero, while the bus potential returns to normal. When CB A recloses, if the fault no longer exists, the line current and voltage traces return to their pre-fault values.

It is instructive to compare the simplified simulation shown in Figure 13.8 with Figure 13.9 which shows a real-life digital recording of a similar successful HSR following a phase-to-ground fault on a 500 kV line. Note that only the line potentials are recorded. The faulted phase current provides a convenient timing reference. Referring to the figure, note that the current trace goes to zero when the line terminal opens and returns to pre-fault level when the remote terminal closes. This allows us to determine the HSR time. In this case, the digital recording informs us that the HSR time is 1.113 s.

Figure 13.9 also shows that the voltage on the de-energized line does not go to zero, after the CB interrupts the fault. The smaller residual voltage is due to the capacitive coupling with the adjacent unfaulted phases which are also disconnected from the power system, but with remaining potential. The remaining potentials on the disconnected phases are slowly reduced as stored energy in the line shunt reactors and shunt capacitances gradually dissipates.

Figure 13.9 also shows that when the line is reclosed, both terminals are not necessarily reclosed at the same time. In this case, one terminal (A, for which the recordings are shown) closes before the other. When the A terminal is reclosed, capacitive charging current can

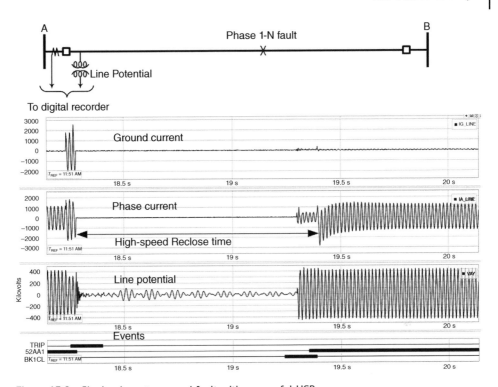

Figure 13.9 Single-phase-to-ground fault with successful HSR.

flow into the line. Only after the remote terminal (B) is closed is the reclosing sequence completed and the load current is restored.

Digital recordings may also be expanded and examined in detail for detailed review of disturbances and performance of equipment and protection. Figure 13.11 shows expanded portions of two especially interesting times during the disturbance.

Figure 13.11a shows the details during fault initiation. The time between fault initiation and protection trip ($t_{Trip} - t_{Fi}$) is found to be 19.5 ms. The time between a trip signal being sent to the CB and the fault being cleared ($t_{Fc} - t_{Trip}$) is found to be 28 ms. The CB auxiliary switch (52 A) changes status to indicate that the CB is open approximately 1.5 ms after the fault current is interrupted. All of these times are within the expected times of the equipment. The phase angle of the pre-fault current with respect to the pre-fault voltage is found to be $0°$, which indicates real power flow out of this terminal. During the fault, the phase current lags the phase voltage, by about $45°$, indicating some significant resistance in the fault.

Figure 13.11b shows the details during reclose. The time between initiating the close signal to the CB and the CB closing ($t_{Rc} - t_{Ri}$) is found to be 28 ms. The time between closing of the first (A) terminal and the second (B) terminal ($t_{Bc} - t_{Rc}$) is found to be 95 ms. The time between the CB main contacts closing and the auxiliary switch indication closed ($t_{CBaux} - t_{Rc}$) is found to be 54.5 ms. All of these times are within the expected times of the equipment. After the first terminal (A) closes and before the second terminal closes, the phase current leads the phase voltage by 90°. Such a phase lead would be expected with the terminal A supplying only line charging capacitive current. After the second (B) terminal closes, the phase current becomes exactly in phase with the voltage showing the return of real power flow.

As is shown in Figures 13.4–13.6, 13.9, and 13.10, during any switching operation and system faults, transients are generally present. The 60 Hz signal must be inferred or extracted by processing the data. In the figures that follow (in Sections 13.3.5–13.3.10), however, we have created illustrative example figures to show phenomena of interest. Unless the transients are part of the analysis, we have eliminated them and, for clarity, show only the fundamental frequency component.

13.3.5 Phase-to-ground fault and unsuccessful HSR

Figure 13.11 shows an example of an unsuccessful reclose. This is similar to the situation as Figures 13.8 and 13.9 except that the fault is not temporary and reappears when the line is reenergized. The reclose is, therefore, unsuccessful and the line trips out again. The traces after the reclose are a repeat of the initial fault. Figure 13.11 shows a typical switching transient in the bus potential trace.

13.3.6 Analyzing fault types

For illustrative purposes, except for Figures 13.4 and 13.6, we have shown only one phase current and voltage in the previous figures. Actually, we want to see, if possible, all three phase currents and voltages in a given line to determine which phases were faulted. This is readily available in a digital relay record. In some legacy station digital fault recorders, this may require more traces than are available, so some compromises were sometimes necessary. Since the voltages during a fault will be the same throughout the station, a common practice was to record a different phase for each line. Also, by recording phases 1 and 3 and the ground current on each line, all fault types can be determined. Modern digital recorders usually measure all three phase currents and voltages.

Example 13.1   Referring to Figure 13.12, note that, when the fault occurs, the voltage traces of phases 1 and 3 are reduced, phase 2 is normal, currents of line AB of phases 1 and 3 are increased, and ground current appears. These are clear indications of a phase-1-to-phase-3-to-ground fault on line AB. If there were no ground current, we would conclude that it was a phase-1-to-phase-3 fault.

13.3.6.1 Circuit breaker restrike

Figure 13.13 shows how to determine that a CB requires maintenance by the fact that it is restriking, that is, the insulating medium is degraded or the contacts are out of adjustment so that current flow is reestablished even though the CB contacts have separated. Another indication, not shown in this figure, might be the presence of high-frequency signals as the contacts begin to close or open. Historically, in the case of an oscillographic recording, depending upon the frequency response of the oscillograph, we might see the actual

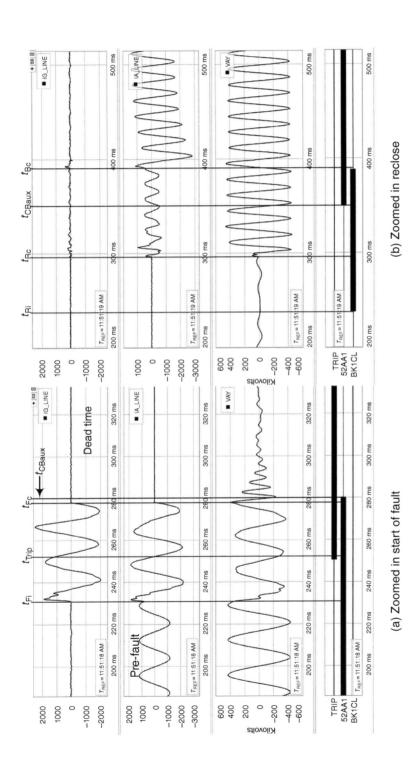

(a) Zoomed in start of fault

(b) Zoomed in reclose

Figure 13.10 (a) and (b) Enlargements of fault start and reclose.

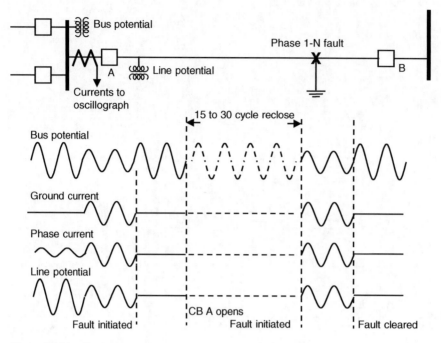

Figure 13.11 Single-phase-to-ground fault with unsuccessful HSR (CB A, circuit breaker A).

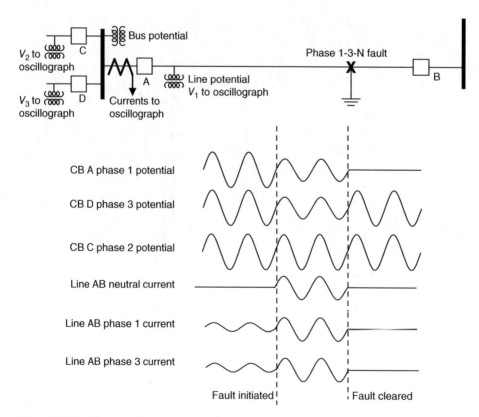

Figure 13.12 Phase-to-phase-to-ground fault.

waveshape or just a fuzzy record, but definitely not a clean 60 Hz trace. This is an indication that an arc is being established or maintained through the insulating medium.

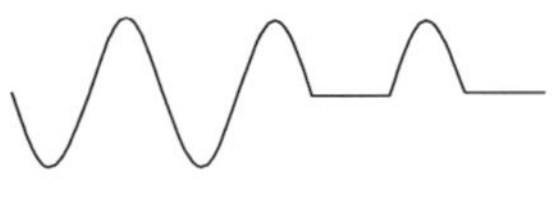

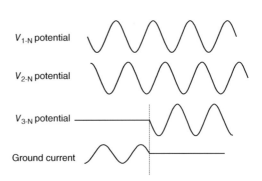

Figure 13.13 Circuit breaker restrike.

13.3.7 Unequal pole closing

Figure 13.14 shows an oscillogram of the three-phase-to-neutral potentials and the ground current associated with closing a CB. There is always a certain amount of lag between the poles of a CB closing. This is normally of the order of milliseconds. It is a function of the CB mechanism and is checked periodically by timing tests. In Figure 13.14, the lag is equal to two cycles. As shown, ground current flows during this time and it is possible that ground relays will operate.

$V_{1\text{-}N}$ potential

$V_{2\text{-}N}$ potential

$V_{3\text{-}N}$ potential

Ground current

Figure 13.14 Circuit breaker unequal pole closing.

13.3.8 CT saturation

Figure 13.15 shows the waveshape of a severely saturated CT. This phenomenon was discussed in Chapter 3. The shaded portion of the current wave is the current delivered to the relay. The remaining portion is shunted to the magnetizing branch of the CT.

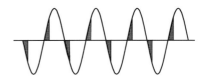

Figure 13.15 CT saturation.

13.3.9 System swing

Figure 13.16 shows the simulated single-phase current and voltage during a system swing as discussed in Section 11.5 (Chapter 11). The periodic oscillations in voltage and current magnitude indicate that various generators in the system are attempting to fall out of step with each other.

Figure 13.17 shows the three phase currents and voltages during an actual system swing [5]. It can be seen that the rate of change of current and voltage gradually increases until an out-of-step condition arises. During the out-of-step condition, the rate of change of currents and voltages may increase further.

13.3.10 Summary

It is, of course, not practical to give examples of all of the information that recordings provide. However, some of the other interesting system parameters that are recorded during system operations and should be studied whenever the opportunity presents itself are:

- Trapped charge oscillations due to discharge of stored energy (referenced in discussion of Figure 13.9).

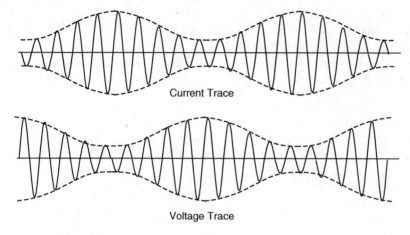

Figure 13.16 Simulated system swing.

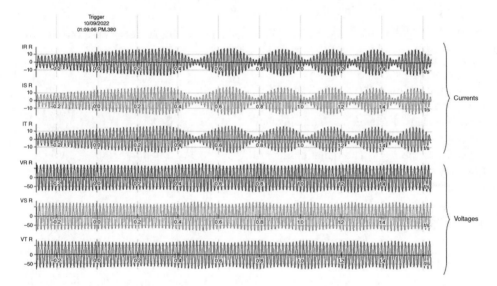

Figure 13.17 Actual system swing.

- Ferroresonance.
- Evolving faults.
- Inter-circuit faults (also known as cross-country faults).

13.4 Synchronized sampling

The digital recorders mentioned in Section 13.1 have become commonplace in modern power systems. All the signals recorded by digital recorders are sampled at the same instant, so that one can obtain a simultaneous snapshot of the recorded event across the complete set of voltages and currents. Analyses of the type described in Section 13.3 are much enhanced when temporal evolution of an event as reflected in all the recorded signals can be analyzed through a sequence of simultaneous samples. Expanding this thought a bit further, it seems desirable that recorded data or, more generally, sampled data from around an entire network simultaneously should also be obtained.

In recent years, a new technology has emerged, which can achieve precise synchronized data sampling across arbitrary distances reliably and economically. This is the technology of synchronizing the sampling clocks used by many (if not all) digital sampled data systems in a power network. The synchronization is very precise, nominally with errors of less than 1 μ s. In more familiar terms, this precision corresponds to errors of less than 0.022° for 60 Hz waveforms. Although several techniques are available for providing the synchronizing clock pulses, the method of choice at present is to use the transmissions of the global positioning satellite (GPS) system [6]. This is a system designed and implemented by the US Department of Defense for use in navigation, consisting of 24 satellites in subsynchronous orbits. At any time, a minimum of four of these satellites are visible from any point on earth. For timekeeping purposes, only one satellite is sufficient, and its basic transmission consists of 1 pulse/s, which on the designated "mark" is within 1 μs of a pulse received by any other GPS receiver at any point on earth.[1] The satellites also transmit the identity of the second, with zero second arbitrarily chosen to be 00:00:00 on 1 January 1900.

A typical substation system for synchronized sampling is shown in Figure 13.18. The analog inputs (secondary voltages and currents) are filtered with anti-aliasing filters [7], and then sampled at the clock pulses generated by a sampling clock which is phase-locked to the GPS receiver signal. The data samples are permanently linked with the identity of the second, as well as the sample number within the second. Thus, the data obtained at any substation can be sent to another site through a modem and a communication channel,

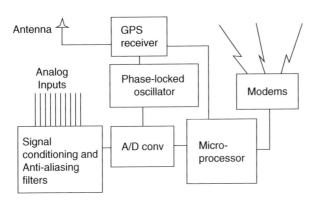

Figure 13.18 Computer-based measurement system for synchronized sampling using GPS transmissions.

---

1 The GPS transmissions are available free of charge, that is, only the receiver must be acquired. There is no usage fee for the transmissions, although there is always the possibility that a charge may be initiated in the future.

and the speed of the channel would not affect the precise specification of the instant when the data were taken. Of course, the system shown in Figure 13.18 is capable of producing far more sophisticated outputs than just the synchronized samples. These outputs are computed from the synchronized data samples in the microprocessor shown, and are time-tagged. This type of hardware is expected to be the basis for most digital substation systems of the future, including digital relays, digital fault recorders, and digital meters.

We will provide a brief catalog of the uses of synchronized sampling systems that are currently under investigation. The reader is referred to the literature [8] for more detailed information on this subject.

13.4.1 Uses of synchronized sampled data

One of the principal uses of synchronized sampled data is to compute phasors from these data (a method of computing phasors from sampled data is given in Section 13.4). The phasors—either phase quantities or symmetrical components of voltages and currents—when measured at the same instant on the power network indicate the true state of the power system at that instant. One could judge the quality of the network operating state, overloads, undervoltages, secure modes of operation, and so on, from knowledge of the synchronized phasors. This type of monitoring activity is common at a power system control center, and synchronized phasors are expected to have a considerable impact on central monitoring and security functions.

Synchronized sampled data could also be used in forming a consistent picture of faults and other transient events as they occur on a power system. Thus, oscillographs obtained from any substation could be correlated precisely, and one would have an outstanding tool for postmortem analysis on a system-wide basis. By making a series of snapshot pictures of unfolding events, one could trace cause-and-effect phenomena accompanying complex system events. Reference [9] provides good insight into the use of wide-area synchronized information for event reconstruction.

Synchronized sampled data collected from different ends of a transmission line could also be used in performing differential relaying (or other advanced relaying functions) digitally, provided dedicated high-speed communication links, for example, fiber-optic links, are available for the protection function. Synchronized measurements have also been considered for adaptive relaying, and improved control of power systems [10]. Fault location using synchronized data from the terminals of a transmission line would be particularly straightforward (as discussed in Section 13.5). We are likely to see many innovative applications of this technology in the coming years.

13.5 Fault location

It is advantageous to determine the location of a fault on a transmission line or a cable, as this is extremely helpful in any maintenance or repair operation that may be necessary in the wake of a fault. It should be recognized that distance relays may set a flag which indicates the zone in which a fault is detected, but such a coarse estimate of fault location is not sufficient to be of help in maintenance work. There are techniques for locating a fault, which depend upon the fault creating a permanent discontinuity on the line, which can then be determined by subjecting the unenergized line to injected traveling waves. By using the time for reflections at the discontinuity, it is possible to determine the fault location. This type of fault-locating technique requires special equipment, and is time-consuming and expensive to apply for all faults. It is usually reserved for underground cables, where faults tend to be permanent, and the use of specialized equipment and the expense are justified.

For overhead transmission lines, an alternative technique is to use the fault-induced currents and voltages to determine the fault location. This is akin to the fault calculation procedure using current and voltage phasors during the fault. Once the fault impedance is determined, the location of the fault can be ascertained from knowledge of the transmission line impedance per mile. This method of fault location calculation does not require any special equipment, nor does it require a special test procedure with the line de-energized. Also, if the fault is not permanent, the fault location can still be determined from the fault current and voltage data, if these latter have been saved.

The fault location task is usually undertaken with the help of saved transient records of currents and voltages during the fault. These records may be analog oscillographic records, digital fault records, or records available from many of the digital computer-based relays. In the case of analog recordings, the calculation is based upon the reading of the current and voltage phasors from the analog tracings, while with the digital records a direct calculation of the phasor quantities could be made from the data samples. A convenient method of finding the fundamental frequency phasor from one cycle of a waveform sampled at N samples per cycle is to use the discrete Fourier transform formula:

$$X = \frac{\sqrt{2}}{N} \sum_{k=1}^{N} x_k \left\{ \cos \frac{2k\,\pi}{N} - j \sin \frac{2k\,\pi}{N} \right\}, \tag{13.1}$$

where X is the phasor representation of the fundamental frequency component of waveform with samples x_k. In determining the phasors of currents in this fashion, it should be remembered that the DC offset (if any is present) in the current must be removed prior to the phasor calculation. When phasors from simultaneously taken samples of several inputs are determined, the resulting phasors are on a common reference, and may be used for fault location calculations.

Example 13.2 Consider the samples obtained from a voltage and a current channel with a digital fault recorder at a sampling rate of 12 samples per cycle. Assume that the samples are obtained from the following inputs:

$$e(t) = 141.42 \cos \left(\frac{377t - \pi}{6} \right),$$

$$i(t) = 1414.42 \cos (377t - \pi) + 1414.42\varepsilon^{-20t}.$$

The 12 samples shown in Table 13.1 are obtained from these two signals. The last column in the table contains the current samples from which the DC offset term has been eliminated.

Table 13.1 Data for the 12 samples of Example 13.2.

t (ms)	e(t)	i(t)	i(t) − 1414.2e^{−20t}
0.0	122.47	0.0	−1414.2
1.388	141.42	150.7	−1224.7
2.778	122.47	630.7	−707.1
4.164	70.71	1301.2	0.0
5.555	0.0	1972.6	707.1
6.944	−70.71	2455.5	1224.7
8.333	−122.47	2611.3	1414.2
9.722	−141.42	2389.0	1224.7
11.110	−122.47	1839.5	707.71
12.499	−70.71	1101.4	0.0
13.889	0.0	364.1	−707.1
15.227	70.71	−182.8	−1224.7

The phasor calculation proceeds as follows:

$$E = \frac{\sqrt{2}}{2} \sum_{k=1}^{12} e_k \left\{ \cos \frac{k\pi}{6} - j \sin \frac{k\pi}{6} \right\} = (86.6 - j50),$$

which is the correct phasor representation of the assumed voltage signal. Also,

$$I = \frac{\sqrt{2}}{12} \sum_{k=1}^{12} i_k \left\{ \cos \frac{k\pi}{6} - j \sin \frac{k\pi}{6} \right\} = (-971.64 - j87.92).$$

If we use the samples of currents from the last column of Table 13.1, the current phasor turns out to be $(-1000 + j0)$, which is the correct value of the phasor for the assumed current waveform. Note that the current phasor obtained without removing the DC offset is in error.

If the phasors are to be calculated from analog oscillographic records, one could follow the same procedure as mentioned in the earlier text, after the waveform samples are first digitized. Alternatively, a less accurate procedure of reading off the phase angles from the zero crossing of waveforms could also be used.

The problem of locating the distance to a fault can be simply expressed in terms of a single-phase circuit. This case will be considered first, and the results extended to the case of a phase-to-ground fault. The Thévenin representation of the sources at the two ends of the transmission line is generators with voltages E_r and E_s behind impedances Z_r and Z_s, respectively. Let the phase angles of E_r and E_s be δ_r and δ_s. The difference between these two angles is primarily responsible for the prefault load flow in the transmission line. Let Z_r represent the impedance of the transmission line (Figure 13.19). Note that R is the unknown resistance in the fault path. The fractional distance to the fault is k, which is to be determined as accurately as possible, knowing only the current I_x and voltage E_x at the relay terminal. If the current in the fault is $I_f = (I_s + I_r)$, the voltage at the relay location is given by

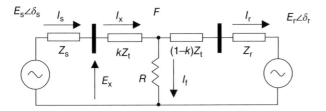

Figure 13.19 Ground fault through a resistance in a single-phase circuit.

$$E_x - kZ_tI_x + RI_f. \tag{13.2}$$

As I_f is unknown, it is sometimes assumed that R is zero, and thus k can be solved directly. As we will see in the following example, this can introduce significant error in k, if R is nonzero and if there is a significant difference between δ_r and δ_s. A much better procedure [11, 12] is to assume that the unknown fault current is proportional to the "change" in the current at the relay location:

$$I_x - I_{x0} \equiv \Delta I_x = d\,I_f, \tag{13.3}$$

where d is known as a distribution factor [13], and describes the superposition of a current at the fault location caused by the presence of the fault. Clearly, d depends upon the net equivalent impedances on the two sides of the fault as seen from the fault location. For the case shown in Figure 13.19,

$$d = \frac{Z_r + (1-k)Z_t}{Z_r + Z_s + Z_t}. \tag{13.4}$$

I_{x0} is the current at the relay location before the occurrence of the fault. Substituting for I_f from Equation (13.3) into Equation (13.2) gives

$$E_x = kZ_tI_x + \frac{R}{d}(I_x - I_{x0}). \tag{13.5}$$

If the distribution factor d is known, then the only unknowns in the abovementioned equation are k and R, both real numbers. Since Equation (13.4) is a complex equation, it can be separated into two real equations, from which the unknowns can be determined. The following example illustrates this procedure.

Example 13.3 Let the prefault data for the system of Figure 13.19 be as follows (all quantities are in pu):

$$E_s = 0.0866 + j0.5,$$
$$E_r = 1.0 + j0.0,$$
$$Z_s = 0.0 + j0.05,$$
$$Z_r = 0.0 + j0.03,$$
$$Z_t = 0.0 + j0.5.$$

The prefault load current is given by

$$I_{x0} = \frac{E_s - E_r}{Z_s + Z_r + Z_t} = 0.8359 + j0.3031,$$

and the prefault voltage at the fault point is

$$\begin{aligned}
E_f &= E_s - [Z_s + 0.3(0.05 + j0.5)]I_{x0}\\
&= (-0.866 + j0.5) - [0.0 + j0.05 + 0.3(0.05 + j0.5)](0.8359 + j0.3031)\\
&= 0.9141 + j0.3283.
\end{aligned}$$

Assume that a fault at a fractional distance of $k = 0.3$ occurs through a fault resistance of 0.3 pu. The Thévenin impedance of the network as seen from the fault point is given by

$$\begin{aligned}
E_{th} &= \frac{[Z_s + kZ_t][Z_r + (1-k)Z_t]}{Z_r + Z_s + Z_t} + R\\
&= \frac{[j0.05 + 0.3(0.05 + j0.5)][j0.03 + 0.7(0.0 + j0.5)]}{j0.05 + j0.03 + 0.05 + j0.5} + 0.3\\
&= 0.3106 + j0.1310.
\end{aligned}$$

The fault current I_f is found by dividing the prefault voltage at the fault point by the Thévenin impedance:

$$I_f = \frac{0.9139 + j0.3288}{0.3106 + j0.1310} = (-2.8768 + j0.1568).$$

The current and voltage at the relay location (again obtained by solving the faulted circuit) are $I_x = 2.7211 + j0.1892$ and $E_x = 0.8755 + j0.3639$. Note that if we assumed that there was no fault path resistance, we would compute the fractional distance to the fault by simply taking the ratio of E_x and I_x, and normalizing it by the total transmission line impedance Z_t:

$$k_{approx} = \frac{E_x}{I_x Z_t} = 0.2847 - j0.6304,$$

which answer, as expected, in serious error. In order to use Equation (13.5), one must know the distribution factor d, which itself, as seen from Equation (13.4), depends upon k. Thus, one must use an iterative procedure to compute d and k simultaneously. However, it will be seen shortly that k is not too sensitive to errors in d. For the moment, let us assume that we know k to be 0.3. Then, the distribution factor d is given by

$$\begin{aligned}
d &= \frac{Z_r + (1-k)Z_t}{Z_r + Z_s + Z_t} = \frac{(0 + j0.03) + 0.7(0.05 + j0.5)}{(0 + j0.05) + (0 + j0.03) + (0.05 + j0.5)}\\
&= 0.6555 - j0.0038.
\end{aligned}$$

Substituting these values of d, I_x, I_{x0}, and E_x in Equation (13.5) gives

$$\begin{aligned}
(0.8755 + j0.3639) &= k(0.05 + j0.5)(2.7211 + j0.1892) + R[(2.7211 + j0.1892)\\
&\quad - (0.8359 + j0.3031)]/(0.6555 - j0.0038)\\
&= k(0.0414 + j1.37) + R(2.8769 - j0.1568).
\end{aligned}$$

Separating the two sides into real and imaginary parts, and eliminating R, k is found to be 0.3, the correct value.

Let us see the effect of approximating the distribution factor d. If we assume d to be 1.0 (a real number) of some arbitrary magnitude, and repeat the calculation given in the earlier text, it is found that k is now 0.3037. Considering the very serious approximation made in d, this result is indeed very accurate. In fact, in the references cited, it has been shown that the correct answer does not depend upon the magnitude of d, but rather upon the angle of d, and, furthermore, it is often accurate enough to assume that the angle of d is 0°.

In a power system, the most common type of fault is the phase-to-ground fault. Furthermore, the fault path resistance is likely to be significant for such faults, and as discussed in Section 5.4.4 (Chapter 5), the presence of the ground-fault resistance introduces an error in the fault distance calculation as shown in Example 13.3. For a phase-a-to-ground fault, the relay location voltage is E_a and the relay location current is I'_a as shown in Equations (5.10)–(5.14). Using these quantities in place of E_x and I_x in Equation (13.5), an accurate estimate of distance to the fault for a ground fault can be determined.

The accuracy of the fault location calculation is affected by several factors. For example, the errors in current and voltage transformers directly affect the distance estimate. Similarly, uncertainties in line constants, effects of untransposed transmission lines, effects of line charging capacitors, and so on are all responsible for some error. In addition, the distribution factor d is seldom known (or constant) for a transmission network, as the system equivalent impedance depends upon the (changing) network configuration. All of these effects have been discussed in the references cited in the earlier text, which may be studied by the interested reader. In passing, we note that if simultaneously obtained data (synchronized phasors as discussed in Section 13.3) from the two terminals of a line are available, one could estimate the fault distance directly without making any assumptions about the distribution factor d. This subject has also been studied, but is beyond the scope of this section.

In addition to the phasor-based methods, availability of precise time discussed in Section 13.4 makes possible the use of traveling wave for fault location. Two methods are available:

I) Using data from a single terminal, the arrival times of the first traveling wave from a fault and the arrival time of its first reflection from the fault can be used to determine the location.

II) Using data from both terminals, the difference in arrival times at the terminals can be determined and the fault location estimated.

Fault location using traveling waves is immune to several of the accuracy limitations noted in the earlier text for phasor-based fault location, and very accurate locations can be estimated. The application of the technique of course needs equipment with precise time (errors of less than 1 μs) and measurement at a very high sampling rate (of the order of 1 MHz). A full discussion of both phasor-based and traveling-wave-based fault location techniques can be found in Reference [14].

13.6 Alarms

13.6.1 Attended stations

Alarms, meters, recorders, and annunciators provide the interface between an operator and the status of the equipment and system being operated. They play an obvious role in attended stations and plants, but are also valuable in unattended locations when relay or operating personnel arrives at the station to determine what has occurred and what steps must be taken to restore service.

Indicating lights, meters, and recorders are used to describe the steady-state condition of electrical equipment such as CBs, motors, transformers, and buses, and mechanical equipment such as valves and pumps. They also are used to display the progress of automatic or semiautomatic processes. Although every utility and industrial plant will have its own standards regarding color-coded lamps, labels, and so on, there are several commonly accepted practices. For instance, a red indicating light almost always indicates that the equipment being monitored is in service. In the case of a CB, this means that the breaker is closed and current and/or voltage is present; for valves, red means the valve is open, and the fluid is flowing. The opposite is true for a green light. The CB is open and the valve is closed. Auxiliary contacts or limit switches can be used to infer the condition of the equipment being monitored, but it must be recognized that this is not a direct indication of the device itself and could lead to an error. For instance, if the connecting rod from the CB main contacts to the auxiliary switch breaks, the auxiliary switch will not reflect the true position of the main contacts. The red light supervising the trip coil has a special application. It means that the breaker is closed "and it will" be tripped if the trip coil is energized. It is not good practice to use auxiliary contacts to energize this red light, since the breaker may be closed as indicated, but the trip coil may be open and incapable of tripping the breaker. An amber light is usually a warning indication that some condition is approaching a level at which some corrective action will be required. White lights are generally simply status lights. Meters and recorders should be sized so that the normal operating point is about three-quarters of full-scale. This allows the operator to monitor an overload or abnormal condition. This is the same practice followed in calculating CT ratios.

The increasing use of digital meters has promoted lively discussions on the relative merits of digital versus analog readouts. There is a tendency to assume that a digital display is inherently more accurate than an analog display of the same quantity simply because there can be a number of digits following a decimal point. This is not true. The accuracy of any reading is the cumulative accuracies of the entire installation, including the transducer (maybe the most critical element of the chain), the burden, the means of communicating the information (hard-wired, high-frequency carrier, microwave, or fiber-optic), and the panel instrument itself. The more important consideration in determining whether to use an analog or a digital meter is its application. A digital meter is the better choice if it is used in a regulating function. The reference is then to a specific number rather than to the relative position of a pointer on an analog scale. However, an operator will often prefer to see the operating value relative to some abnormal value and the absolute numbers are of little importance. For instance, an analog ammeter in a motor circuit is preferable to a digital meter. Many times the operator will mark the normal, alarm, and tripping currents on the ammeter face and will operate accordingly, without any interest or concern with the actual motor current.

In addition to the normal complement of meters and recorders, the use of annunciators has become standard practice in power plant and system control rooms. The design details of annunciators reflect the particular preferences of the user, but there are general practices that are common to all. The annunciator is primarily used to alert the operator to a change in the operating mode of a piece of equipment or operating parameter. A window with the appropriate message is illuminated when a condition goes "off-normal." This is usually accompanied by flashing the window light and sounding an audible alarm. The operator then silences the alarm and stops the flashing light. If the condition is off-normal, the window stays lit. The same sequence occurs when the condition returns to normal, only now the light stays out. The operator knows the status of the equipment at all times by reviewing the lit annunciator windows. On a 1300 MW unit, there may be between 1000 and 1500 annunciator windows, and for any trip-out, several hundred alarms will be actuated. SERs are used to supplement the annunciator displays and provide the operator with a more meaningful record of each event. As computers are being introduced into power plant control rooms and system dispatching centers, the annunciator displays are modified so that only the initiating causes are shown initially. The complete status of all equipment can then be displayed on demand.

13.6.2   Unattended station

The same complement of meters, recorders, indicating lights, and annunciators is used at unattended stations to facilitate maintenance and troubleshooting. However, the change of status of equipment or the occurrence of any trip-outs or overload must be immediately reported to a central location for proper operation of the system. Before the widespread use of digital communication systems using microwave or fiber-optic cable, relatively simple alarm systems were used. Basically, two types of alarm were sent from the station to the dispatching center. Variously called "critical/noncritical," "operation/maintenance," or some other descriptive terms, the purpose was to have the proper switch, relay, or maintenance personnel sent to the station to correct the problem, begin the investigation, and restore service. As communications became more reliable and sophisticated, specific details were sent, eventually resulting in very complete data acquisition systems (DASs). A DAS can now incorporate both normal and off-normal values, equipment condition, and specific operation or maintenance messages. In addition, with the advent of digital relays and oscillographs, a remote terminal unit (RTU), which is the communication interface between the DAS and the remote dispatching or engineering center, can be used to access both devices.

13.7   COMTRADE and SYNCHROPHASOR standards

Two recent industry standards are of particular interest for protection systems. These are COMTRADE (Institute of Electrical and Electronics Engineers [IEEE] standard and a parallel International Electrotechnical Commission [IEC] standard) and SYNCHROPHASOR (IEEE standard). We will offer brief overviews of the subject matter of these standards.

13.7.1 COMTRADE

This standard defines a standard format for data files created by computer relays, digital fault recorders, or other substation-based systems [15]. The original idea for this standard evolved from the need for interchangeability of data files for various user groups. Figure 13.29 shows on the left-hand side the various groups who may produce transient data from system voltages and currents. The potential users of the transient data are shown on the right-hand side. It became clear that various data producers were using proprietary file structures, and users had to have access to interpreters of data from those file sources. By defining a common file structure, a great stride has been made in opening up data access to a much wider audience.

The initial impetus for this work came from a Conférence Internationale des Grands Réseaux Électriques à Haute Tension (CIGRE) working group working on a document to offer an overview of the computer relaying technology in the 1990s. A working group of the IEEE Power System Relaying and Control Committee took on the task of producing a standard based upon the CIGRE report. This standard was later revised (1999), and at the same time a parallel working group of the IEC successfully adapted the standard as an IEC standard. This is now the worldwide accepted standard for storing and using transient data obtained from power systems, simulators, relays, and other recording devices.

The standard calls for three main files: a header file, a configuration file, and a data file. The first two help describe in detail the contents of the data file. The data file contains sample values of voltage and current measurements, as well as status information of various contacts available to the recording device. Each record is time-stamped to fractions of a microsecond.

The interested reader is referred to the standard document for additional details. The value of using a common format for all data collected from substations became particularly evident when attempts were made to reconstruct the sequence of events during major power system disturbances. There is a strong push in the electric power industry to make sure that all files stored and used for analysis are in COMTRADE format.

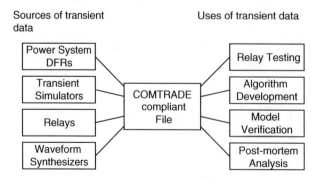

Figure 13.20 Community of producers and users of transient data produced in electric power systems. This concept emphasized the need for the creation of the COMTRADE standard.

13.7.2 SYNCHROPHASOR

Two standards are relevant to the application of synchronized phasor measurements provided by phasor measurement units (PMUs). Reference [16] defines the structure of the output files. Reference [17] standardizes the transfer of data from the PMU to another location. The IEC recently adopted [16] to create a harmonized (joint) standard.

The rationale for creating these standards is similar to that described in the earlier text for the COMTRADE standard. As PMUs began to be manufactured by different manufacturers, it became imperative that their output files should conform to a common format so that interoperability of PMUs from different manufacturers can be assured. The standards are also fashioned after the COMTRADE standard in their structures. There are header, configuration, and data files in the SYNCHROPHASOR standards. The data files carry time stamps correlated with GPS transmissions. With the available timing precision of 1 μs from these systems, synchronized data from various substations of the power system can be obtained, providing measurements upon which real-time monitoring, control, and protection systems can be based.

With the recurring major blackouts around the world, it became clear that data with precise synchronization are extremely valuable in determining the sequence of events and contributing causes to a catastrophic power system failure. This realization has led to a worldwide move toward installing PMUs on power grids. Many applications of these measurements are being investigated by researchers around the world, and widespread use of this standard and the PMU data is expected to be a common feature of power system monitoring, control, and protection systems of the future.

13.8   Summary

In this chapter, we have presented examples of power system fault recording devices and the analysis of typical oscillograms of records associated with faults and other operating situations. We have described how to analyze faults, determine if equipment needs maintenance, and determine if the protection scheme is working as intended. We have discussed the concept of synchronized sampling which allows us to correlate the records from several devices and locations. We have described various calculating procedures to determine the location of a fault even if information from only one terminal is available. We have described typical alarm and annunciator schemes used in power plants and substations. We have described the technology of synchronized sampling using the GPS system and have shown a typical substation system.

Problems

13.1   Given the oscillograph voltage trace shown in Figure 13.21 for a 345 kV bus, determine the normal and fault phase-to-neutral and phase-to-phase primary voltage. The potential transformer ratio is 3000:1, the galvanometer calibration is 4 V mm^{-1} and connected phase-to-neutral.

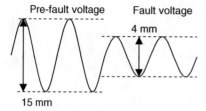

Figure 13.21 Oscillograph voltage trace for Problem 13.1.

13.2 Given the oscillograph current traces of Figure 13.22, determine the normal and fault phase and ground primary current. The CT ratio is 240:1 and the galvanometer calibration is 1 A mm^{-1} for both traces. Why is there any ground current prior to the fault?

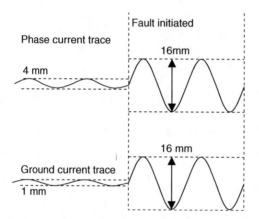

Figure 13.22 Oscillograph voltage trace for Problem 13.2.

13.3 From the oscillograph record of Figure 13.23 determine what type of fault has occurred (i.e., phase-a-to-ground, phase-a–b, etc.). Is there a reclose? Is it successful or unsuccessful? How many cycles before the fault was cleared and how long was the line de-energized?

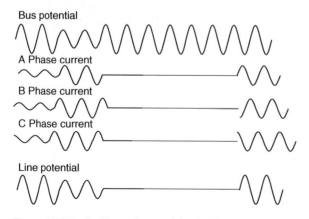

Figure 13.23 Oscillograph record for Problem 13.3.

13.4 From the system one-line diagram and the oscillograph record of Figure 13.24, describe the sequence of operation. For example, at $T = 0$–1.5 cycles—no fault, voltage normal, current = load current, current = zero. (Note the voltage E_a on line 1 when $i_a = 0$. What is it?)

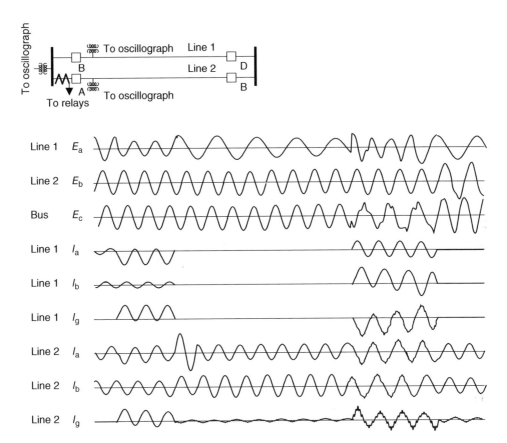

Figure 13.24 System one-line diagram and oscillograph record for Problem 13.4.

13.5 Derive equations similar to Equations (13.4) and (13.5) for a b–c fault and an a–g fault on a three-phase system. Is it necessary to use distribution factors for the positive, negative, and zero networks for these cases? (Hint: use the fact that in the "fault current," the positive-, negative-, and zero-sequence currents are all equal to each other.)

13.6 Assume that the distribution factor d used in fault location equations (Equation (13.4), etc.) has a phase angle error of $5°$. Assuming that the rest of the data are as in Example 13.3, determine the error in fault location as a function of the load angle, that is, as a function of the difference between the phase angles of the sending and receiving-end voltages. Vary the load angle in steps of $10°$ from $-30°$ to $+30°$.

References

1 Steinmetz, C.P. (1914) Recording devices. AIEE Midwinter Convention, February 27, 1914.

2 IEEE Power System Relaying Committee (1987) Application of Fault and Disturbance Recording Devices for Protective System Analysis, IEEE special publication 87TH0195–8-PWR.

3 IEEE Power System Relaying Committee Report (2006) Considerations for Use of Disturbance Recorders, available on line at http://www.pes-psrc.org/kb/published/reports/C5-Final%20Report.pdf

4 IEEE Power System Relaying Committee Report (2016) Analysis of System Waveforms and Event Data available on line at http://www.pes-psrc.org/kb/published/reports/0-0-Working%20Group%20I21-Disturbance-Events-Paper-Final-May-09-2016.pdf

5 Torres, F., Jácome, Y., and Henville, C. (2007). An Out-of-Step Event in the Peruvian Power System. *Western Protective Relaying Conference*, Spokane, WA.

6 Phadke, A.G. (1993). Synchronized phasor measurements in power systems. *IEEE Comput. Appl. Power* **6** (2): 10–15.

7 Phadke, A.G. and Thorp, J.S. (1988). *Computer Relaying for Power Systems*. Ltd: Research Studies Press/Wiley.

8 Phadke, A.G. and Thorp, J.S. (1991). Improved control and protection of power systems through synchronized phasor measurements. In: *Control and Dynamic Systems*, vol. **43** (ed. C. T. Leondes), 335–376. San Diego, CA: Academic Press.

9 IEEE Committee Report (2008) Event Reconstruction Using Data from Protection and Disturbance Recording Intelligent Electronic Devices available on line at http://www.pes-psrc.org/kb/published/reports/I11_Report_FINAL_080409.pdf

10 Phadke, A.G. and Horowitz, S.H. (1990). Adaptive relaying: myth or reality, evolution or revolution. *IEEE Comput. Appl. Power* **3** (3): 47–51.

11 Phadke, A.G. and Xavier, M.A. (1993) Limits to fault location accuracy. Proceedings of the 1993 Fault and Disturbance Analysis Conference, April 14–16, 1993, College Station, TX.

12 Novosel, D., Phadke, A.G., and Elmore, W.A. (1993) Improvements in fault location estimate. Proceedings of the 1993 Fault and Disturbance Analysis Conference, April 14–16, 1993, College Station, TX.

13 Takagi, T., Yamakoshi, Y., Yamaura, M. et al. (1982). Development of a new type fault locator using the one-terminal voltage and current data. *IEEE Trans. PAS* **101** (8): 2892–2898.

14 IEEE (2015). *IEEE Std C37.114-2014 IEEE Guide for Determining Fault Location on AC Transmission and Distribution Lines*. New York, NY: IEEE.

15 IEEE (2013). *IEEE Std C37.111-2013 (IEC 60255-24 Edition 2.0 2013-04) IEEE/IEC Measuring relays and protection equipment – Part 24: Common format for transient data exchange (COMTRADE) for power systems*, 1. Piscataway, NJ: 73, Standards Association of IEEE.

16 *IEC/IEEE 60255-118-1:2018 International Standard - Measuring relays and protection equipment - Part 118-1: Synchrophasor for power systems – Measurements*, IEC Central Office, Geneva, Switzerland, and IEEE New York, NY

17 *IEEE Std C37.118.2-2011 (Revision of IEEE Std C37.118-2005) IEEE Standard for Synchrophasor Data Transfer for Power Systems*, IEEE New York, NY

14

Improved Protection with Wide-Area Measurements (WAMS)

14.1 Introduction

The technology of wide-area measurements (WAMS) on power systems using synchronized phasor measurement units (PMUs) has matured to the point that most power systems around the world are currently installing these devices in their high-voltage substations and evolving toward an integrated WAMS. References [1–4] will provide detailed information about this technology, and include additional references. The worldwide WAMS-related activities were well summarized in a number of papers [4] in a 2008 issue of the "Power & Energy Magazine."

The basic building block of WAMS is the PMU. The hardware of a PMU is akin to that of a computer relay (see Section 2.6, Chapter 2). The principal difference is that the sampling clock of a PMU is synchronized to a universally available signal—at present the global positioning system (GPS) is the preferred source of such signals. By using a common synchronization, signal measurements made at different substations in the power system are assured to be simultaneous. A GPS signal provides synchronization accuracy of better than 1 μs, which is equivalent to a phase angle error of less than $0.021°$. The PMUs measure positive-sequence and individual-phase voltages and currents of the feeders in a substation, as well as local frequency and rate of change of frequency. They also provide the status of selected switches and circuit breakers. The measurement accuracy requirements and the communication facilities needed to transfer the data collected by PMUs are described in five recent Institute of Electrical and Electronics Engineers (IEEE) standards [5–9].

14.2 WAMS organization

The number of PMUs placed on a power system depends upon the stage of WAMS development at a particular power company, as well as upon the intended use of the WAMS system. In most cases, the final WAMS configuration assumes that every extra-high-voltage (EHV) substation will be equipped with PMUs, and that bus voltages and feeder currents available in that substation will be measured by the PMUs. The intermediate stages of

Power System Relaying, Fifth Edition. Stanley H. Horowitz, Arun G. Phadke and Charles F. Henville.
© 2023 John Wiley & Sons Ltd. Published 2023 by John Wiley & Sons Ltd.

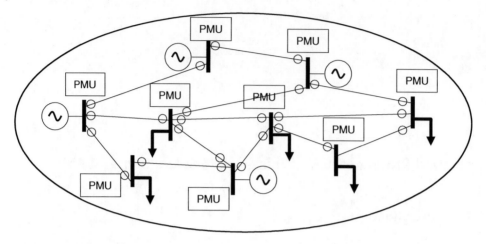

Figure 14.1 PMU installations in a generic power system. Bus voltages and feeder currents are measured by the PMUs.

WAMS implementation seek to determine the optimum number of PMUs and their locations to achieve some specific objective such as observability of the network state [10].

Figure 14.1 is a conceptual representation of a power transmission system equipped with PMUs placed at all of its substations. It is assumed that bus voltages as well as feeder currents are being measured by the PMUs. Although only one PMU is shown at each substation, in reality there may be more than one PMU if required to accommodate all the signals required to be measured.

Figure 14.2 illustrates the architecture of WAMS that is representative of the current industry practice. The PMUs transfer the synchronized measurements of voltages and

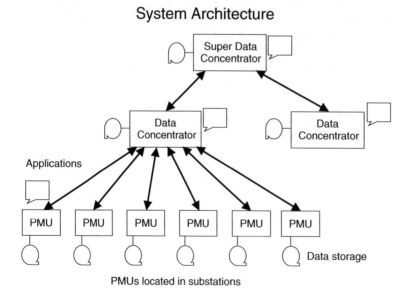

Figure 14.2 PMU and PDC architecture in a WAMS installation.

currents to a first-level phasor data concentrator (PDC) [9, 10], and several such PDCs transfer their collected data to the PDC at the next level of the hierarchy. This PDC is often termed the "super PDC." At the PMU as well as the PDC levels, appropriate applications of PMU data available at that level are implemented.

14.3   Using WAMS for protection

There are certain considerations regarding the use of WAMS and phasor measurements in protection systems. The first consideration is that of the latency of measurements. Phasor measurements are obtained from sampled data collected over some duration—typically one cycle of the nominal power system frequency. The data are communicated to the next level of hierarchy of every n cycle. The recommended values for "n" are given in the IEEE standard [5]. Depending upon the filters and processing algorithms used in phasor estimation, phasors will have a latency of the order of $n + 1$ cycles. In addition, there are communication delays involved in getting the data to a remote site where protection decisions are made. There is an industry consensus that total delay from the instant when an event takes place on the power system to the instant when the measurement is available for use at a remote site is of the order 40–100 ms, depending upon the communication system in use. It is therefore clear that measurements provided by WAMS cannot be used in high-speed protection functions.

Data storage is available at various levels of the architecture, and appropriate applications of the collected data may exist at all levels.

Another consideration is that the phasors provide an estimate of the fundamental frequency component of the system currents or voltages. Thus, components associated with transients—such as the direct current (DC) offset in the currents, or harmonics in voltages and currents—are not available in WAMS measurements. Consequently, protection functions that depend upon the transient components or those requiring waveform sample data cannot benefit from WAMS measurements.

This leaves the slower protection functions for consideration for improvement using WAMS data. This is indeed a very fortunate circumstance, since it is exactly the slower responding protections that depend upon an assessment of the system state. As the power system state continues to evolve over time, these slower protection systems may not be appropriate for the prevailing conditions. Many of the power system cascading failures and blackouts have resulted because of the mismatch between the settings of these protection functions and what would have been appropriate for the prevailing system state. The slower protection functions often isolate larger parts of the system, which is the key reason why their misoperation often leads to cascading failures.

Ideally, the protection settings of these slower functions should be revised periodically in order to make sure that they are compatible with the evolving power system. However, many power companies do not have the manpower to handle the relay settings revision task on a regular basis. At the minimum, the WAMS systems could be used to alarm the relay engineers that a setting currently in use may be inappropriate, and should be looked at. It may also be possible to supervise or alter some settings if the WAMS system indicates this is necessary. Both of these approaches will be considered here.

The protection functions that have been studied [3, 11] with a view to using WAMS to improve them are:

a) Supervision of backup zones of distance relays.
b) Alarming for distance relay characteristic penetration.
c) Adaptive out-of-step relaying.
d) Adaptive loss-of-field relay.
e) Intelligent load shedding.
f) Adjusting balance between security and dependability.
g) Improved remedial action schemes (RASs) and system integrity protection schemes.

These functions and the technique of using WAMS for their improvement are in the following sections.

Reference [12] sheds additional light on the relaying performance improvement using WAMS. Some of other exiting applications of WAMS noted in Reference [3] are listed here, but not discussed further in this chapter:

a) Improved selectivity for automatic reclosing on mixed cable/overhead line transmission circuits using WAMS for fault location.
b) Power system analysis.
c) Synchrophasor-assisted black start.
d) Distributed generation anti-islanding.
e) Automatic generator shedding.
f) Communications channel analysis.
g) Verifying voltage and current phasing.

Reference [3] also describes several other possible future applications.

14.4 Supervising backup protection

In most cascading failures of power systems, it has been noted that unwanted operation of backup zones (particularly the third zone of distance relays) is a prominent contributing factor. The third zone (and to a lesser extent the second zone) of distance relays is designed to back up the protection systems of neighboring transmission lines and is designed to overreach the neighboring circuits by 30–50% of the line lengths (see Chapter 5). It is sometimes argued that the third zone should be eliminated to avoid such misoperations, although on further thought it is found that third zone is required and presents a desirable option in case of certain failures in primary protection of neighboring circuits [13]. When a third zone of protection is deemed necessary at a critical location in the network, it is possible to use WAMS to devise a supervision scheme that would prevent misoperation of the third zone. Additional information on supervisory protection principles will be found in Reference [14].

Consider a portion of the power system shown in Figure 14.3. The distance relay at terminal A of the transmission line AB is provided with a third zone function, and this location is considered to be critical for the security of the power system, so that it is required that this

Figure 14.3 WAMS-based supervision of the third zone of a distance relay at terminal A of line AB. The third zone function is deemed to be necessary for this relay, although the line is critical from the point of view of system security so that an unwanted trip of this relay would lead to cascading failures of the power system.

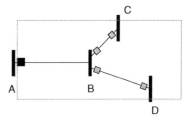

relay does not operate unless it is operating to back up the neighboring circuits in case their primary protection systems fail.

We assume that the third zone of the relay at A has picked up, and it will trip the circuit breaker shown in solid black in zone three times (typically of the order of 1 s). If this is a genuine backup operation, there must be a zone one fault seen by the distance relays at stations B, C, and D corresponding to circuit breakers shown by dotted squares. In this case, PMUs situated at stations B, C, and D can be programed to determine the apparent imped-ance seen at those locations by dividing the positive-sequence voltage at the buses by the positive-sequence currents of the lines BC and BD. If none of these apparent impedances are less than the corresponding line impedances, the zone three operation of relay at loca-tion A is not justified and should be blocked. Such a supervisory signal can be transmitted to the relay at station A, and the zone three operations blocked. It should be noted that these actions are taken only if the negative-sequence current at all the terminals is negligible, rul-ing out an unbalanced fault. The inappropriate operation of the third zone at station A should only be blocked when the line currents are balanced, and the supervisory signal from PMUs at B, C, and D, indicating that a blocking of the third zone is appropriate.

The logic of this supervision can be summarized by the flowchart shown in Figure 14.4.

It should be noted that this supervision function is provided by PMUs that are close to the critical station A, and thus the amount of data as well as the communication delay involved should be quite small. One should take such supervisory actions only at critical locations, and wholesale supervision of all third zones is not contemplated. The tolerance "ε" in

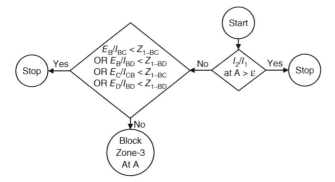

Figure 14.4 Supervision of zone three of the relay at station A based upon the input from PMUs at stations B, C, and D.

Figure 14.4 should be around 0.2 in order to make sure that there is no unbalanced fault detected at station A.

14.5 Impedance excursions into relay settings

Another cause of relay misoperation during cascading power failure events is the inappropriate setting of various relays for the prevailing power system state. The settings are not such that the protection system misoperates during normal operation at the prevailing state, but once another event (such as a fault followed by a line trip) occurs, the setting of the relay in question may not be able to handle the change in power flows resulting from the event. The famous 1965 blackout in Northeastern United States was caused by just such a circumstance. The offending relays had been set in 1956, and over the intervening 9 years, the system loading changed so that the relay settings of the critical lines from Niagara Falls to Ontario were dangerously close to the prevailing loading levels [14].

A relatively simple task for PMUs at critical locations in the power systems is to monitor the trajectories of apparent impedance (ratio of positive-sequence voltage to positive-sequence current in a feeder) and alarm the relay engineer if the apparent impedance comes near a tripping zone of a relay. The monitoring should follow load excursions during normal and abnormal system states, as well as during generator swings. A conceptual view of this task is shown in Figure 14.5.

Line AB in Figure 14.5 is determined to be critical for the reliability of the power system shown. The tripping zones of the relays at stations A and B are as shown. Although the zone shapes are shown to be circles, the same considerations will apply for other zone shapes. The results of two system events are depicted in the figure: one is the load change due to some switching operation in the system, and the other is the swings of the generators in the system due to another event. In both cases, the apparent impedance seen by the

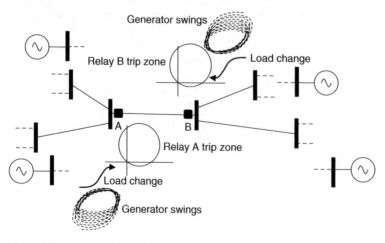

Figure 14.5 Apparent impedance trajectories seen at critical relay locations in relation to the relay tripping zones. An alarm is generated for the relay engineer if the trajectories approach within a preset distance of the relay characteristic.

PMU at stations A and B is determined and checked whether or not they approach the tripping characteristic of the relays in question. As indicated in the figure, the relay at B is approached quite closely because of both these events, and consequently an alarm will be generated for the relay engineer to check the settings of the relay at B, whether or not it is considered to be appropriate. It may well be that the engineer will decide that the settings should remain as they are, or it may be concluded that the settings must be revised in order to avoid unnecessary trips in the future.

14.6 Stability-related protections

There are several stability-related functions that could benefit from WAMS measurements. In this section, five types of stability-related protection functions will be considered. These are:

1) Loss-of-field relaying that deals with steady-state stability of a generator.
2) Out-of-step relaying that deals with transient stability of the system.
3) Small oscillation stability.
4) Intelligent load shedding.
5) Voltage stability.

14.6.1 Loss-of-field relaying

The principles of loss-of-field relaying are discussed in Section 10.3 (Chapter 10). As noted there, the loss-of-field relay has a circular characteristic denoting the limiting condition either because of steady-state stability limit or undervoltage reactive power limit. There is a second concentric circle that is used to provide an alarm to the generator control room operator.

The steady-state stability limit is dependent upon the generator reactance and the Thévenin impedance of the connected network at the connection point. It is clear that the system Thévenin impedance will depend upon the network configuration near the generator. Consider the power system shown in Figure 14.6.

When both lines connecting buses A and B are in service, the system Thévenin impedance is small leading to a smaller relay characteristic. During the course of system operation, if one of the lines goes out of service, the Thévenin impedance of the system increases, thereby increasing the size of the circle representing the stability limit. If in this case should a loss-of-field condition occur, the actual stability limit will correspond to the larger circles, leading to an unstable condition even before an alarm is produced if the characteristic is not altered corresponding to the loss of line. Such a situation would be encountered if the power system is undergoing a cascading event causing loss of several transmission lines sequentially.

The WAMS system can provide an online determination of line outages as they occur, and require the generator protection system to change the loss-of-field relay setting to correspond to the prevailing Thévenin impedance. It should be noted that only the lines in the immediate neighborhood of the generator will affect the Thévenin impedance

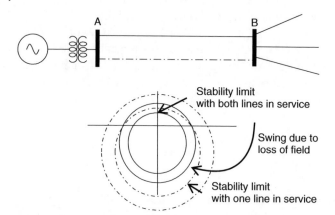

Figure 14.6 Power system configuration in the neighborhood of a generator. As the network changes due to line outages, the Thévenin impedance of the system changes, affecting the stability limit of the generator.

significantly, and only a relatively small number of lines need to be monitored by the WAMS for this application.

14.6.2 Out-of-step relaying

Out-of-step relaying has been discussed in Section 11.5 (Chapter 11). There, it was noted that the settings of these relays are based upon simulations of transient stability and instability events for various assumed system configurations and loadings. Often the system assumes configurations that have not been foreseen, or unanticipated disturbances lead to unexpected outcomes. It has been observed in practical systems that the settings of these relays are often inappropriate for prevailing system conditions, and this leads to the motivation to investigate out-of-step relays that are based on real-time measurements provided by WAMS.

Equal-area criterion (see Example 10.1, Chapter 10) can be utilized to determine the outcome of an evolving transient stability event when the power system behaves like a two-machine system (Figure 14.7). This is often the case when the region in question is peninsular with interconnection to the main grid over a few tie lines. It is also the case when a remote plant is connected to a large network over long transmission lines. Using PMU data to track the behavior of the angle difference between the two centers, a real-time estimate of the outcome of the evolving transient can be made with a high degree of certainty. As this technique depends upon real-time measurements and not any precomputed system performance, the results of the estimation tend to be more reliable. Such a protection scheme was implemented on a system in United States and the results reported in a publication [15].

For power systems that are more tightly integrated and are not reducible to a two-machine equivalent, the problem of detecting instability in real time is far more difficult. A number of approaches including Lyapunov techniques, energy function methods, extended equal-area criterion, and faster than real-time simulation of transient swings have been discussed in the technical literature. One approach that has been somewhat successful

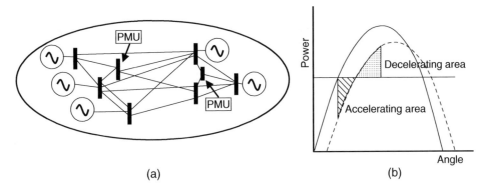

Figure 14.7 (a) Power system exhibiting the behavior of a two-machine-like system for transient stability oscillations. PMUs are placed in central buses of the two regions, and the angle movement is tracked. (b) Equal-area criterion can be applied to determine the stability of an evolving event.

has been to track in real time the movement of rotors over a certain period after the transient event commences, and identify groups of machines that form coherent groups. Once the coherent groups have been identified, each group (usually only two coherent groups are formed) is replaced by a single equivalent machine. One can now apply the equal-area criterion to the two equivalent machines or project the angular separation between the two equivalent machines with a time series and then determine whether or not the predicted angular separation indicates a stable or an unstable swing. These ideas are illustrated in Figure 14.8.

14.6.3 Small oscillation stability

From time to time, most power systems exhibit sustained oscillations with low damping. The oscillations have come to be noticed when PMUs with fast response have been placed at network buses. The oscillations are in power flows, voltage, and current magnitudes, and in voltage and current phase angles (Figure 14.9).

These oscillations, if unchecked, can cause damage to the power system. There could be undesirable relay operations due to the oscillations in power flows, and also there could be damaging torsional vibrations in large generator rotor systems.

It has been shown in several publications that these oscillations could be damped by judicious use of power system stabilizers (PSSs), appropriately tuned excitation systems of the generators, or by modulating flexible AC transmission system (FACTS) devices [16, 17]. In all these cases, it is advantageous to use WAMS in the form of phasor measurement feedback. Most of these studies have been analytical with relatively few practical implementations. It remains a fertile and important area of research in making use of real-time phasor measurements as controller feedback signals.

14.6.4 Intelligent load shedding

Underfrequency load shedding has been used by most power systems in order to balance the load shedding and generation in an islanded system. It is often desirable to take load action

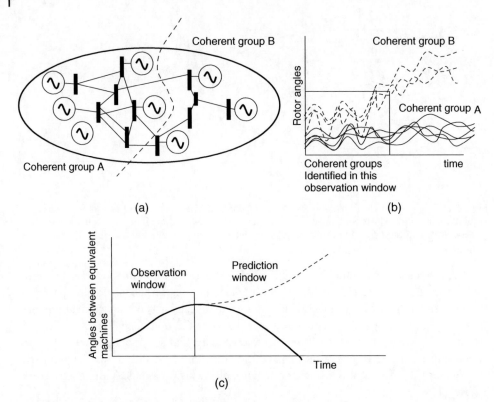

(a)

(b)

(c)

Figure 14.8 (a) Network rotor angles monitored to detect coherency. (b) Coherent groups converted to two equivalent machines. (c) Using polynomial curve fitting to rotor angle difference between the two equivalent machines, a prediction is made to determine whether the swing is stable (solid line) or unstable (dotted line).

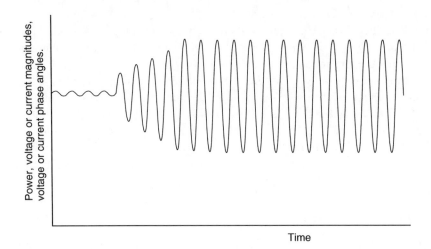

Figure 14.9 Sustained oscillations in voltage, current, and power uncovered by PMUs.

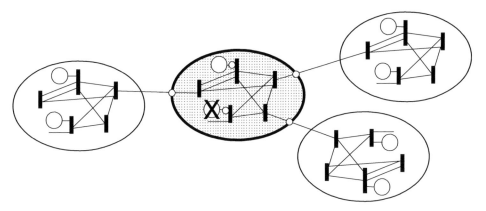

Figure 14.10 Intelligent load shedding for shaded area (which is presumed to have a loss of generation) to be based on real-time measurement of tie-line power flows and generation outputs.

in order to avoid the formation of an island—which is often the precursor to a cascading outage event. Consider the interconnected grid consisting of multiple control areas as shown in Figure 14.10. The shaded area is assumed to have a loss of generation at the location denoted by "X." As a result, the tie-line power flows, as well as the output of the remaining generators in the area changes. The net change in tie-line power flows is similar to the area control error (ACE) [18] used in automatic generation control (AGC) systems commonly used in controlling the generation in an area. Of course, since the system is still interconnected, there is no contribution from frequency deviation to the ACE.

The net tie-line power flow and internal generator output changes provide a sound measure of the amount of load that could be shed and thus restore the system to an approximate state that existed before the generation outage. This is to be confined to the shaded area, and if accomplished with sufficient speed would prevent a system split and the formation of the island.

14.6.5 Voltage stability

Voltage stability of a network has been discussed extensively in the technical literature. In recent years, some papers have addressed the problem of estimating voltage stability margin with phasor measurements [19–21]. Much of the research in this area remains speculative, and the reader is referred to the publications mentioned in the earlier text for additional information on the progress in this direction. It is to be expected that further research in this important area will be undertaken and a practical method of assessing voltage stability margin at an operating point determined using WAMS data.

14.7 SIPS and emergency control with WAMS

RAS or special protection systems (SPSs) have been in use in power networks for many years in order to take corrective actions in anticipation of a catastrophic event from occurring. In 2010, IEEE Power System Relaying and Control Committee (PSRC) recommended that

Critical
Inputs

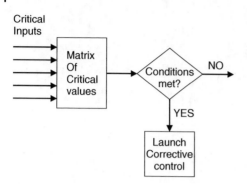

Figure 14.11 A generic representation of a SIPS logic.

these acronyms be included in a more expanded term: system integrity protection scheme (SIPS). The basic concept behind such systems is to determine through simulations of the power network for various contingencies if a certain sequence of events are likely to lead to major disturbances and possible cascading failures. It is also verified through simulations that such conditions can be detected by taking real-time measurements of critical power flows, circuit breaker status on key transmission lines, levels of generation and load, and so on. In certain hierarchical SIPS, critical measurements are brought to a central controller and each of these variables is compared with predetermined limits in order to determine if a catastrophic event is imminent. A set of precalculated corrective measures—such as tripping of loads and generation, modifying settings of controllable devices—are then initiated. A generic view of a hierarchical SIPS system is shown in Figure 14.11. The critical inputs are brought to a decision matrix and compared with their critical values. If certain predetermined patterns of these measurements are met, then an automatic control action is initiated to launch a corrective measure.

In many power systems, a SIPS is introduced when a particularly troublesome contingency condition is uncovered. Over several years, there may well be a large number of SIPS systems installed, each one dedicated to deal with its specific problem. A natural concern is that are these SIPS systems appropriate when the nature of the power system has changed due to network-switching events or loading and generation pattern changes? In particular, is there any danger that the controls delivered by some SIPS systems may be deleterious to other SIPS?

The idea behind using WAMS-based SIPS is to use the WAMS-based estimate of the actual needs of the power system in real time and then take those corrective actions that would match the prevailing conditions. Research in this area is currently being done, and it can be expected that in coming years, a WAMS-based SIPS strategy would evolve. References [22–25] shed further light on WAMS influence on SIPS.

14.8   Summary and future prospects

This chapter provided an overview of the technology of WAMS using real-time phasor measurements to improve the performance of certain protection systems. This field is currently being studied by several organizations around the world and it is likely that we will begin to see many applications of this kind being implemented on power networks. The main candidates for these improvements are currently limited to slower protection functions that can tolerate the delays involved in getting the WAMS data to a decision center, and also because these slower operating protection functions cause major disruptions to the power network when they fail to operate correctly. Using real-time data to modify or supervise these slower protection functions makes them appropriate for the prevailing condition of the power networks. One should also note that WAMS provides significant improvements to many other power system functions including state monitoring and control.

References

1 Phadke, A.G. and Thorp, J.S. (2009). *Computer Relaying for Power Systems*, 2e. Chichester: Wiley.

2 Phadke, A.G. and Thorp, J.S. (2008). *Synchronized Phasor Measurements and Their Applications*. Berlin: Springer.

3 IEEE PES Power System Relaying Committee Report (2013). *Use of Synchrophasor Measurements in Protective Relaying Applications* On Line at http://www.pes-psrc.org/kb/published/reports/Use%20of%20Synchrophasor%20Measurements%20in%20Protective%20Relaying%20Applications_final.pdf

4 Phadke, A.G. (Guest Editor)(2008). *Power & Energy Magazine*, Special Issue: "Shelter in a storm" **6** (5).

5 *"IEEE Standard for Synchrophasor Measurements for Power Systems"* C37.118.1, 2011.

6 *"IEEE Standard for Synchrophasor Measurements for Power Systems – Amendment 1: Modification of Selected Performance Requirements"* C37.118.1a, 2014.

7 *"IEEE Standard for Synchrophasor Data Transfer for Power Systems"*, C37.118.2, 2011.

8 *"IEEE Guide for Phasor Data Concentrator Requirements for Power System Protection, Control, and Monitoring"*, C37.244, 2013.

9 *"IEEE Standard for Phasor Data Concentrators for Power Systems"*, C37.247, 2019.

10 Nuqui, R.F. and Phadke, A.G. (2005). Phasor measurement placement techniques for complete and incomplete observability. *IEEE Trans. Power Syst.* **20** (4): 2381–2388.

11 Phadke, A. and Novosel, D. (2008). *Wide Area Measurements for Improved Protection Systems*. Croatia: CIGRE Colloquium in Cavtat.

12 Phadke, A.G., Wall, P., Ding, L., and Terzija, V. (2016). Improving the performance of power system protection using wide area monitoring systems. *J. Modern Power Syst. Clean Energy* **4** (3): 319–331.

13 Horowitz, S.H. and Phadke, A.G. (2006). Third zone revisited. *IEEE Trans. Power Deliv.* **21** (1): 23–29.

14 Phadke, A.G. (1980). Recent Developments in Digital Computer Based Protection and Control in Electric Power Substations, 'State of the Art Lecture'. *Conference on Power System Protection*, The Institution of Engineers (India), Madras, India.

15 Centeno, V., Phadke, A.G., Edris, A. et al. (1997). An adaptive out-of-step relay for power system protection. *IEEE Trans. Power Deliv.* **12** (1): 61–71.

16 Pal, A. and Thorp, J.S. (2012). *Co-ordinated Control of Inter-area Oscillations using SMA and LMI*. Innovative Smart Grid Technologies (ISGT) IEEE PES.

17 Ota, H., Ukai, K., and Nakamura, H. (2002). *Fujita PMU based midterm stability evaluation of wide-area power system*. In: *IEEE/PES Transmission and Distribution Conference and Exhibition*, vol. **3**.

18 Cohn, N. (1971). *Control of Generation and Power Flow on Interconnected Systems*. New York: Wiley.

19 Begovic, M.M. and Phadke, A.G. (1990). Voltage stability assessment through measurement of a reduced state vector. *IEEE Trans. Power Syst.* **5** (1): 198–203.

20 H. Innah and T. Hiyama. *Neural Network Method Based on PMU Data for Voltage Stability Assessment and Visualization*. TENCON 2011–2011 IEEE Region 10 Conference, p. 822–827.

21 Milosevic, B. and Begovic, M. (2003). Voltage-stability protection and control using a wide-area network of phasor measurements. *IEEE Trans. Power Syst.* **18** (1): 121–127.

22 Biswal, M., Brahma, S.M., and Cao, H. (2016). Supervisory protection and automated event diagnosis using PMU data. *IEEE Trans. Power Deliv.* **31** (4): 1855–1863.

23 Kundu, P. and Pradhan, A.K. (2016). Enhanced protection security using the system integrity protection scheme (SIPS). *IEEE Trans. Power Deliv.* **31** (1): 228–235.

24 Madani, V., Novosel, D., Horowitz, S. et al. (2010). IEEE PSRC report on global industry experiences with system integrity protection schemes (SIPS). *IEEE Trans. Power Deliv.* **25** (4): 2143–2155.

25 Yang, Q.; Bi, T.; Wu, J., (2007). *WAMS Implementation in China and the Challenges for Bulk Power System Protection*, 2007 IEEE Power Engineering Society General Meeting.

15

Protection Considerations for Renewable Resources

James K. Niemira, P.E.

Power Systems Solutions – Engineering Services, S&C Electric Company

15.1 Introduction

Application of renewable resources is becoming more prevalent and can be expected to have a significant impact on the design of the power system in coming years. Within the United States, many states have passed legislations mandating the use of renewable energy sources. There has been a push to have as much as 20% of electricity generation sourced by wind or other renewable sources in the near future [1]. At the time of this writing (2013), wind generation accounts for 3.5% of electricity generation and solar generation accounts for approximately 0.1% of electric energy production in the United States [2]. The Canadian and US governments as well as various state legislatures and provincial governments have offered production tax credits (PTCs) and other tax rebates, special pricing, and other incentives to encourage the installation and adoption of renewables. Continuation or expansion of such programs is tenuous in times of economic hardship and government deficit spending. Also, the recent technological development of hydraulic fracturing now being used in the production of natural gas and oil has resulted in decreasing prices of these resources due to the ability to access previously uneconomical sources of supply and has put ever more down ward price pressure on alternative resources. The goal of 20% renewable by 2030 may not be met, but use of renewables is and will be an important source of generation in the future.

15.2 Types of renewable generation

Renewable generation includes wind, solar photovoltaic (PV), solar heat engines, geothermal, and biogas. Any of these sources may be used for generators ranging in size from several kilowatts up to a few megawatts. Generators may be distributed throughout the power system and commingled among the connected loads (known as "distributed generation" or "DG") or may be aggregated together at one location to provide a generation plant with total output capacity of 20–100 MW or even larger. Each of these types of connections to the power system will have different effects on the power system and consequently will have their own special considerations for protection of the power system and equipment.

Power System Relaying, Fifth Edition. Stanley H. Horowitz, Arun G. Phadke and Charles F. Henville.
© 2023 John Wiley & Sons Ltd. Published 2023 by John Wiley & Sons Ltd.

15.2.1 Heat engines

The biogas and geothermal generation plants are generally smaller-scale plants that use the fuel gas or naturally occurring heat source to heat a working fluid to drive a heat engine connected to conventional induction or synchronous generators. Solar concentrator heat engines also use mirrors to focus the incident sunlight on a thermal collector to provide a high-temperature heat source to produce steam or heat some other working fluid, such as hydrogen for use in a Stirling engine, which again is used to drive a conventional induction generator or synchronous generator. Because these are standard electric machines, the behavior under faulted circuit condition of these generators is well known and the fault current contribution can be predicted with good accuracy.

15.2.2 Wind turbines

Wind turbines use the power of the wind to drive a fan, propeller, or blade, which in turn is used to drive a generator. A gearbox is most often used to convert the relatively slow speed of the turning blades to a higher speed more efficient for the generation of electricity. The mechanical system of the blades capturing the wind may be either stall-regulated or pitch-regulated.

Stall regulation is a robust design, wherein the shape of the airfoil limits the amount of energy capture at high wind speed. Due to the design and shape of the blades, the turbulence increases and the blades stall out at high speed, which limits the power input to the mechanical system. These designs are intended to operate at nearly a constant rated speed over a wide range of input wind speeds. Pitch-regulated designs actively control the pitch of the blades to more efficiently capture mechanical energy. Some pitch-regulated designs will even use individual pitch control of each blade to smooth the power input throughout the rotation, even with differing wind speed at the top and the bottom of the swept area.

Presently, the industry broadly classifies turbines into five basic types, which is described in further detail in the following text: Type 1 uses a simple induction generator (SIG) also known as a squirrel cage induction generator (SCIG); Type 2 uses a wound rotor induction generator (WRIG) with variable resistance in the rotor circuit to provide some speed and torque control over the generator; Type 3 is a doubly fed asynchronous generator (DFAG), also known as doubly fed induction generator (DFIG), which uses a wound rotor machine with a power electronic converter in the rotor circuit; Type 4 uses a full-power converter/inverter to convert the electric power of the generator to synchronous voltage and current required by the power system; Type 5 uses a hydraulic fluid coupling to convert variable speed/variable torque input from the turbine blades to a constant speed/variable torque output to drive a standard synchronous generator at constant speed [3].

15.2.2.1 Type 1: Squirrel Cage Induction Generator

The Type 1 wind turbine generator is shown schematically in Figure 15.1. This design uses a squirrel cage induction machine to generate electric power by driving the machine's rotor faster than the synchronous speed. These machines have the advantage of simple and rugged design. Input speed and power are controlled either using a stall-regulated design or in some designs using active pitch control of the blades. Precise control of the speed is not required, although most designs operate within a narrow range close to a rated design speed. When the blades are driven faster, the slip increases and more power is produced.

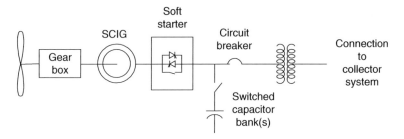

Figure 15.1 Type 1 WTG, squirrel cage induction generator.

   The disadvantage of the Type 1 design is that magnetizing power must be supplied by the power system. That is, reactive power (vars) is drawn from the power system. Switched capacitors are usually included to provide the required reactive power to operate at near-unity power factor. Another disadvantage is the high inrush current that may occur when the machine is initially switched on. A soft starter is often included to reduce the inrush current. The soft starter uses power electronic switches to reduce the amount of voltage applied to the machine until the rotor is up to speed and magnetized. This is accomplished by gating the switches at precisely controlled switching angles on the voltage waveform until the rotor is up to speed, and then closing the switch to connect the machine across the line. A step-up transformer, usually mounted on a pad next to the base of the turbine tower, is used to connect the turbine output to the collector system or local utility distribution system. Designs of wind turbines evolved from commonly used industrial machinery. Typical design voltage for the generator ranges from 575 to 690 V.

15.2.2.2 Type 2: Wound Rotor Induction Generator
The Type 2 wind turbine generator is shown schematically in Figure 15.2. This design uses a wound rotor induction machine which, like the Type 1, is driven above the synchronous speed to act as a generator. By the use of a variable resistance in the rotor circuit, slip and torque can be controlled over a wider range than is possible with the Type 1 machines. The variable resistance may be externally mounted and connected to the rotor circuit using slip rings, or, in some designs, the resistors are mounted on the rotor along with power electronic switching circuits. With the rotor-mounted switches and resistors, optical couplers are used to transmit switching commands to the rotor. This design eliminates the slip rings and associated maintenance. Speed control of Type 2 machines in the range of about 10% is common.

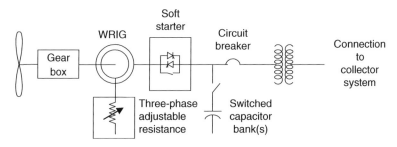

Figure 15.2 Type 2 WTG, wound rotor induction generator.

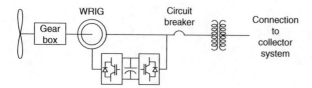

Figure 15.3 Type 3 WTG, doubly fed asynchronous generator.

15.2.2.3 Type 3: Doubly Fed Asynchronous Generator

The Type 3 wind turbine generator, shown schematically in Figure 15.3, uses a wound rotor machine similar to that used in the Type 2; however, in the Type 3 design, the rotor magnetic field is not produced using the principle of induction. Instead, a power electronic converter is used to actively control the rotor field. This design is referred to as "doubly fed" because power connections are made to both the stator circuit and to the rotor circuit of the machine. The power electronic converter can either draw power from the rotor or deliver power to the rotor, depending on the switching angles of the power electronic switches.

Low-frequency alternating currents (AC) are supplied to the three-phase rotor circuit in such a manner that the magnetic field on the rotor may be controlled to rotate relative to the rotor itself. A much wider range of speed control is possible, up to about 50%, with allowable operation both above and below the synchronous speed. The converter is sized based on the rotor power it is required to handle, which is typically about 30% of the total power rating of the full machine.

When the rotor is spinning faster than synchronous speed, power can be drawn from the rotor circuit. The effect is similar to the power dissipated in the external resistors of the Type 2 design, but when the Type 3 is operating in this mode, the converter delivers the power to the power system instead of just dissipating the energy as heat in an external resistor. Efficiency is increased and more power can be delivered to the power system as compared to the Type 2 design.

When the rotor is spinning at exactly synchronous speed, direct currents (DCs) are fed to the rotor windings and the machine has the effect of operating as a synchronous generator. Alternatively, the converter can deliver power to the rotor. In this mode of operation, the Type 3 machine operates with the rotor spinning at slower than synchronous speed while still delivering active power to the power system from the stator, that is, while still acting as a generator.

In any of these modes of operation, rotor excitation can be adjusted to produce or consume reactive power. The control scheme may allow fast response of reactive compensation to respond to changing power system conditions.

15.2.2.4 Type 4: Full-Scale Converter

The Type 4 wind turbine generator uses a full-scale power electronic converter, as shown schematically in Figure 15.4. This type requires a larger converter than the Type 3; the converter must be sized for the full output of the machine instead of just the rotor power. Use of the full-scale converter mechanically decouples the input from the output. Generated power may be at any frequency; so, this allows a wide range of input speed.

The generator in the Type 4 wind turbine may be an induction generator, permanent magnet synchronous generator, or adjustable field synchronous generator. Some turbine designs

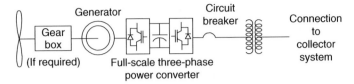

Figure 15.4   Type 4 WTG, full-scale power electronic converter.

of this type use slow-speed generators with a large number of poles and by doing so are able to eliminate the need for a gear box, which has historically been one of the chief reliability concerns of wind turbine generators. The Type 4 design can also be controlled to produce or consume reactive power, similar to the function of a static compensator (STATCOM).

15.2.2.5   Type 5: Synchronous Generator with Variable Speed Input Torque Converter
The Type 5 wind turbine generator, as shown schematically in Figure 15.5, uses a variable input speed torque converter to drive a synchronous generator at fixed constant speed.

The generator is an adjustable field synchronous machine with medium-voltage output, typically in the range of about 12–25 kV. Voltage must be specified at the time of purchase to match the intended application. These generators may be installed individually or in small groups as distributed resources connected to the utility distribution system. A step-up transformer is not normally required because the generator output voltage is selected to match the distribution system to which it will connect.

15.2.3   Wind turbine behavior during faults

Behavior of wind-powered generation plants under faulted circuit conditions has been the subject of much study in the last decade [4]. Behavior of the Type 1, Type 2, and Type 5 generators is that of the well-known induction and synchronous machines. However, most large-scale plants have used the Type 3 and more recently the Type 4 machines. Because of the use of power electronics and active control systems applied to the Type 3 and Type 4, the behavior of these machines under faulted circuit condition is not well known, and neither can general statements necessarily be made.

Before the widespread use of wind-powered generation, the general practice was to simply trip the machine during fault conditions. Then, after detecting the return of the utility

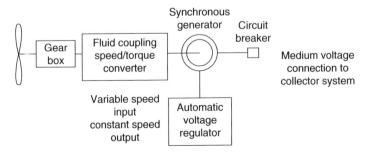

Figure 15.5   Type 5 WTG, synchronous generator with speed/torque converter.

source and after a significant intentional time delay, the turbine would be restarted and reconnected to the line. This is still the practice with small-scale plants (less than 10 MW). Larger plants cannot be allowed to trip in this manner. The loss of generation can lead to system unbalance between load and generation, which could result in system instability and blackout. Thus, larger plants have a requirement for low-voltage ride-through (LVRT): the plant must not trip during a system fault unless the location of the fault is such that it is necessary to trip the plant in order to clear the fault.

With the use of the power electronics in the Type 3 and Type 4 machines, the behavior of the machine depends most predominantly upon the control algorithm and programing developed by the manufacturer. As such, the behavior may vary from one model to the next. The Type 3 machines may use a "crowbar" circuit to short the rotor circuit and protect the power electronics. Under the crowbarred condition, the Type 3 machine will behave as a standard induction machine. After the crowbar is turned off, the power electronics will control the machine output, often switching to produce reactive power to support the system voltage during the fault condition.

Some manufacturers have reported that their Type 4 machines will behave as a positive-sequence current source during fault condition [4]. The amount of current produced is thus only slightly more than the rated current and may not be detectable by overcurrent relays that are typically set with some margin above the expected maximum load current. The Type 4 machines will also have LVRT capability and should also produce reactive power during faulted circuit conditions.

15.2.4 Photovoltaic generating systems

PV cells convert the incident sunlight directly into a DC source. Individual solar cells are combined in series and parallel combinations and encapsulated by the manufacturer to form a module, and several modules together form a PV panel. The panel is the basic building block for the PV array. The DC current from the array is connected to an inverter and converted to AC for connection to the power system. Figure 15.6 shows an operational schematic of a PV-generating system.

In the installation, sufficient panels are connected in series to obtain the desired input voltage for connection to the inverter. Then, a sufficient number of series strings of panels

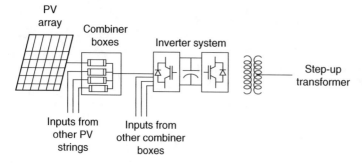

Figure 15.6   Photovoltaic system schematic.

are connected in parallel to provide the desired input power to the inverter. The total installation of the series- and parallel-connected panels forms the array.

The current–voltage relationship for a PV module is shown in Figures 15.7 and 15.8 for various operating conditions. Figure 15.7 shows generated current and voltage for various illumination levels at constant temperature, 25 °C. Figure 15.8 shows the current–voltage relationship at full illumination with varying temperature.

It can be seen in Figure 15.7 that the generated current varies with the amount of illumination of the panel and with the terminal voltage. At any illumination level, the short-circuit current is only slightly higher than the normal operating current and the open-circuit voltage changes only slightly with illumination level.

In Figure 15.8, it can be seen that current at full illumination varies only slightly with temperature, but that open-circuit voltage increases at lower temperature. These characteristics must be considered when deciding on the number of panels to be series-connected to assure that the inverter input voltage will not be exceeded and that the wiring and other components will have sufficient insulation- and voltage-withstanding capability under all operating conditions.

A common control scheme is to apply maximum power point tracking. That is, the controller adjusts the input voltage presented to the array by the inverter to the value that captures maximum power from the array and the controller continuously adjusts to changing illumination conditions to maintain the maximum power capture.

Installations of solar PV systems are addressed in Article 690 of the National Electrical Code (NEC) [5].

Under faulted-circuit conditions, behavior of the inverters for PV systems will depend upon the control algorithms and programing used, but typically the output current will only slightly exceed rated current during fault conditions. Smaller installations complying with the Institute of Electrical and Electronics Engineers (IEEE) Standard 1547 [6] and/or Underwriters' Laboratories (UL) Standard 1741 [7] will trip as soon as there is a disturbance on the utility grid. Larger installations connecting to the utility transmission system or subtransmission system will have a LVRT requirement and will have to stay on line during short-duration disturbances.

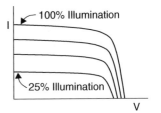

Figure 15.7 Solar PV module *I–V* characteristic: variation with illumination at constant temperature, 25 °C.

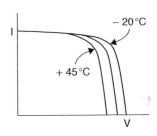

Figure 15.8 Solar PV module *I–V* characteristic: variation with temperature at full illumination.

15.3 Connections to the power grid and protection considerations

Connections of the renewable generation to the power system may be made at the usage voltage level, at the utility distribution system level, or at the utility transmission level. Where the connection is made depends on the location and size of the generation plant.

Rules will vary slightly by jurisdiction, but are generally similar from one location to another. Connections at the usage level and on the distribution system level are smaller installations and are also known as "distributed generation" because the generation is dispersed and intermingled with the connected loads. Connections at the transmission voltage level will generally be larger plants, at least 10 MW and up to about 100 MW or larger.

Small residential installations up to about 10 kW or so can be made at the usage voltage level and are usually "net-metered." That is, the generation is connected on the customer side of the revenue meter, the power generated is used to offset the amount purchased from the utility company, and any surplus generation can be allowed to flow "backward" through the meter back to the utility and can be used as credits toward future purchases at a different time of day or on a different day. Accounts are reconciled at the end of the month or end of the year, depending on the specific rules. If there is a surplus of generation, the customer may be allowed to roll credits over to the next month, or may be paid for the excess credits or may lose the excess credits.

Larger installations at industrial sites may be 250 kW up to 10 MW. These installations may also be net-metered or may be separately metered and compensated at the feed in tariff rate. There may be economic advantages, tax rebates, or other incentives to choose separate metering over net metering. The utility will have to conduct studies on the proposed installation to determine the impact of the generation on the existing facilities and what system upgrades will have to be done to safely connect the generation. Facility upgrades may include basic infrastructure (e.g., wire and cables) to carry the power, upgrades to the relaying systems to detect faults, and additional communication infrastructure to assure tripping of the remote generation when required.

Plants larger than 10 MW in size require special study and consideration. Such large plants are most often connected at the subtransmission or transmission system level.

15.3.1 Distributed generation

In the United States, installations of distributed generation up to 10 MW in size are usually governed by IEEE Standard 1547 [6] or rules similar to it adopted by the local authority having jurisdiction (AHJ). The electric utility distribution system was originally designed with expected power flow occurring in only one direction. The energy source was the transmission grid, power is stepped down to the distribution system at the distribution substation, and several distribution feeders carry the power to pole-top or pad-mounted transformers where voltage is stepped down to the final customer-usage level. The feeder-protective relays are usually simple overcurrent and time–overcurrent relays and often with an automatic reclosing relay. On more recent installations, microprocessor-style multifunction relays may be applied, but there is a large installed base of existing equipment using electromechanical relays that is still in service. With the addition of generation on the feeders, several problems can occur: the new generation source will increase the available short-circuit current on the power system; reach of existing relays may be affected; current flow from unexpected directions may cause misoperation of existing relays; unintended islanding operation may occur; and depending on the type of interconnection transformer, the new generator may be an ungrounded source during islanding operation [8]. These concerns are illustrated in Figure 15.9 and discussed in the following text.

Figure 15.9 One-line diagram of typical utility distribution system with distributed generation and illustrating faults at various locations.

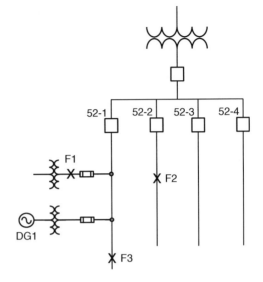

Consider the distribution substation shown in Figure 15.9 with the distributed generation DG1 installed on one of the feeders. For a fault anywhere on the distribution system, the additional generation provides additional fault current that may result in existing equipment on the power system being subjected to faults in excess of their ability to interrupt. For example, for the fault at location F1, the fuse must interrupt not only the fault current from the substation bus, but also the additional fault current supplied from the DG1 source. If the fuse had been selected based on the fault current contribution from the substation alone, it may be inadequately rated for the total fault current when the DG1 contribution is included. As a result, it may be necessary to replace the fuse with a device having higher ratings before allowing the DG1 to be brought on line. All other equipment must be similarly evaluated.

For the fault at location F2 in Figure 15.9, the generator DG1 will also contribute fault current. In addition to the possibility of exceeding the fault-interrupting capacity of the circuit breakers, the backfeed of fault current may result in unnecessary trip of the feeder circuit breaker 52-1 if the current contribution from DG1 exceeds the trip setting of the feeder overcurrent relays.

For a fault at location F3, the DG will contribute fault current, and the infeed from the DG source may result in a reduced fault contribution from the substation bus. Thus, the ability of the feeder relays to detect this fault may be reduced.

Also, once the feeder circuit breaker has tripped, an unintended island may form. If the DG1 source generation approximately matches the connected load on the feeder at the time of the trip, the DG can maintain the voltage on the feeder and continue to supply power to the connected loads. This could interfere with automatic circuit reclosing, which is commonly applied on distribution feeders. While islanding, the frequency and phase angle are not perfectly maintained in synchronism with the larger power system. So, the reclosing may occur with a large phase difference between the islanded feeder and the larger main power system. Such out-of-phase switching can result in large inrush current transients and damage to equipment, including the DG or other equipment or machines.

Damaging overvoltages may occur on the power system following a feeder trip if the DG is not also promptly tripped. For example, if the interconnection transformer is delta-connected on the feeder side, it will not be able to source current to a ground fault on the feeder, but will try to maintain line-to-line voltage. As a result, the unfaulted phases will have their voltage elevated during such a fault—the voltage on the faulted phase is shorted to ground by the fault, but the voltage on the unfaulted phases is elevated above ground to the full line-to-line system voltage. If the loads on those phases are only intended for line-to-ground connection (the most common practice in North America), then they may be damaged by sustained overvoltage if the DG is not tripped promptly.

The rules for connection of small DG require the DG to automatically trip within 2 s following the formation of an unintentional island, or within 0.16 s (10 cycles at 60 Hz) if the voltage or frequency deviates substantially from normal [6]. If the DG is substantially smaller than the minimum connected feeder load, the generator will not be able to provide the required power to maintain the frequency and should trip in less than 10 cycles. Utility companies often do not have historical information about the minimum connected load, but interconnection rules usually allow a fast-track process for connection of smaller generators, provided the aggregate generation capacity including the newly proposed generation does not exceed 15% of the maximum feeder load [9]. If the proposed addition will result in exceeding 15% of the feeder maximum load, then the addition may still be allowed, but only after detailed studies are performed to determine the impacts on the system and the required mitigation means that must be implemented. If the DG is large enough to supply the connected load, it may be necessary to install communications equipment to provide direct transfer trip from the distribution substation so as to assure trip of the generation when it is necessary to trip the feeder to clear a fault. The island must not be allowed to persist because it may result in reclosing out of phase when automatic reclosing attempts to restore the feeder following an overcurrent trip.

15.3.2 Connection to the transmission system

Large wind or solar-generating plants can be up to about 100 MW and it is possible that the total output of several such plants may be combined at one point of interconnection to the transmission system. Individual generators are typically rated up to about 2 MW, with even larger units up to 5 MW available for offshore applications. The output from the individual generators is stepped up to the collector system voltage, usually 34.5 kV, and brought to the collector substation by the collector system feeders. Feeders are usually underground cables, but overhead systems are also sometimes used. The collector substation usually has about four feeders, each feeder having approximately 12 turbines connected. There may be additional substation feeders for connection of capacitor banks and other reactive power compensation systems. The collector substation transformer, also referred to as the main power transformer, steps the voltage up to the transmission system level. The high side of the station may be the point of interconnection, but often an interconnection transmission line is required to carry the power from the high-voltage side of the collector substation to the interconnection substation. If the interconnection substation is nearby, such as adjacent to the collector substation, the "transmission line" may just be rigid bus over the fence between the facilities.

Reliability of the transmission system is a very important consideration for any facilities connected to it. The transmission interconnection rules are intended to provide fair and open access for all generators while assuring continued reliability of the transmission system. The transmission provider may require installation of a three-terminal switching substation at the point of interconnection. The three-terminal station may be three circuit breakers arranged as three taps from a common bus or as a three-breaker ring. Examples are shown in Figure 15.10. The ring bus arrangement uses the same number of circuit breakers, but allows more operating flexibility. With the ring bus, any circuit breaker can be taken out of service for maintenance while still feeding through and keeping all lines in service. It is also common practice to design the ring bus to accommodate future expansion to a breaker-and-a-half scheme.

The three-terminal interconnection switching station splits the existing transmission line into two lines and connects the new generation between the two sections via another spur line to the collector station. This arrangement avoids the complications of attempting to protect "three-terminal" transmission lines and is fair to any future generation providers that may also want to connect to the same transmission line—at present there is no technologically feasible means to protect transmission lines with more than three terminals. In addition, connection to the generation will be more reliable. If a three-terminal line were to be allowed, the generation would have to be tripped for a fault anywhere within the three terminals of the line. By sectioning the existing line into two lines, there are two paths for power flow from the generation to the transmission network, and the generation will only be tripped in the event of a fault on the spur line to the generating collector substation.

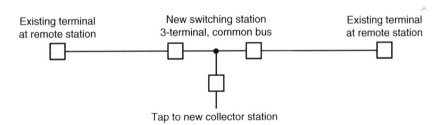

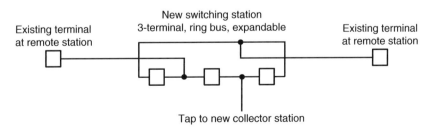

Figure 15.10 Interconnection switching station examples.

A fault on either of the other two lines, or elsewhere on the transmission network, should not result in tripping of the generation.

Plants connected to the transmission system will usually also have a requirement for LVRT. For a fault outside the plant, the plant must not trip unnecessarily while the fault is cleared by the appropriate protective system. That is, for a fault on a transmission line just outside the plant, the plant must not trip during the time it takes for the line protection to operate and clear the fault. Of course, if the location of the fault is such that the generation must be tripped in order to clear the fault, then loss of the generation is unavoidable in this instance. But any unnecessary loss of generation must be avoided. Otherwise, more serious system issues may occur because of unbalance between load and generation.

15.3.3 Typical protection schemes on wind plant collector system

Collector systems of large wind-powered generation plants in North America are typically at 34.5 kV, which is the highest voltage of commercially available distribution equipment. A typical collector system is shown schematically in Figure 15.11.

Due to fault current considerations, the maximum connected generation on any one collector bus is limited to approximately 100 MW. Substation circuit breakers and air-insulated disconnect switches are available for fault currents up to 40 kA, but the insulated separable connectors commonly used with the underground cables (also known as "dead-break elbows") are limited to only 25 kA. Underground cables are the most commonly used for transporting the power back to the collector station, but overhead lines are also sometimes used.

The generator step-up transformer at the base of the turbine tower or in the nacelle is most often a delta connection on the collector system side. As a result, these transformers will not stabilize the collector system voltages with respect to ground if the feeder circuit breaker should trip. For this reason, grounding transformers are usually used on the feeder

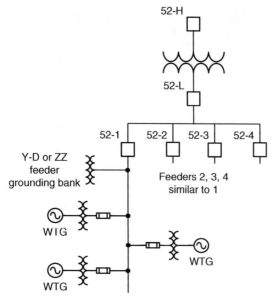

Figure 15.11 Typical large wind-generation plant's one-line diagram.

side of the circuit breaker at the substation. Some installations do not use feeder grounding transformers, but instead use a specially designed circuit breaker having an integral grounding switch. When the feeder circuit breaker trips, it immediately applies a three-phase ground to the feeder circuit. This effectively limits the feeder voltage and saves the expense of a grounding transformer bank, but may result in increased stress on the generators by applying a three-phase-to-ground fault when the feeder trips. The remainder of this section highlights the most commonly applied protection schemes used on large wind plants. Details of the protection schemes for the various types of equipment are discussed in the preceding chapters. It is important to note also that microprocessor-type relays are used because of the availability of this technology. Through appropriate control logic programing, the microprocessor-type relays allow the creation of sophisticated schemes, utilizing many protection elements, with different levels and time delays, and even communication-assisted tripping schemes at reasonable cost and size.

Main power transformer protection depends upon the size and cost of the transformer and the importance of the transmission system to which it connects. Transformers larger than about 20 MVA will usually utilize a current differential relay (87T)[1] with backup overcurrent relay (50/51) as well as mechanical relays for sudden pressure (63), winding temperature (49), liquid temperature (26), and liquid level (71). Redundant relaying may be used on larger more expensive transformers, or may be a requirement for reliability of the transmission system to assure prompt disconnection of a faulted unit. The transformer differential zone usually includes the high-side circuit breaker and the medium-voltage-side main circuit breaker, if there is one. If there are only one or two collector feeders, the medium-voltage side bus may also be included in the transformer differential zone. However, it may be advisable to use a separate protection zone for the medium-voltage side bus from the point of view of service restoration following a fault. Whenever a transformer differential trip occurs, it is advisable to test the transformer to assure there is no internal problem before reenergizing the transformer. Transformer testing can be costly and time-consuming, but not nearly so much so as costs and delays that can occur when energizing a faulted transformer. If a bus fault should cause operation of the transformer differential, delays could be experienced while it is confirmed that it is safe to restore the transformer to service.

Main collector bus protection may be a bus differential relay (87B), or may be included in the transformer differential zone, or may be protected with time-delayed overcurrent relays (50TD/51) in a "fast bus" scheme. As explained earlier, except for small stations where the cost may be prohibitive, it is usually advisable to provide a separate bus protection zone, so that in case of a fault, the fault location can be ascertained with certainty and unnecessary tripping of the transformer differential can be avoided. A high-impedance bus differential relay provides secure and reliable detection of faults in the bus zone, but requires a dedicated set of current transformers at each terminal of the bus—every feeder circuit breaker and the transformer main medium-voltage side circuit breaker, or transformer medium-voltage side bushings if there is no main medium-voltage side circuit breaker. A "fast bus" scheme can be created using a time-delayed overcurrent (50TD) element set for sensitive, fast response to bus faults with operation of this element restrained by instantaneous overcurrent (50) relay elements applied on the feeders. The 50TD element detects overcurrent on the

---

1 Numbers in parentheses are IEEE standard device function numbers: Appendix A and Reference [10].

transformer medium-voltage side main terminals or medium-voltage side main circuit breaker. The 50TD element has an intentional time delay of a few cycles to allow time for a blocking signal to be received and processed from the feeder relays. Pickup of the feeder relay bus blocking element (50), indicating that the fault is outside the bus zone, can be communicated to the fast bus relay either through a physical contact wired to an input on the relay or through a communication circuit [11]. A time–overcurrent element (51) is also included in the bus protection and coordinated with the feeder relays as backup to the fast bus scheme or if a fast bus scheme is not used.

Collector feeder protection typically uses overcurrent and time–overcurrent relay elements (50/51). Because all of the feeders have connected generation, a fault on one feeder will be supplied from adjacent feeders as well as the utility source. For a fault on a collector feeder, it is important to assure coordination so that the faulted feeder trips and clears prior to tripping of the adjacent feeders or the connection to the transmission utility. The utility source usually will supply much more fault current than one collector feeder; so, it is usually not difficult to select pickup settings and appropriate time delays that will assure coordination. Additionally, directional overcurrent relay elements (67) may also be used to help discriminate between a faulted feeder, where the current flow is toward the feeder, from a fault on an adjacent feeder or a fault on the transmission system, where the fault current flow is from the feeder toward the substation collector bus. For faults on the transmission system, the collector feeders should restrain and allow clearing of the fault by the transmission protection. In such a situation, the collector feeders only operate as backup protection.

Collector feeder grounding banks, if used, are usually included in the collector feeder protection zone. Feeder settings should be set appropriately to provide protection of the feeder grounding bank; directional overcurrent relay elements are sometimes useful in providing the required sensitivity. A neutral current transformer is also sometimes used and wired to an overcurrent relay element (51G or 50TDG) of the feeder relay to provide backup protection of the grounding bank. Ground fault detection should not be set too sensitively nor with too short a time delay because the grounding bank must be allowed to source current to a ground fault, even faults on adjacent feeders, without tripping unnecessarily.

Turbine protection will depend on the design of the turbine and location of the step-up transformer. If the design uses a pad-mounted step-up transformer installed next to the base of the turbine tower, internal fuses are usually installed on the transformer medium-voltage side and low-voltage switchgear is installed inside the base of the turbine tower. If the design uses a step-up transformer at the top of the tower, either inside the nacelle or on a platform near the top, a medium-voltage fault interrupter with overcurrent relay (50/51) is usually installed in the tower base.

15.3.4 Transmission system protection

The connection to the transmission system may include a short transmission line from the collector substation to the substation at the point of interconnection to the existing transmission system. Usually, this line will be several miles in length, and optical fiber cables can be included for secure and reliable communications between the stations. Relaying for this interconnecting line usually employs a line differential (87L) and may also include impedance relay elements (21) and overcurrent relays (50/51) for backup. Redundancy may also be a requirement for reliability of the transmission system.

If the collector substation happens to be built next to the existing transmission line, then the interconnecting "line" may be a section of rigid bus between the two substations. In this situation, the two substation ground mats will be tied together and the two substations are effectively two sections of one large substation, although the two sections may be owned and operated by different companies. Protection of the connecting bus may be accomplished with a bus differential relay (87B).

As described earlier, it is usually not possible to simply tap into the transmission line with a single circuit breaker at the point of interconnection. The resulting "three-terminal" transmission line may be difficult or impossible to protect. Existing high-speed impedance elements may have to be delayed and the reach reduced to avoid overreaching due to the inflow of fault current from the third terminal. However, when the third terminal is out of service or not generating, for maintenance or otherwise, fault clearing may be delayed on a section of the line as a consequence of maintaining selectivity when all terminals are in service. Because of these difficulties and the requirement for reliability of generation, it is usual to construct a switching substation with three line terminals at the point of interconnection to the transmission grid [12]. The switching substation may be a ring bus arrangement for increased operating flexibility.

By the addition of the switching substation, the existing transmission line is split into two lines and a third terminal supplies the spur line to the collector substation. At the new switching station, the relaying must be compatible with the existing relaying at the remote terminals. Otherwise, it may be necessary to upgrade and replace the remote terminals as well. If specialty relaying was used on the existing line, it may be necessary to relocate the existing relaying from one of the remote stations to the new switching station, keeping the two existing line terminals as a matched set on one segment of the line, and supply new relaying for both terminals of the new line segment.

15.4 Grid codes for connection of renewables

Interconnection requirements or "grid codes" are intended to maintain the safety and reliability of the power system and to provide consistent rules that are fair to all participants. The requirements usually include construction and operation requirements consistent with good utility practice, generation requirements for the range of reactive power that must be supplied, voltage operating ranges, response of generation during a system fault, LVRT requirements, and response to frequency variations.

For connections to the distribution system, the requirements are usually consistent with IEEE 1547 [6]. These requirements include operation at unity power factor, not actively regulating the voltage or frequency, and isolating promptly from the grid during system disturbances. Once disconnected, the DG must not reconnect until the system voltage has returned and has been confirmed to be stable for a specified time, generally about 5 min.

If there is a risk of operation as an unintended island, the utility may require additional protective equipment such as communications equipment to facilitate transfer trip. Such means are avoided if possible because of the increase of the system cost, but may be required for the safe operation of the distribution system.

For connection to the transmission system, the generation will be required to supply or consume reactive power along with the active power supplied and will have to remain on line during temporary system disturbances.

Reactive power requirements generally fall into one of two basic requirements as illustrated in Figure 15.12.

Either there is a range of power factor specified that the generating plant must be capable of maintaining regardless of power output [13], illustrated by the area inside the *V* curve in Figure 15.12, or there is specified range of reactive power or var capability based on the plant-rated output, and the plant must be capable of producing reactive power anywhere within this capability range regardless of the active power generation of the plant down to a specified minimum output [14], as illustrated by the rectangular region in the figure.

Meeting the reactive power requirements may require the addition of switched capacitor banks, switched reactor banks, and possibly the use of fast-responding dynamic reactive power sources. Type 3 and Type 4 wind turbines include dynamic var capability, but often this requirement is met with a separate device such as a distribution static compensator (DSTATCOM) or other power-electronics-controlled device. A complete var compensation system usually controls the switched reactive devices (capacitors and reactors) and automatically fine-tunes the total system output moment to moment by adjusting the power electronics device within its operating range. Figure 15.13 shows an example reactive compensation system. The reactive compensation system may be required to supply additional reactive current and support the system voltage during faults. Some systems have significant short-time overload capability that facilitates meeting the temporary performance requirements during system fault conditions, whereas other systems may require a somewhat larger rated continuous capability in order to have sufficient capacity to meet temporary requirements.

LVRT requirements are a means of specifying the behavior of the generation during system faults to avoid unnecessarily tripping the generation. An example LVRT curve is shown in Figure 15.14 [15]. Requirements usually include withstanding a three-phase fault that may occur on the transmission system just outside the plant (i.e., 0 V on the high-voltage side of the interconnect transformer) for a duration of up to nine cycles, which is a time duration consistent with high-speed detection and clearing of such a fault. As can be seen

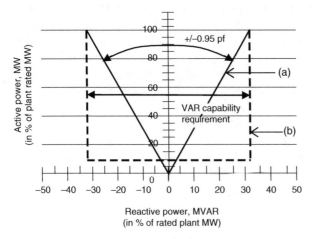

Figure 15.12 Typical reactive power-generation requirements: (a) constant power factor *V* curve and (b) reactive power range requirement based on total plant size.

Figure 15.13 Reactive compensation system rated 78/12 megavolt ampere of reactive power (MVAR) including three stages of switched capacitors, 22 MVAR each, and dynamic capability of 12 MVAR continuous and 33 MVAR short time. *Source:* Courtesy of S&C Electric Company.

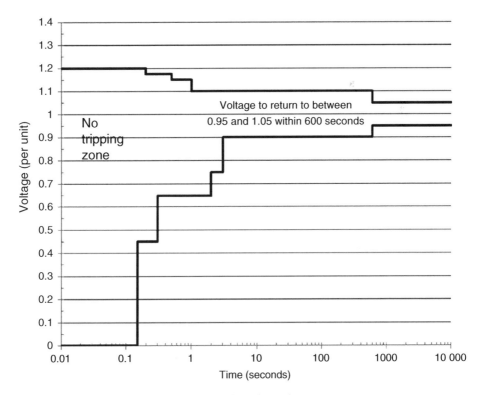

Figure 15.14 Example of low-voltage ride-through requirements.

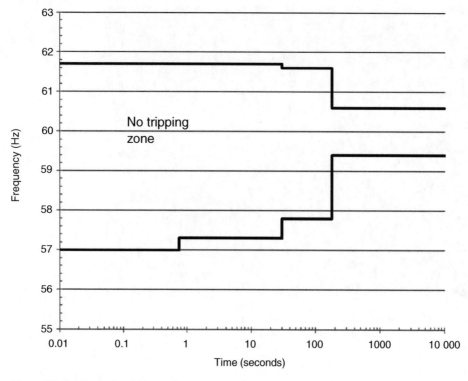

Figure 15.15 Example of frequency variation requirements.

in the figure, the required withstand time duration is longer for voltage sags that are less severe. This requirement is consistent with faults occurring further out on the transmission system; for such remote faults, current flow through the intervening impedance between the plant and the fault location will support the voltage at the plant, and the response of the relay systems may be intentionally delayed in responding to remote faults (i.e., Zone 2 response).

Transmission-connected generation must also withstand frequency variations during transient events. Figure 15.15 shows example requirements [15]. As with the voltage requirements, the requirement to ride-through frequency variations is longer for less severe deviations from normal, and the intent of this requirement is to avoid cascading effects and possible system blackout due to loss of generation.

15.5 Summary

Application of renewable energy resources to the electric power system is increasingly more prevalent and growth is expected to continue. The flexibility of power electronics devices allows means of capturing energy from sources previously untapped. But the behavior of the power electronics devices during system disturbances is different from that of classical machines, depending more upon the programed control algorithms decided upon by the designer rather than the inherent physics of devices themselves. Detailed application studies using accurate device models are often required to determine the behavior during faulted circuit conditions and to develop protective relay schemes that will adequately detect the faulted conditions. Application of renewable generation to distribution systems presents challenges to the existing protection systems that did not consider the possibility of such generation sources at the time the power system was designed and constructed. For new construction, the application of modern microprocessor-based relaying systems allows sophisticated protection schemes to be employed using programed logic to develop fault detection and control schemes that would be cost-prohibitive and impractical using previous generations of technology.

References

1 *US Department of Energy Report* (2008). *"20% Wind Energy by 2030: Increasing Wind Energy's Contribution to U.S. Electricity Supply"*. DOE/GO-102008-2567, http://www1.eere.energy.gov/wind/pdfs/41869.pdf.

2 Data from US Energy Information Administration. *All Sectors Wind Generation* divided by *All Sectors Total Generation* and *All sectors Solar Generation* divided by *All Sectors Total Generation*, http:// http://www.eia.gov/forecasts/steo/query/index.cfm? periodTypeNNUAL&startYear 2009&endYear 2014 &formulas x146x200m8.

3 Camm, E.H., Behnke, M.R., Bolado, O., et al. (2009). *Characteristics of Wind Turbine Generators for Wind Power Plants. IEEE PES Wind Plant Collector System Design Working Group, Power & Energy Society General Meeting, PES* 09 July 26–30, 2009, Calgary. IEEE, New York.

4 Miller, D.H., Walling, R., Harley, R., *et al.* (2013). *Fault current contributions from wind plants*, Joint Working Group on Fault Current Contributions from Wind Plants. PES-TR26, IEEE, New York, IEEE report.

5 NFPA-70: National Electrical Code, NEC, published by the National Fire Protection Association, Quincy, Massachusetts, USA.

6 IEEE 1547 – 2003. (2003). *IEEE standard for interconnecting distributed resources with electric power systems*, Institute of Electrical and Electronics Engineers, New York.

7 UL 1741 (2010). *Standard for Inverters, Converters, Controllers and Interconnection System Equipment for Use with Distributed Energy Resources*, (standard), 2e. Northbrook, Illinois, USA: Underwriters' Laboratories, Inc.

8 Seegers, T.T., Birt, K., Beazer, R., et al. (2004). *Impact of Distributed Resources on Distribution Relay Protection. A Report to the Line Protection Subcommittee of the Power System Relay Committee of The IEEE Power Engineering Society prepared by Working Group D3*, http://www.pes-psrc.org/kb/published/reports/wgD3ImpactDR.pdf, IEEE, New York, IEEE report.

9 See for example, *North Carolina Interconnection Procedures, Forms, and Agreements for State-Jurisdictional Generator Interconnections Effective June 16, 2009, Docket No. E-100, Sub 101*, http://ncuc.commerce.state.nc.us/cgi-bin/webview/senddoc.pgm?dispfmt &itype Q&authorization &parm2 JAAAAA77190B&parm3 000122299.

10 IEEE Std C37.2-2008. (2008). (Revision of IEEE Std C37.2-1996) *IEEE standard for electrical power system device function numbers*, acronyms, and contact designations, IEEE, New York.

11 Pickett, B., Sidhu, T., Anerson, S., et al. (2009). *Reducing outage durations through improved protection and auto restoration in distribution substations*, Working Group K3 Report to the Subcommittee of the Power System Relay Committee, http://www.pes-psrc.org/Reports/WG_K3_draft%20full_paper-13-2.pdf, IEEE report.

12 Miller, D.H. and Niemira, J.K. (2011). *Fault Contribution Considerations for Wind Plant System Design and Power System Protection Issues. Power and Energy Society General Meeting*, July 24–29, 2011. New York: IEEE.

13 United States of America, Federal Energy Regulatory Commission (2005). 18 CFR Part 35, (Docket No. RM05- 4-001; Order No. 661-A), Interconnection for Wind Energy. http://www.ferc.gov/EventCalendar/Files/20051212171744-RM05-4-001.pdf.

14 ERCOT (2012). *Nodal Protocol Revision Request NPRR Number 452, Clarification of Reactive Power Requirements*. http://www.ercot.com/content/mktrules/issues/nprr/451-475/452/keydocs/452NPRR-01_Clarification_of_Reactive_Power_Requirements_0326.doc.

15 North American Electric Reliability Corporation (NERC), *NERC PRC-024-1, Proposed LVRT Requirements from Attachment 2 and WECC Frequency Withstand Requirements from Attachment 1*, http://www.nerc.com/docs/standards/sar/PRC-024-1_clean_third_posting_2012Feb22.pdf (accessed 24 June 2013).

16

Solutions

*{**Note:** The solutions provided here have been checked for accuracy. However, it is possible that some errors remain. If the user finds such errors, a note about this to the authors would be welcome.*

If any instructor wishes to assign homework for students, it is of course not possible to use the problems in the book as their solutions are given here. However, what we have done in our own classes is to change some of the data in the problems and assign them as homework.}

16.1 Chapter 1 problems

(1.1) Assume that the fault is near bus 5.

	System normal		Stub end	
Line out	Fault current	Current in B_1	Fault current	Current in B_1
None	7.8083	2.7074	3.5136	3.5136
1–2	5.3242	1.9136	2.9543	2.9543
2–6	7.7747	2.7066	3.5136	3.5136
2–5	5.3639	3.1840	3.5136	3.5136
2–4	7.5616	2.2852	2.8960	2.8960
3–5	6.2659	2.9795	3.5136	3.5136
3–6	6.2659	2.9795	3.5136	3.5136
4–6	7.2264	2.0550	2.8060	2.8060
6–7	5.1207	1.7067	2.7895	2.7895

Power System Relaying, Fifth Edition. Stanley H. Horowitz, Arun G. Phadke and Charles F. Henville.
© 2023 John Wiley & Sons Ltd. Published 2023 by John Wiley & Sons Ltd.

(1.2) For b–c fault, all the currents in problem 1.1 are multiplied by $\sqrt{3}/2$.

	System normal			Stub end	
Line out	Fault current	Current in B_1		Fault current	Current in B_1
None	6.7620	2.3446		2.0843	2.0843
1–2	4.6107	1.6606		1.5352	1.5352
2–6	6.7329	2.3439		2.0843	2.0843
2–5	4.6451	2.7573		2.0843	2.0843
2–4	6.5483	1.9789		1.7960	1.7960
3–5	5.4262	2.5802		2.0843	2.0843
3–6	5.4262	2.5802		2.0843	2.0843
4–6	6.2580	1.7796		1.7224	1.7224
6–7	4.4345	1.4087		1.4748	1.4748

(1.3)

		B_2		B_3		B_4		B_5		B_6
Fault at	I_a	E_a at B1	I_a	E_a at B2	I_a	E_a at B3	I_a	E_a at B4	I_a	E_a at B5
2	12.851	1.0	0	0	0	0	0	0	0	0
3	6.3381	1.0	6.3381	0.51	0	0	0	0	0	0
4	4.2319	1.0	4.2319	0.67	4.2319	0.33	0	0	0	0
5	3.1130	1.0	3.1130	0.76	3.1130	0.50	3.1130	0.26	0	0
6	2.8170	1.0	2.8170	0.78	2.8170	0.55	2.8170	0.33	2.8170	0.09

(1.4) Assume the fault to be near bus 5.

	With T_2		Without T_2	
Breaker	I_b	E_{bc}	I_b	E_{bc}
B_1	4.24	1.73	8.81	1.73
B_2	5.06	1.65	4.83	1.61
B_3	5.06	0.87	4.83	0.87

(1.5) All elements are assumed to be closed. In sequence, perform the following operations to remove B_1 from service.

Open B_1. Open S_{11} and S_{12}. Close B_1 (for safety to the breaker).

To return to service, Open B_1. Close S_{11} and S_{12}. Close B_1.

For Figure 1.7a, follow the same procedure as mentioned in the earlier text.

For Figure 1.7b,c, any lines on the transfer bus (1) are transferred to the main bus (2). This is accomplished by first opening the line breaker, opening the switch to bus

1, closing the switch to bus 2, and closing the breaker. At this point, all the lines are on the main bus (2).

To maintain a breaker, the breaker is opened, its switch to bus 2 is opened, and the breaker is closed for safety. The bypass switch to the transfer bus is closed, and the bypass breaker switches and the breaker are closed. At this stage, the breaker to be maintained is isolated, and the line is energized through the transfer bus and the transfer breaker. The procedure is reversed to return the breaker to service.

For Figure 1.7d, follow the same procedure as for 1.7a,c. A subsequent fault may split the bus.

For Figure 1.7e, follow the same procedure as mentioned in the earlier text. Any subsequent fault should not split the bus.

(1.6) The load at the remote terminal is lost or not as follows.

Fault on	(a)	(b)[a]	(c)	(d)	(e)
Line	lost	lost	lost	lost	lost
Breaker	lost	no	no	no	no
Switch (line side)	lost	lost	lost	lost	lost
Switch (bus side)	lost	no	no	no	no
Bus	lost	no	no	no	no

[a]Note: In the main and transfer bus scheme, the remote end power will be lost in all cases. Except for the cases indicated by "no," where it is restored after the bus reconfiguration with the transfer bus is completed.

The conclusion is that the single-bus arrangement is the least preferable, and that the double-bus or the breaker-and-half arrangement is the most preferred.

(1.7) (a) Response is correct and appropriate.
(b) B_2 has apparently failed. Responses of R_1, B_1, R_2, R_3, B_3, R_4, and B_4 are correct and appropriate.
(c) R_1, B_1, R_2, B_2 operated correctly and appropriately, and R_5 and R_6 apparently operated incorrectly and tripped breaker B_5. This is also an inappropriate response.
(d) R_2 has failed to operate: this is incorrect, a loss of dependability. R_5 and R_6 operated in remote backup mode and took out their associated circuit breakers. This response is appropriate from the point of view of the power system.

(1.8) In Figure 1.21a,b, connect the CT on the right to the bus protection, and the CT on the left to the line protection (see Figure 16.1).

Figure 16.1 For Problem 1.8.

(1.9)   The breaker and relay operations are to be scheduled as shown in the following text (see Figure 16.2), so that for the slowest combination of relay and breaker operation times, appropriate margins are maintained.

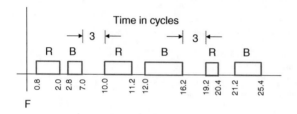

Figure 16.2 For Problem 1.9.

(1.10) (a) Fault is on the line connecting B_1 and B_2.

(b) Fault is in the overlap between the zones of the bus in the middle, and the line connecting breakers B_3 and B_4.

(c) Transformer fault.

(d) Fault on the bus in the middle.

16.2   Chapter 2 problems

(2.1)   This is to be considered a single-phase system. (The student may solve it as a three-phase system, in which case positive-sequence data must be given, and phase and ground faults considered separately. This is covered in Chapter 4.) It is sufficiently accurate to neglect all resistances. The fault at far right produces a fault current through the relay of magnitude $1.0/(0.3 + 0.15) = 2.22$ pu. This is the minimum fault current. The maximum load is 1.0 pu. An overcurrent relay pickup between 1.0 and 2.22 would be appropriate. Perhaps a setting of 1.5 pu could be used. However, as seen in Chapter 4, this is not sufficiently secure.

The smallest voltage under load would be approximately 1.0 pu. (The small voltage drop in the series impedances during load flow is negligible.) The largest voltage at the relay location during fault is $1.2 \times 0.3/0.45 = 0.8$ pu. Thus, a setting for the undervoltage relay between these two extremes, say around 0.9 pu, would be adequate. As in case of the overcurrent relay, this is not sufficiently secure.

(2.2) The delta side of the transformer is bus 1. Bus 2 is the wye side of the transformer. We will assume another branch between the wye side of the transformer and the transmission line, which connects buses 3 and 4. The branch
between buses 2 and 3 is needed, so that we may put a reverse fault at bus
2. A reverse fault at the line terminal will produce zero voltage, and the phasor
relationship for directionality will be indeterminate. (Of course, an actual fault
at the line terminal will use the memory voltage for directionality). The source
at the receiving end is at bus 5. All positive-sequence impedances will be
assumed to be $j0.1$, and all zero-sequence impedances will be assumed to be
$j0.3$. The actual values of the impedances do not matter, as the main interest
is in determining the phase relationships between various voltages and currents.
The receiving-end source will be assumed to be grounded. The delta winding
will of course have an open circuit in the zero-sequence circuit. The phase shift
in the positive- and negative-sequence circuits introduced by the wye–delta
transformer will be omitted.

The symmetrical component representation for a b–c fault in the forward direction at bus 4 is shown in the abovementioned diagram (see Figure 16.3). The phase
voltages and currents at the relay locations are

$$E_a = 0.666 + 0.333 = 1.0,$$

$$E_b = \alpha^2 0.666 + \alpha 0.333 = -0.5 - j0.2884,$$

$$E_c = \alpha 0.666 + \alpha^2 0.333 = -0.5 + j0.2884,$$

$$I_a = -j1.667 + j1.667 = 0,$$

$$I_b = \alpha^2(-j1.667) + \alpha(j1.667) = -2.8867,$$

$$I_c = \alpha(-j1.667) + \alpha^2(j1.667) = +2.8867.$$

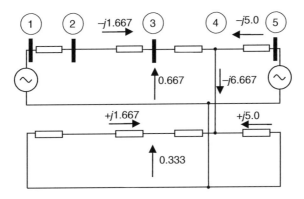

Figure 16.3 For Problem 2.2.

The appropriate relaying quantities for this fault are $E_{bc} = E_b - E_c$ and $I_{bc} = I_b - I_c$, which are

$$E_{bc} = -j5.768,$$

$$I_{bc} = -5.7734.$$

The phasor diagram which follows shows their relationship.
Consider now the following symmetrical component representation of a reverse b–c fault at bus 2 (see Figure 16.4).

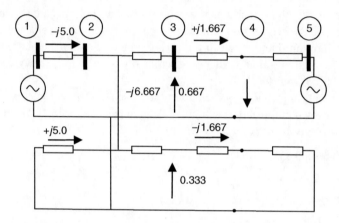

Figure 16.4 For Problem 2.2.

The symmetrical components of voltages and currents, and b–c voltages and currents are

$$E_b = \alpha^2 0.666 + \alpha 0.333 = -0.5 - j0.2884,$$
$$E_c = \alpha 0.666 + \alpha^2 0.333 = -0.5 + j0.2884,$$

$$I_b = \alpha^2(j1.667) + \alpha(-j1.667) = 2.8867,$$

$$I_c = \alpha(j1.667) + \alpha^2(-j1.667) = -2.8867,$$

$$E_{bc} = -j5.768,$$

$$I_{bc} = -5.7734.$$

The phasor diagrams for the forward and reverse faults are shown in the following text (see Figure 16.5). It follows that the b–c voltages and currents provide appropriate inputs for a directional determination, and hence E_{bc} and I_{bc} should be used for directional determination.

The symmetrical component circuit for a–g fault at bus 4 (forward) is shown in the following text (see Figure 16.6). Note that the zero-sequence circuit has an open circuit on the delta side. The phase voltages and currents at the relay locations are

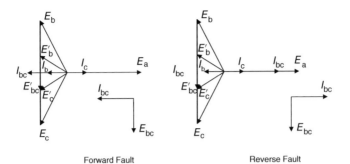

Forward Fault   Reverse Fault

Figure 16.5   For Problem 2.2.

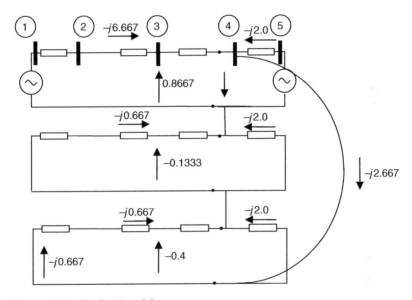

Figure 16.6   For Problem 2.2.

$$I_a = I_0 + I_1 + I_2 = 3(-j0.6667) = -j\,2.0,$$

$$I_b = 0,$$

$$I_c = 0,$$

$$E_a = E_0 + E_1 + E_2 = 0.8667 - 0.1333 - 0.4 = 0.333,$$

$$E_b = E_0 + \alpha^2 E_1 + \alpha E_2 = -0.7667 - j0.8667,$$

$$E_c = E_0 + \alpha E_1 + \alpha^2 E_2 = -0.7667 + j0.8667.$$

It is clear from the b–c fault calculations that the reverse fault will produce the same voltages, and the same current as these, except that the direction of the currents at the relay locations will be reversed. These relationships are shown in the following phasor diagram (see Figure 16.7).

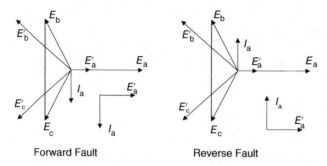

Forward Fault   Reverse Fault

Figure 16.7 For Problem 2.2.

It can be seen from the phasor diagrams that the phase a voltage and the phase a current provide a secure direction determination for the phase to ground fault. It is also clear that the current in the transformer neutral $I_n = 3I_0 = -j2.0$, whether the fault is in the forward or reverse direction as seen by the relay. Thus, it too is a secure polarizing signal for determining the direction of the fault. Of course, when the neutral current is used, the forward fault is indicated with an angle of $0°$ between the phase a current and the neutral current, and the reverse fault is indicated when the angle is $180°$.

As will be seen in Chapter 5, the operating current for ground distance relays is not I_a but a compensated phase current: $I_a' = I_a + I_0[(Z_0 - Z_1)/Z_1]$, where Z_0 and Z_1 are the zero- and positive-sequence impedances of the transmission line. The phase angle of I_a' is the same as that of I_a, and hence the directional determination is not affected whether the compensated or uncompensated phase current is used.

(2.3) $L = \dfrac{\mu_0 \pi d^2 N^2}{4\left(x + \dfrac{gd}{4a}\right)}$, $W' = -\dfrac{1}{2} I^2 \dfrac{\partial L}{\partial x}$,

$$W' = -\frac{1}{2} I^2 \frac{\mu_0 \pi d^2 N^2}{4\left(x + \dfrac{gd}{4a}\right)}. \text{ Therefore, } K = -\frac{1}{2} \frac{\mu_0 \pi d^2 N^2}{4}.$$

(2.4) A computer program written to solve the differential equation

$$0.005 \frac{d^2 x}{dt^2} = -0.001 \left\{ \frac{1.05^2}{(0.05 + x)^2} \right\} I^2 - 1.$$

From $t = 0$ to sometime T, such that with $x = 1.0$, the final x is less than 0.7 produced the following results for different values of I:

$I = 1.1$		$I = 1.5$		$I = 3.03$		$I = 10.0$		$I = 20.0$	
x	t	x	t	x	$t\times10^{-3}$	x	$t\times10^{-4}$	x	$t\times10^{-5}$
1.	0.	1.	0.	1,	0.	1.	0.	1.	0.
0.9894	0.0001	0.9901	0.0001	0.9960	0.005	0.9956	0.005	0.9965	0.01
0.7847	0.0021	0.7958	0.0011	0.9168	0.105	0.9074	0.105	0.9295	0.21
0.5921	0.0041	0.6076	0.0021	0.8379	0.205	0.8193	0.205	0.8554	0.41
				0.7593	0.305	0.7311	0.305	0.7848	0.61
				0.6809	0.405	0.6431	0.405	0.7143	0.81
								0.6473	1.0

Using interpolation, the time at which $x = 0.7$ is calculated at each value of the current I. The results are as follows:

I in pu	$t\times10^{-4}$
1.1	29.8
1.5	16.1
3.0	1.006
10.0	0.3403
20.0	0.085

The result can be plotted on a log–log scale as follows (see Figure 16.8):

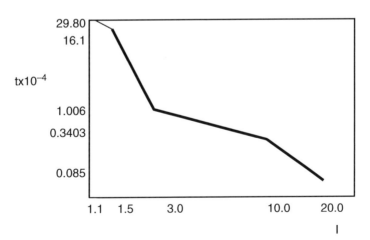

Figure 16.8 For Problem 2.4.

(2.5) From Equations (2.3) and (2.4), $I_d/I_p = (x_0 + gd/4a)/(x_1 + gd/4a)$. Let the coil inductances be L_0 and L_1 when the plunger position corresponds to x_0 and x_1, respectively. Then, by Equation (2.1), $L_0/L_1 = (x_0 + gd/4a)/(x_1 + gd/4a) = I_d/I_p$. Hence, $I_d/I_p = 2.0/2.05 = 0.9756$.

(2.6) $\tau = KI_{m1}I_{m2}(\cos\omega t\ \sin(\omega t + \theta) - \cos(\omega t + \theta)\sin\omega t)$.

Using trigonometric identities

$$\tau = KI_{m1}I_{m2}(\cos\omega t\ \sin\omega t\ \cos\theta) - \cos^2\omega t\ \sin\omega t\ \cos\theta + \sin^2\omega t\ \sin\theta)$$
$$= KI_{m1}I_{m2}\sin\theta.$$

The torque is zero if either I_{m1} or I_{m2} is zero, or if the angle θ is zero.

(2.7) Let the polar coordinates $Z\angle\theta$ be converted to rectangular coordinates R and X:

$$R = Z\ \cos\ \theta\ \text{and}\ X = Z\ \sin\ \theta.$$

The equation $Z\ \sin\ (\theta + \phi) = 0$ becomes

$$Z\ \sin\ \theta\ \cos\ \phi + Z\ \cos\ \theta\ \sin\ \phi = 0,$$

or $X\cos\ \phi + R\sin\ \phi = 0,$

or $X = -R\ \tan\ \phi,$

which is an equation of a straight line passing through the origin and making an angle of $-\phi$ with the R axis.

If the spring torque is not negligible, equation for this case becomes

$$K_3IV\ \sin\ (\theta + \phi) = \tau_s\ \text{or},$$
$$V/I = \{K_3V^2/\tau_s\}\ \sin\ (\theta + \phi).$$

If we let $\{K_3V^2/\tau_s\} = Z_r$ a constant which depends upon the voltage, we get

$$Z = Z_r\ \sin\ (\theta + \phi).$$

This is an equation of a circle in polar coordinates passing through the origin, with a diameter of Z_r, and making an angle of $-\phi$ with the vertical axis.

In more familiar rectangular coordinates, multiply the two sides by $\cos\ \theta$ and $\sin\ \theta$, respectively:

$$Z\ \cos\ \theta = R = Z_r\ \cos\ \theta\ \sin\ (\theta + \phi),$$
$$Z\ \sin\ \theta = X = Z_r\ \sin\ \theta\ \sin\ (\theta + \phi),$$

or $R - (Z_r\ \sin\ \phi)/2 = \{Z_r\ \cos\ \theta\ \sin\ (\theta + \phi) - \sin\ \phi/2\},$

$X - (Z_r\ \cos\ \phi)/2 = \{Z_r\ \sin\ \theta\ \sin\ (\theta + \phi) - \cos\ \phi/2\}.$

Squaring and adding the two expressions, and using trigonometric identities,

$$\{R - (Z_r\ \sin\ \phi)/2\}^2 + \{X - (Z_r\ \cos\ \phi)/2\}^2 = \{Z_r/2\}^2,$$

which is an equation of a circle with its center at $\{(Z_r\ \sin\ \phi)/2, (Z_r\ \cos\ \phi)/2\}$ and a radius of $Z_r/2$. This is the same circle as that expressed in the earlier text in polar coordinates.

(2.8) Let K_4 have a projection along the R axis of R_0, and a projection along the X axis of X_0. The polar coordinate Z has the projections of R and X. The equation $|Z + K_4| = \sqrt{\{K_1/K_2\}}$ is thus equivalent to $\{(R + R_0)^2 + (X + X_0)^2\} = K_1/K_2$, which is the equation of a circle.

(2.9) $|Z| = (K_3/K_2) \sin (\theta + \phi) = Z_r \sin (\theta + \phi)$.

The rest of the derivation is as in Problem (2.7) mentioned in the earlier text.

(2.10)   Fault at C produces a fault current of $(1.0)/(0.1 + 0.2 + 0.2) = 2.0$ pu. The pickup setting for the relay is $2.0/3 = 0.666$ pu. The dropout current is $1/10$ of this value, or 0.0666 pu. This is the maximum load current that the relay at A will tolerate before it trips falsely due to a fault at C.

(2.11) $Z\text{-}IZ_r \sin (\theta + \phi) = 0$.
Set $IZ_r = E_r$. The equation then becomes

$$E = E_r \sin (\theta + \phi).$$

The demonstration that this is a circle follows the arguments in Problems (2.7) and (2.9). The diameter of the circle is $|E_r|$, and the diameter makes an angle of $\pi/2 - \phi$ with the horizontal axis.

16.3   Chapter 3 problems

(3.1)   The effective turns ratio of the CT is

$$N' = I_1'/I_2,$$

where I_2 is the current in the burden, and I_1' is the current in primary winding. The actual turns ratio n is used to obtain the primary current as reflected in the secondary.
    Hence, $n = I_1'/I_1$. Thus $n' = (I_1'I_1)/(I_1I_2) = nI_1/I_2$. Therefore, by definition R, the ratio correction factor is $R = I_1/I_2$. Thus,

$$R = I_1/I_2; \text{and } \varepsilon = (I_1\text{-}I_2)/I_1 . \text{Hence } R = 1/(1\text{-}\varepsilon).$$

R can be less than 1.0 only if I_m is in phase opposition to I_2. This in turn can happen only if the burden is completely capacitive. This is not very likely.

(3.2)   Since $R = (I_2 + I_m)/I_2$ and since $I_2(Z_b + Z_1) = I_mZ_m$, where Z_b is the burden imped-ance and Z_1 is the secondary leakage impedance, it follows that

$$R = 1 + [(R_b + R_1) + j(X_b + X_1)]/(R_m + jX_m).$$

    A program was written to calculate R for given values of Z_b, Z_1, and Z_m. Substitut-ing the values from Example 3.1, that is, $R_1 = 0.01$, $X_1 = 0.1$, $R_m = 4$, $X_m = 15$, and $Z_b = 2\angle\theta$, where θ varies between $-90°$ and $90°$ in steps of $10°$ produced these results (see Figure 16.9)
:

θ	|R|	∠R
−90	0.8825	−2.09
−80	0.8912	−3.42
−70	0.9037	−4.62
−60	0.9196	−5.61
−50	0.9382	−6.39
−40	0.9588	−6.93
−30	0.9807	−7.23
−20	1.0031	−7.29
−10	1.0254	−7.13
0	1.0469	−6.77
10	1.067	−6.23
20	1.0852	−5.53
30	1.1011	−4.69
40	1.1142	−3.75
50	1.1244	−2.79
60	1.1313	−1.64
70	1.1348	−0.52
80	1.1348	0.61
90	1.1314	1.73

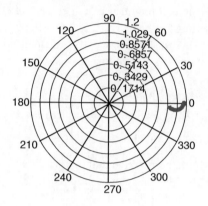

Polar plot

Figure 16.9   For Problem 3.2.

θ	|R|	∠R
−90	0.8825	−2.09
−80	0.8912	−3.42
−70	0.9037	−4.62
−60	0.9196	−5.61
−50	0.9382	−6.39
−40	0.9588	−6.93
−30	0.9807	−7.23
−20	1.0031	−7.29
−10	1.0254	−7.13
0	1.0469	−6.77
10	1.067	−6.23
20	1.0852	−5.53
30	1.1011	−4.69
40	1.1142	−3.75
50	1.1244	−2.79
60	1.1313	−1.64
70	1.1348	−0.52
80	1.1348	0.61
90	1.1314	1.73

The same program was run with θ fixed at 0, and the Z_b magnitude varied between 0 and 2 in steps of 0.2. The results are shown in the following text (see Figure 16.10):

Z_b	$|R|$	$\angle R$
0	1.0064	0.059
0.2	1.0098	−0.647
0.4	1.0133	−1.35
0.6	1.017	−2.05
0.8	1.0208	−2.74
1.0	1.0248	−3.42
1.2	1.0289	−4.10
1.4	1.0332	−4.78
1.6	1.0376	−5.45
1.8	1.0422	−6.114
2.0	1.0469	−6.772

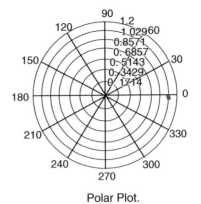

Polar Plot.

Figure 16.10 For Problem 3.2.

Z_b	$|R|$	$\angle R$
0	1.0064	0.059
0.2	1.0098	−0.647
0.4	1.0133	−1.35
0.6	1.017	−2.05
0.8	1.0208	−2.74
1.0	1.0248	−3.42
1.2	1.0289	−4.10
1.4	1.0332	−4.78
1.6	1.0376	−5.45
1.8	1.0422	−6.114
2.0	1.0469	−6.772

(3.3) The current in the burden is varied in steps of 10 A. The resulting R is obtained as follows: $|R| = |I_1/I_2|$.

I_2	E_m	I_m	I_1	$|I_1|$	$|R|$
	0	0	0	0	1.000
10	20	$0.086\angle -60°$	$10.04 + j0.07$	10.04	1.004
20	40	$0.136\angle -60°$	$20.065 - j0.1126$	20.065	1.003
30	60	$0.26\angle -60°$	$30.1 - j0.173$	30.100	1.003
40	80	$0.46\angle -60°$	$40.2 - j0.3464$	40.20	1.005
50	100	$2.06\angle -60°$	$51 - j1.732$	51.03	1.021

For a constant Z_m, R is given by

$$R = 1 + I_m/I_2 = 1 + 2/(300\angle 60^0).$$

Since at 60 V, the magnetizing impedance is $300\angle 60°$. Hence, the magnitude of R is 1.033 for all burden currents. As can be seen from the abovementioned table, this is a good approximation for most of the range of currents.

(3.4) See Figure 16.11. This nonlinear problem must be solved iteratively.

ITERATION (i): Assume 0 magnetizing currents. Then $e_{m1} = 100$ and $e_{m2} = 110$. For these values of the voltages, the magnetizing current magnitudes are $i_{m1} = 2$ and $i_{m2} = 10.0$. The corresponding magnetizing impedances are $Z_{m1} = 50\angle 60°$ and $Z_{m2} = 11\angle 60°$. The magnetizing impedance Z_{m1} is paralleled with the burden to produce $Z_{eq} = 0.9898 + j0.017$. The Kirchhoff's voltage equation for the inner impedance loop is $I(0.2) + (I + 50)Z_{eq} + (I - 50)Z_{m2} = 0$. This leads to $I = 50(Z_{m2} - Z_{eq})/(0.2 + Z_{eq} + Z_{m2}) = 44.51 + j7.57$. Solving for the voltages across the magnetizing branches, $e_{m1} = 93.86$ and $e_{m2} = 102.87$. As these are different from the voltage values with which we began the iteration, we will continue with the next iteration.

ITERATION (ii): Re-solving the same equations as before, $I = 47.282 + j5.02$, $e_{m1} = 96.46$ and $e_{m2} = 105.96$.

ITERATION (iii): $i = 47.32 + j4.6855$, $e_{m1} = 96.48$ and $e_{m2} = 105.99$. These values are close enough to the starting values of the two voltages, and we may consider the iterations to have converged.

The current in the secondary of the first CT is $CT1 = i = 47.32 + j4.685$, the current in the secondary of the second CT is $CT2 = 50 - i_{m2} = 48.93 + j1.8684$, and the total current in the burden is the sum of these two currents: $96.26 + j6.554$.

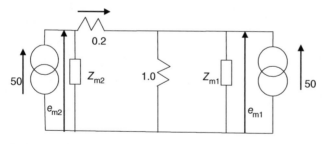

Figure 16.11 For Problem 3.4.

(3.5) Consider a burden voltage of 200 V. The 10C200 CT has error of 10% or less at this voltage. The 10C400 CT has a somewhat less error, 5% or less. As the behavior of the error in this region is highly nonlinear, a slight increase above 200 V at the secondary will cause a disproportionate increase in the CT error. It is therefore safe to rate the combination at 10C200 which is that of the poorer quality CT.

(3.6) In the case of (a), the two secondary currents of 50 and 25 A add in the burden, creating a voltage of $75 \times 1 = 75$ V. In case (b), the two currents subtract, and the voltage at the burden $25 \times 1 = 25$ V.

(3.7) In (a), the secondary currents are $I_1 = -8.333\angle 0°$ and $I_2 = -8.333\angle 180°$ in the directions indicated in the figure. Since there is no current in the primary of phase 3, the current I_3 is 0. Consequently, $I_0 = I_1 + I_2 + I_3 = 0$. In (b), the currents inside the delta in the three phases are $16.666\angle 180°$, $-8.333\angle 0°$, and $-8.333\angle 0°$, respectively. The currents outside the deltas in the direction are therefore: $I_1 = -16.666\angle 180° - 8.333\angle 0° = 25\angle 0°$; $I_2 = -8.333\angle 0° - (-8.333\angle 0°) = 0$; and $I_3 = -8.333\angle 0° - (-16.666\angle 180° = 25\angle 180°)$.

(3.8) Substituting from Equation (3.9) into Equation (3.12), we get

$$\lambda \frac{R_b + R_c}{R_b R_c I_{max} \cos\theta} = \left[\frac{1}{s + \frac{1}{\tau}}\right]\left\{\frac{s}{s^2 + \omega^2} - \frac{T}{1 + sT}\right\} + \tan\theta \left[\frac{1}{s + \frac{1}{\tau}}\right]\left\{\frac{\omega}{s^2 + \omega^2}\right\}.$$

Working on the three terms on the right-hand side,

#1:

$$\left[\frac{1}{s + \frac{1}{\tau}}\right]\left\{\frac{s}{s^2 + \omega^2}\right\} = -\frac{\tau}{1 + \omega^2\tau^2}\left[\frac{1}{s + \frac{1}{\tau}} - \frac{s}{s^2 + \omega^2} - \frac{\omega^2\tau}{s^2 + \omega^2}\right],$$

which, after taking the inverse transform and using trigonometric identities, with $\omega\tau = \tan\phi$:

$$L^{-1}\left[\frac{1}{s + \frac{1}{\tau}}\right]\left\{\frac{s}{s^2 + \omega^2}\right\} = -\tau \cos^2\phi \, \varepsilon^{-\frac{t}{\tau}} \cos\phi \cos(\omega t - \phi).$$

#2:

$$\left[\frac{-1}{s + \frac{1}{\tau}}\right]\left\{\frac{T}{1 + sT}\right\} = -\left[\frac{1}{s + \frac{1}{\tau}} - \frac{T}{1 + sT}\right]\frac{\tau T}{\tau - T},$$

$$L^{-1}\left[\frac{-1}{s + \frac{1}{\tau}}\right]\left\{-\frac{T}{1 + sT}\right\} = \frac{\tau T}{T - \tau}\varepsilon^{-\frac{t}{\tau}} - \frac{\tau T}{T - \tau}\varepsilon^{-\frac{t}{T}}.$$

#3:

$$\tan\theta \left[\frac{1}{s + \frac{1}{\tau}}\right]\left\{\frac{\omega}{s^2 + \omega^2}\right\} = -\frac{\omega\tau^2}{1 + \omega^2\tau^2}\tan\theta \left[\frac{1}{s + \frac{1}{\tau}} - \frac{s}{s^2 + \omega^2} + \frac{\frac{1}{\tau}}{s^2 + \omega^2}\right],$$

substituting for $\omega\tau$, and rearranging, the inverse transform becomes

$$L^{-1}\left[\tan\theta\left[\frac{1}{s+\frac{1}{\tau}}\right]\left\{\frac{\omega}{s^2+\omega^2}\right\}\right] = \frac{1}{\omega}\sin^2\phi\tan\theta\left[\varepsilon^{-\frac{t}{\tau}} + \frac{\sin(\omega t - \phi)}{\sin\phi}\right].$$

Combining the three results, and rearranging the result, we get

$$\lambda\frac{R_b + R_c}{R_b R_c I_{max}\cos\theta} = -\tau\cos^2\phi\,\varepsilon^{-\frac{t}{\tau}} + \tau\cos\phi\cos(\omega t - \phi)$$

$$+ \frac{\tau T}{T-\tau}\varepsilon^{-\frac{t}{\tau}} - \frac{\tau T}{T-\tau}\varepsilon^{-\frac{t}{T}} + \frac{1}{\omega}\sin^2\phi\tan\theta\left[\varepsilon^{-\frac{t}{\tau}} + \frac{\sin(\omega t - \phi)}{\sin\phi}\right],$$

which upon further use of $\omega\tau = \tan\phi$ and trigonometric identities leads to Equation (3.15).

Equation (3.16) is obtained by differentiating Equation (3.15) with respect to t, and dividing the result by R_b.

When the load is purely inductive, R_b is 0. For this case, from Equations (3.10) and (3.11), it follows that

$$i_f = \frac{R_c}{L_m R_c + L_b R_c + sL_m L_b}i_1 \text{ and } \lambda = L_m i_f.$$

Hence,

$$\lambda = L_m\frac{R_c}{L_m R_c + L_b R_c + sL_m L_b}i_1 = R_c\frac{1}{s+\frac{1}{\tau}}i_1$$

with

$$\tau = \frac{L_m L_b}{R_c(L_m + L_b)}.$$

As the form of the expression for λ is the same as that for the previous case, the form of the result for λ and i_2 is also identical. The only difference is the difference in the value of τ, and the multiplier $R_b R_c/(R_b + R_c)$ is now replaced by R_c.

(3.9) This is clearly a more advanced problem. It may be set as an undergraduate project problem. Any simulation program can be used to develop the desired output. The CT model in the EMTP has been developed as a part of the Ph.D. dissertation of A.K.S. Chaudhary at Virginia Tech, and a paper based on the work was presented at the Summer Power Meeting of the IEEE Power Engineering Society in Vancouver, BC in July 1993.

(3.10) If the burden is $Z(\cos\phi + j\sin\phi)$, and the leakage impedance is $(1 + j6)$ ohms, the burden voltage is given by

$$E_2 = E_1\{Z(\cos\phi + j\sin\phi)/[(1 + Z\cos\phi) + j(6 + Z\sin\phi)]\},$$

where E_1 is the primary voltage referred to the secondary. The ratio error is $R = E_2/E_1$, which has a magnitude and angle, which has been calculated by a computer program for three values of Z (50, 100, 200 Ω), and at variable angles ranging between 0° and 340° in steps of 20°.

The results are shown in the following table.

ϕ	Z = 50 Magnitude R	Z = 50 Angle R	Z = 100 Magnitude R	Z = 100 Angle R	Z = 200 Magnitude R	Z = 200 Angle R
0	0.9737	−6.71	0.9884	−3.40	0.9946	−1.71
20	0.9389	−5.71	0.9697	−2.94	0.9849	−1.49
40	0.9130	−4.14	0.9551	−2.16	0.9772	−1.10
60	0.8971	−2.19	0.9459	−1.16	0.9723	−0.59
80	0.8915	−0.06	0.9427	−0.03	0.9705	−0.02
100	0.8965	2.08	0.9456	1.10	0.9721	0.56
120	0.9119	4.04	0.9545	2.11	0.9769	1.08
140	0.9372	5.64	0.9688	2.91	0.9845	1.48
160	0.9716	6.67	0.9873	3.38	0.9940	1.70
180	1.0128	6.98	1.0083	3.47	1.0046	1.73
200	1.0570	6.43	1.0293	3.12	1.0148	1.54
220	1.0977	4.98	1.0476	2.37	1.0235	1.16
240	1.1273	2.76	1.0601	1.30	1.0293	0.63
260	1.1385	0.07	1.0648	0.03	1.0314	0.02
280	1.1284	−2.62	1.0606	−1.23	1.0295	−0.60
300	1.0097	−4.87	1.0484	−2.32	1.0238	−1.13
320	1.0593	−6.37	1.0304	−3.09	1.0153	−1.52
340	1.0152	−6.97	1.0094	−3.46	1.0051	−1.72

It should be noted that the error goes down as the burden impedance increases. This is of course opposite to the effect of load impedance in a CT secondary. This is because in the VT the burden is a parallel load, while in a CT it is a series load. Also note that these errors are unrealistically high, as the leakage impedance in proportion to the burden impedance is quite high. In a well-designed secondary circuit, the errors would of course be much smaller.

(3.11) Substituting from Equation (3.21) into Equation (3.23), we get

$$e_2'(s) = -E_{max}\frac{\cos\theta}{\tau'}\times\left\{\frac{s^2 - s\omega\tan\theta}{(s^2+\omega^2)(s^2+s/\tau'+\omega^2)}\right\}.$$

Using a table of Laplace transforms (for example, "Laplace Transform Theory and Electric Transients" by Stanford Goldman, Dover Publications, p. 422), the time domain solution becomes

$$e_2'(t) = -E_{max}\frac{\cos\theta}{\tau'}\times\frac{1}{\omega}\left\{\frac{\omega^4 + \omega^4\tan^2\theta}{[1/4\tau'^2 + \tan^2\phi/4\tau'^2 - \omega^2]^2}\right\}^{\frac{1}{2}}\sin(\omega t + \psi_1)$$

$$+\frac{2\tau'}{\tan\phi}\left\{\frac{\left[1/4\tau'^2 - \tan^2\phi/4\tau'^2 + \frac{\omega\tan\theta}{2\tau'^2} + \tan^2\phi/4\tau'^2[\omega\tan\theta - 2/2\tau']^2\right]}{[1/4\tau'^2 + \tan^2\phi/4\tau'^2 - \omega^2]^2 + 4\omega^2/4\tau'^2}\right\}^{\frac{1}{2}},$$

$$\times\varepsilon^{-t/2\tau'}\sin(\omega t + \psi_2)$$

where $\psi_1 = \tan^{-1}(\tan\theta) - \tan^{-1}\dfrac{2\omega/2\tau'}{1/4\tau'^2 + \tan^2\phi/4\tau'^2 - \omega^2}$

and

$$\psi_2 = \tan^{-1}\frac{\tan\theta/2\tau'\,[-\omega\tan\theta - 2/2\tau']}{\left[1/4\tau'^2 - \tan^2\phi/4\tau'^2 + \omega\tan\theta/2\tau'\right]} - \tan^{-1}\frac{-2(1/2\tau')(\tan\phi/2\tau')}{1/4\tau'^2 + \tan^2\phi/4\tau'^2 - \omega^2}.$$

When $2\omega\tau'$ is set equal to $\sec\phi$, and trigonometric identities are used, $\psi_1 = \theta - \pi/2$ and $\psi \equiv \psi_2 = \tan^{-1}[-\sin\theta/(\cos\phi + \tan\theta)]$.

Similarly, the denominators in the two expressions on the right-hand side become

$$1/4\tau'^2 - \tan^2\phi/4\tau'^2 + \omega\tan\theta/2\tau' = \sec^2\phi/4\tau'^4.$$

The numerator of the first term:

$$\frac{1}{\omega}\left\{\omega^4 + \omega^4\tan^2\theta\right\}^{\frac{1}{2}}\sin(\omega t + \psi_1) = [\sec\phi\sec\theta/2\omega]\cos(\omega t + \theta).$$

And the numerator of the second term:

$$\frac{2\tau'}{\tan\phi}\left\{\left[1/4\tau'^2 - \tan^2\phi/4\tau'^2 + \frac{\omega\tan\theta}{2\tau'^2} + \tan^2\phi/4\tau'^2[\omega\tan\theta - 2/2\tau']^2\right]\right\}^{\frac{1}{2}}$$

$$= \frac{1}{2\tau'}\left\{[\cot\phi - \tan\phi + \csc\phi\tan\theta]^2 + [2 + \sec\phi\tan\theta]^2\right\}^{\frac{1}{2}}.$$

Dividing by the denominator, $\sec^2\phi/4\tau'^4$, the second term becomes

$$\left[\csc^2\phi + 2\tan\theta\cos\phi\csc^2\phi + \csc^2\phi\tan^2\theta\right]^{1/2},$$

which upon further simplification becomes

$$\left[1 + (\cot\phi + \csc\phi\tan\theta)^2\right]^{1/2},$$

and finally putting all the pieces together,

$$e_2'(t) =$$

$$-E_{max}\left\{\cos(\omega t + \theta) - \cos\theta\left[1 + (\cot\phi + \csc\phi\tan\theta)^2\right]^{\frac{1}{2}}\varepsilon^{-\omega t\cos\phi}\sin(\omega t\sin\phi + \psi)\right\},$$

which with the value of ψ given in the earlier text is Equation (3.24).

(3.12) If C_2 is the capacitance between the tap point and the ground, then by the voltage divider equation,

$$\frac{C_1}{C_1 + C_2} = \frac{4\sqrt{3}}{138}; \text{where } C_1 \text{ is } 0.005\,\mu\text{F. Hence, } C_2 = 0.094593\,\mu\text{F}.$$

$C_1 + C_2 = 0.099593\,\mu\text{F}$, hence the tuning inductance must be

$$L = \frac{10^6}{(377)^2 \times 0.099593} = 70.646 \text{ Henrys}.$$

If the burden is $300\,\Omega$ resistive on the secondary side, as seen on the primary side of the transformer, where the voltage is $4\,\text{kV}$ when the secondary voltage is $69\,\text{V}$, the burden appears to be

$$300 \times \frac{4000^2}{69^2} = 1.0082 \times 10^6 \, \text{ohms.}$$

The ratio correction factor $R = E_2/E_1$, where E_2 is the voltage across the burden, and E_1 is the voltage of the source:

$$R = \frac{1.0082 \times 10^6}{1.0082 \times 10^6 + j\{2\pi f \times 70.646 - 10^6/[2\pi f \times 0.099593]\}}$$

$$= \frac{1}{1 + j0.02641\{f/60 - 60/f\}}.$$

The following values for the magnitude and angle of R are obtained as the frequency is changed between 58 Hz and 62 Hz.

Hz	Magnitude R	Angle R (degrees)
58	0.9999984	0.13
59	0.9999996	0.05
60	1.0	0
61	0.9999996	−0.05
62	0.9999985	−0.10

16.4 Chapter 4 problems

(4.1) See Figure 16.12.

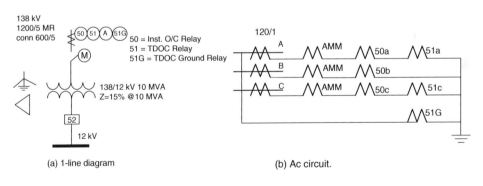

(a) 1-line diagram

(b) Ac circuit.

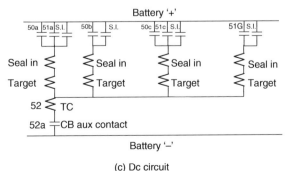

(c) Dc circuit

Figure 16.12 (a)–(c) For Problem 4.1.

(4.2) (a) $I_{3ph-A} = 100/10 = 10$ amperes

$I_{3ph-B} = 100/11 = 9.09$ amperes $^\cdot$

(b) $I_{3ph-A} = 100/1 = 100$ amperes

$I_{3ph-B} = 100/11 = 9.09$ amperes $^\cdot$

(4.3) Setting rules:

1) Instantaneous relay 50 A must not pickup for faults at bus B.

Set 135% I_{ph-B}

2) Check for pickup for faults at bus A to see if there is good protection for close-in faults.

(a) $I_A = 100/10 = 10$ amperes

$I_B = 100/11 = 9.09$ amperes $^\cdot$

Set relay at 12.27 A (use 12 A tap).

Relay 50 A will not pick up for I_{3ph}. There is no application for an instantaneous relay.

(b) $I_A = 100/1 = 100$ amperes

$I_B = 100/11 = 9.09$ amperes $^\cdot$

Set relay at 12.27 A (use 12 A tap).

For faults at bus A, pickup is $100/12 = 8.33$ times pickup. This is a good application.

(4.4) $I_{3ph-A} = 100/5 = 20$ amperes

$I_{3ph-B} = 100/10 = 10$ amperes $^\cdot$

Instantaneous relay pickup $= 1.35 \times 10 = 13.5$ A.

Protection coverage at bus A $= 20/13.5 = 1.48 \times$ pickup—good.

Good protection at bus A, will not overreach bus B.

(4.5) (a) 5 A pickup

T.D. = #3

$I_F = 15$ A

$X_{pickup} = 15/5 = 3$

Curve (a) 0.9 s

Curve (b) 2.0 s

Curve (c) 1.5 s

(b) 5 A pickup

$I_F = 15$ A

Time = 1.5 s

Pickup $= 15/5 = 3 \times$

Curve (a) T.D.is#5

Curve (b) T.D.is#2.1

Curve (c) T.D.is#3

(c) I_F = 20 A

> Time = 2.0 s
>
> For pickup of 5 A, 20/5 = 4 ×
>
> Curve (a) T.D.is#8
>
> Curve (b) T.D.is#5
>
> Curve (c) T.D.is#6

(4.6) Instantaneous relays
At station B
Set at 135% maximum fault at transformer.

> 1.35 X 3000 amperes = 4050 amperes primary.

Set pickup at 4050/20 = 202.5 A secondary.

> Check pickup at bus B for minimum fault 2000/20 = 100 amperes secondary
>
> 100/202.5 = 0.494 X pickup – relay will not pickup.

Check pickup at bus B for maximum fault 5000/20 = 250 A secondary
250/202.5 = 1.23 × pickup—marginal value for the instantaneous relay.
Check @2 times load

> 200/20 = 10 amperes secondary
>
> 10/202.5 = 0.494 – no problem

At station A
Set at 135% Maximum fault at station B.

> 1.35 X 5000 amperes = 6750/30 = 225 amperes secondary.

Check pickup at station A for minimum fault

> 3000/30 = 100 amperes – will pick up.

Check for maximum fault

> 9000/30 = 300 amperes secondary.
>
> 300/225 = 1.33 X pickup – OK for close-in faults.

Check @2 times load

> (2 X 150)/30 = 10 amperes secondary
>
> 10/225 = 0.0444 no problem.

Time-delay overcurrent relays
At station B
The TDOC relay should have 3× pickup for minimum fault at the transformer.

> 1000/20 = 50 A secondary. 50/3 = 16.67 A. Use 16 A tap.

Check for 2 times load

(2× pickup 100)/20 = 10 A secondary. Pickup is 10/16 = 0.625 – OK.
No coordination is necessary; use #1/2 T.D.
Setting is 16.0 @1/2.

At station A

The TDOC relay at station A should back up TDOC at station B.
1000/30 = 33.33 A secondary
33.33/3 = 11.11 A—use 11.0 A tap.

Check @ 2 times load 150 × 2 = 300/30 = 10 A. Close but acceptable.
The time for the TDOC relay at station B for the maximum fault is

5000/(16 × 20) = 15.625 pickup
@1/2 TD = 0.15 s
Add 0.3 s coordinating time

Time for relay at station to operate = 0.45 s.

5000(16 × 20) = 15.625 pickup @0.45 s.
From Figure 3.5, TD = 2.3
Setting is 11 A pickup @2.3 TD.

(4.7) Instantaneous relays
At Station B

For a maximum fault current of 4050 A primary, use a CT that gives no more than 100 A secondary.

From Table 3.1, select 600/5 MR connected 200:5. Instantaneous pickup = 4050/40 = 101.25 A. Close, but acceptable.
Check pickup at bus B for a minimum fault 2000/40 = 50 A secondary.

50/101.25 = 0.5 pickup – will not pick up.

Check pickup for maximum fault 5000/40 = 125 A secondary.

125/101.25 = 1.23 pickup – marginal.

Check @ 2 times load

200/40 = 5 amperes secondary
5/101.25 = 0.04 – no problem.

At Station A

6750/100 = 67.5. Use 600/5 MR, connected 400/5 (80/1) CT from Table 3.1.
Instantaneous pickup = 6750/80 = 84.37 A secondary.
Bus A minimum fault = 3000/80 = 37.5 A—relay will not pick up.
Bus A maximum fault = 9000/80 = 112.5 A
112.5/84.37 = 1.33 pickup—OK.
Time-delay overcurrent relay

At Station B

> $1000/40 = 25$ A secondary. $25/3 = 8.33$ A—use 8.0 A tap.
> Twice load $= 200/40 = 5$ A secondary—OK will not pick up.
> Setting is 8.0 tap—T.D. #1/2.

At Station A

> $1000/80 = 12.9$ A secondary. $12.9/3 = 4.17$ A—use 4.0 A tap.
> Twice load $= 300/80 = 3.75$—OK will not pick up.
> Time for TDOC at station B to operate
> $5000/8 = 625. 625/40 = 15.625$ pickup
> @T.D. #1/2 time is 0.15 s.
> Add 0.3 s coordinating time.
> Time for TDOC at station A to operate is 0.45 s.
> $5000/(80 \times 4) = 15.625$ times pickup
> At 0.45 s $=$ T.D. #2.15
> Setting is 4.0 tap – #2.5 T.D.

(4.8) This example is intended to illustrate the need for directionality. It is not intended to show how to set overcurrent relays on a networked system. Assume this is a single-phase circuit, pickup settings are based on the fault locations as shown, the right-hand generator is also 100 V and is located at breaker 6.

Relay	I_{F1}	pickup	I_{F2}	pickup
Breaker 3	3.7 A	1.2 A	5.2 A	1.2 A
Breaker 4	3.7 A	1.2 A	10 A	3.33 A
Breaker 5	3.7 A	1.2 A	10 A	3.33 A
Breaker 6	50 A	3.33 A	10 A	3.33 A

For faults at F1, relays at breaker 3 are set with enough multiples of pickup to operate and can coordinate with relays at breaker 5 by appropriate time delay.

For faults at F2, relays at breaker 3 will see more current and protect its line properly.

For faults at F1, relays at breaker 4 see the same current as relays at breaker 5 and would have to be set longer to coordinate. This would delay tripping for faults at F2, its primary zone of protection. Therefore, it must be set directional to see only faults to its left.

However, if we move the faults to other line sections, the same reasoning applies, and there all relays at breakers 2, 3, 4, and 5 must be directional.

(4.9) Fault calculations

(a) System normal S/N (1.5x. 5)/2.0 + 0.8 = 1.175

$$I_F = 70/1.175 = 59.6 \text{ amps}$$

$$I_1 = (0.5/2) \text{ x } 59.6 = 14.89 \text{ amps}$$

$$I_3 = (1.5/2) \text{ x } 59.6 = = 44.68 \text{ amps}$$

(b) 1–2 out of service

$$I_3 = I_5 = 70/(.5 + .8) = 53.85 \text{ amps}$$
$$I_1 = 0$$

(c) 3–4 out of service

$$I_1 = I_5 = 70/(1.5 + .8) = 30.43 \text{ amps}$$
$$I_1 = 0$$

These are three phase fault currents. Minimum fault currents would be phase–phase, or $0.866 \times I_{3ph}$. Assume fault is at end of line.
Relay settings
Breaker 5

Instantaneous relay

Maximum current is S/N

$$I = 1.35(59.6) = 80.46 \text{ amperes}$$

Time-delay overcurrent relay

Minimum current is with 3–4 out of service. Set for phase-to-phase fault.

$$\text{Pickup} = (30.43 \times 0.866)/3 = 8.78 \text{ A}$$

Use 8.0 A tap

No coordination required—set TD at #1/2
Breaker 1

Instantaneous relay
Must not overreach breaker 5.

Set $135\%I_{max} = 1.35(70/1.5) = 63$ amps.

Time-delay overcurrent relay

$$I_{max} = 30.43 \text{ amps}$$
$$I_{min} = 14.89 \text{ amps}$$
Set $(14.89 \times 0.866)/3 = 4.3$ amps. Use 4.0 amp tap.

Determine relay time at breaker 5 for $I_{F1max} = 30.43/8.0 = 3.8 \times$ pickup.

Time at #1/2 time dial $= 0.2$ s from Appendix Iva
Add   0.3 coordinating time
0.5 s

At breaker 1 $30.43/4.0 = 7.6\times$ pickup @0.5 s #2 time dial

Setting is 4.0 A, tap #2 T.D.

Breaker 3

Instantaneous relay

$$I_{max\ 3ph} = 53.85 \text{ amperes}$$
$$I_{min\ 3ph} = 44.68 \text{ amps}$$

Must not overreach breaker 5.

Set $1.35 \times (70/0.5) = 189$ amps

Time-delay overcurrent relay

$$I_{min} = 44.68 \times 0.866 = 38.69 \text{ amps}$$
$$38.69/3 = 12.9 \text{ amps}$$

Set at 12 A pickup
I_{max} of breaker 5 at $S/N = 59.6$ A. For this maximum current condition, the current through breaker 3 is 44.68 A.

Relay time at breaker 5 is $59.6/8 = 7.45 \times$ pickup $= 0.18$ s
Add coordinating time $= \underline{0.3 \text{ s}}$
0.48 s.

At breaker 3, relay TD is $44.68/4 = 11.17 \times$ pickup @0.48 s $=$ #1.5 from Appendix Iva.

Setting is 8.0 A, tap #1.5 TD

(4.10) See Figure 16.13.
It is necessary to consider only the system to the left of F1

$$X_H = \frac{X_{HL} + X_{HT} - X_{LT}}{2} = 0.15,$$

$$X_L = \frac{X_{HL} + X_{LT} - X_{HT}}{2} = -0.05,$$

$$X_T = \frac{X_{HT} + X_{LT} - X_{HL}}{2} = 0.25.$$

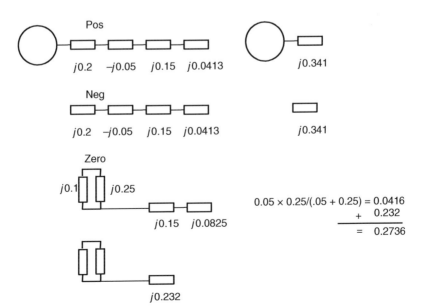

Figure 16.13 For Problem 4.10.

For line-to-ground fault at F_1

$$0.341 + 0.341 + 0.2736 = 0.956,$$

$$Z_{ohm} = \frac{10 \times 95.6 \times (220)^2}{10000} = 462 \text{ ohms},$$

$$I_F = \frac{220000}{\sqrt{3} \times 462} = 274.9.$$

For $I_N = 0, I_L = I_H,$

$$I_L = I_H xk(220/110),$$

$k = 0.5$, that is, the currents through the two legs of the zero-sequence equivalent are equal and so the reactances of each leg are equal:

$$X_r + 0.1\text{-}0.05 = 0.25,$$
$$X_r = 0.2 = 20\%@110 \text{ kV} = 24.2 \text{ ohms}.$$

For faults on the low-voltage side of the transformer, the current in the neutral will always be up from the ground. The zero-sequence equivalent for a low side fault is:

$$I_L = I_H + I_T \text{ i.e.it is always greater than } I_H.$$

$$I_H = 3(I_H\text{-}I_L) \text{ – always up the neutral}.$$

See Figure 16.14.

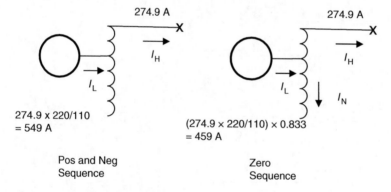

274.9 A

I_H

I_L

274.9 x 220/110
= 549 A

Pos and Neg
Sequence

274.9 A

I_H

I_L

I_N

(274.9 × 220/110) × 0.833
= 459 A

Zero
Sequence

Figure 16.14 For Problem 4.10

16.5 Chapter 5 problems

(5.1) $I_{load} = \left(120x10^6 x\sqrt{3}\right)/\left(3x138x10^3\right) = 502 \text{ amps}.$

Select a CT ratio of 500:5 or 100:1.

The CVT ratio is $138 \times 10^3/120 = 1150:1.$
Hence, the ratio (secondary impedance:primary impedance) $= 11/1150 = 0.08696.$

The secondary values for the three elements are:

Line AB $(3 + j40) : (0.26 + j3.478),$

Line BC $(2 + j50) : (0.174 + j4.348).$

Transformer BD $(0 + j9)$: $(0 + j0.783)$.

Hence, the Zone-1 setting for line AB is $0.9 \times (0.26 + j3.478) = 3.14\angle 85.7°$.

Zone-2 is $1.2 \text{x} (0.26 + j3.478) = 4.185\angle 85.7°$.

Taking the element with the larger impedance beyond bus B (which is line BC), Zone-3 setting is $(0.26 + j3.478) + 1.5 \times (0.174 + j4.348) = 10.01\angle 87°$.
Selecting from the available impedance settings, the chosen settings are:

Zone-1 $= 3.1 \, \Omega$,

Zone-2 $= 4.2 \, \Omega$,

Zone-3 $= 10.0 \, \Omega$.

Select a maximum torque angle of 80°.

(5.2) $I_{load} = \left(10\text{x}10^6 \text{x} \sqrt{3}\right) / \left(3\text{x}13.8\text{x}10^3\right) = 418.37 \text{ amps}$.
Select a CT ratio of 500:5 or 100:1.

CVT ratio is $13800/120 = 115$.

The impedance ratio is $100/115 = 0.8696$.

The line impedance is $0.8696 \times (4\text{:}j40) = 3.478 + j34.78$. This would require a Zone-1 setting of $0.9 \times (3.478 + j34.78) = 3.13 + j31.3 \, \Omega$ secondary. And zone-2 setting would be $1.2 \times (3.478 + j34.78) = 4.17 + j41.7 \, \Omega$ secondary. If the available relay can be set for Zone-1 at a maximum value of $10 \, \Omega$, the above settings could not be achieved.

There select a CT ratio of 2000:5. This will produce an impedance ratio of 0.2174, and the zone settings of

Zone-1 $= 0.78 + j7.82 = 7.85\angle 84°$,

Zone-2 $= 1.04 + j10.42 = 10.47\angle 84°$.

Set the maximum torque angle at 80°, and using the actual ohm settings available in the relay, the zone settings become

Zone-1 $= 7.8 \, \Omega$,

Zone-2 $= 10 \, \Omega$.

(5.3) The primary load impedance corresponding to 120 MVA is $138^2/120 = 158.7 \, \Omega$. The load impedance as viewed from the secondary side is $158.7 \times 0.08696 = 13.8 \, \Omega$. A lagging power factor of 0.8 plots in the first quadrant of the R–X diagram at an angle of 36.9°. This is beyond the setting of Zone 3 of the line in Problem 5.1. The relay is not in any danger of tripping at this load. The R–X diagram of the three zones and the load is shown in the figure in the following text. Both directions of the load flow are shown. If the load is considered to be too close to the Zone-3 boundary (for example, if significant load growth in future is anticipated), one could use a mho characteristic as shown (see Figure 16.15).

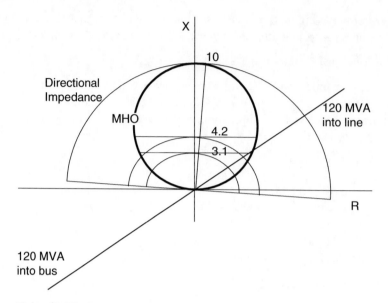

Figure 16.15 For Problem 5.3.

(5.4) a–b fault

Using the result obtained in Example 5.4, for the a–b fault the condition is $I_a = -I_b$ and $I_c = 0$. Also, $I_0 = 0$. Hence, $\alpha I_1 + \alpha^2 I_2 = 0$. Thus,

$$I_1 = -\alpha I_2 = 88.18\angle\text{-}84.92°.\text{Hence } I_2 = 88.18\angle\text{-}24.92°.$$

Thus, $I_a = I_1 + I_2 = 88.18\angle\text{-}84.92° + 88.18\angle\text{-}24.92° = 97.78\text{-}j124.98$ and $I_b = \text{-}I_a$.

Therefore, $I_{ab} = I_a - I_b = 175.56\text{-}j249.96 = 305.45\angle\text{-}54.9°$.

At the relay location, $E_1 = 7967.4 - j0.5 \times (88.18\angle - 84.92°) = 7528.33\angle - 0.3°$. And $E_2 = -j0.5 \times (88.18\angle - 24.92°) = 440.9\angle245.08°$. Thus, $E_a = E_1 + E_2 = 7342.45 - j439.27$, and $E_b = \alpha^2 E_1 + \alpha E_2 = -3359.08 - j6460.88$. $E_{ab} = 10\,701.37 + j6021.61 = 12\,279.21\angle29.37°$.

The impedance seen by the a–b relay is $Z_{ab} = E_{ab}/I_{ab} = (12\,279.21\angle29.37°)/(305.45 \angle - 54.9°) = (4 + j40)$.

b–g fault

For this case, the boundary conditions are $I_a = I_c = 0$. This leads to $I_0 = \alpha^2 I_1 = \alpha I_2$. Thus, $I_1 = 41.75\angle - 84.92°$, $I_2 = \alpha I_1 = 41.75\angle35.41°$, and $I_0 = 41.75\angle155.41°$. $I_b = I_0 + \alpha^2 I_1 + \alpha I_2 = 41.75\angle155.41° + 41.75\angle155.41° + 41.75\angle155.41° = 125.25\angle155.41°$. Also $mI_0 = 1.253\angle - 1.13° \times 41.75\angle155.41°$. Hence, $I'_b = I_b + mI_0 = 177.5\angle155.08°$. The voltages at the relay location are $E_1 = 7759.58 - j19.68 = 7759.60\angle - 0.14°$, $E_2 = -j5 \times 41.75\angle35.41° = 208.75\angle - 54.92°$, and $E_0 = -j10 \times 41.75\angle155.41° = 417.5\angle65.41°$. The b phase voltage is $E_b = E_0 + \alpha^2 E_1 + \alpha E_2 = 417.5\angle65.41° + 7759.60\angle - 0.14° + 208.75\angle - 54.92° = -3634.52 - j6141.56 = 7136.42 \angle - 120.62°$. The impedance seen by the phase relay is $Z_b = (E_b/I'_b) = (7136.42 \angle - 120.62°)/(177.5\angle155.08°) = (4 + j40)$.

(5.5) The symmetrical component circuit for a b–c–g fault is shown in the following text (see Figure 16.16). The negative- and zero-sequence circuits are in parallel.

The equivalent impedance of the two parallel branches is $Z_{20} = [(4 + j45)(10 + j100)]/(4 + j45 + 10 + j100) = 2.863 + j31.04$. This is in series with the positive-sequence impedance. Hence, the net impedance in the fault circuit is $(4 + j45 + 2.863 + j31.04) = 6.863 + j76.04 = 76.35\angle 84.84°$.

The symmetrical component currents in the fault are

$$I_1 = 7967.4/76.35\angle 84.84° = 104.35\angle\text{-}84.84°,$$

$$I_2 = I_1(10 + j100)/14 + j145) = \text{-}71.99\angle\text{-}85.03°, \text{and}$$

$$I_0 = \text{-}(I_1 + I_2) = 32.36\angle 95.58°.$$

The phase currents are

$$I_a = 0,$$

$$I_b = I_0 + \alpha^2 I_1 + \alpha I_2 = 160.67\angle 167.46°,$$

$$I_c = I_0 + \alpha I_1 + \alpha^2 I_2 = 159.84\angle 22.76°.$$

The relay currents for the b–c, b–g, and c–g relays are:

$$I_{bc} = I_b - I_c = 305.43\angle\text{-}174.9°,$$

$$I_b{}' = I_b + mI_0 = 160.67\angle 167.46° + 1.253\angle\text{-}1.13°\text{x}32.36\angle 95.58° = 176.78\angle 154.82°,$$

$$I_c{}' = I_c + mI_0 = 159.84\angle 22.76° + 1.253\angle\text{-}1.13°\text{x}32.36\angle 95.58° = 176.81\angle 35.55°.$$

The symmetrical components of relaying voltages are:

$$E_1 = 7967.4 - j5\text{x}104.35\angle\text{-}84.84° = 7447.92\angle\text{-}0.36°,$$

$$E_2 = \text{-}j5\text{x}(\text{-}71.99\angle\text{-}85.03°) = 359.95\angle 4.97°,$$

$$E_0 = \text{-}j10\text{x}(32.36\angle 95.58°) = 323.6\angle 5.58°.$$

The corresponding phase voltages at the relay locations are

$$E_a = E_0 + E_1 + E_2 = 8128.44 + j15.72,$$

$$E_b = E_0 + \alpha^2 E_1 + \alpha E_2 = 7108.04\angle\text{-}120.88°,$$

$$E_c = E_0 + \alpha E_1 + \alpha^2 E_2 = 12279.55\angle\text{-}90.63°,$$

$$\text{And } E_{bc} = E_b - E_c = 12279.55\angle\text{-}90.63°.$$

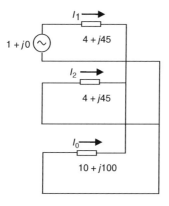

Figure 16.16 For Problem 5.5.

The apparent impedance seen by the three distance relays can now be calculated:

$$Z_{ab} = E_{ab}/I_{ab} = 12279.55\angle\text{-}90.63°/305.43\angle\text{-}174.9° = \mathbf{4 + j40},$$

$$Z_b = E_b/I_b' = 7108.04\angle\text{-}120.88°/176.78\angle154.82° = \mathbf{4 + j40},$$

$$Z_c = E_c/I_c' = 12279.55\angle\text{-}90.63°/176.81\angle35.55° = \mathbf{4 + j40}.$$

(5.6) b–c relay

$I_c = 0$, hence $I_{bc} = I_b - I_c = I_b = 152.72\angle125.1°$. $E_b = -3359.08 - j6460.88$. $E_c = \alpha E_1 + \alpha^2 E_2 = 7528.33\angle119.7° + 440.90\angle125.08° = -3983.36 + j6900.15$. Therefore, $E_{bc} =$

$$E_b - E_c = (-3359.08\text{-}j6460.88 + 3983.36\text{-}j6900.15) = 13375.6\angle\text{-}87.32°.$$

$$Z_{bc} = E_{bc}/I_{bc} = 13375.6\angle\text{-}87.32°/152.72\angle125.1° = 87.58\angle\text{-}124.84°.$$

c–a relay

$$I_{ca} = I_c - I_a = \text{-}I_a = 152.72\angle125.08°. E_{ca} = E_c - E_a = \text{-}3983.36$$
$$+ j6900.15 + j6900.15 - 7342.45 +$$
$$j439.27 = 13495.96\angle147.05°.$$

$$Z_{ca} = E_{ca}/I_{ca} = 13495.96\angle147.05°/152.72\angle125.08° = 88.37\angle21.97°.$$

a–g relay

$$Z_b = E_b/I_b' = E_b/I_b = (-3359.08\text{-}j6460.88)/(87.78 + j124.98) = 47.7\angle117.45°.$$

In general, the other relays see larger impedances than that seen by the appropriate (a–b) relay. The ground relays see quite a small impedance, although the angles are quite different from the angle of the impedance seen by the a–b relay.

(5.7) Zone 1:

All infeeds are removed. The three line segments to be protected are AC: $(4 + j30)$, AD: $(3 + j35)$, AB:$(9 + j85)$. Select the smallest of these to avoid overreach. Hence, zone-1 setting is $0.9 \times (4 + j30) = 3.6 + j27$ ohms primary.

Zone 2:

This should be for the longest segment, with all infeeds in place. Segment AB with infeeds will appear to be $(1 + j10) + (2 + j15) \times (1 + 200/600) + (6 + j60) \times [1 + (200 + 300)/600] = 14.667 + j140$ ohms primary. Zone-2 setting of $1.2 \times (14.667 + j140) = 17.6 + j168$ ohms is a possibility. However, this setting may not coordinate with whatever is beyond C and D, which should be checked.

More importantly, this Zone-2 setting may conflict with Zone-1 settings of BG and BH. Consider the possibility of all the infeeds being out of service. In that case, the proposed Zone-2 setting of $17.6 + j168$ ohms will reach beyond bus B by

$$(17.6 + j168) - (9 + j85) = 8.6 + j83 \text{ ohms}.$$

This excursion beyond B will certainly go beyond the zone setting of BH, which is shorter of the two lines originating at bus B. Zone 1 for BH would be set at

$0.9 \times (4 + j40) = 3.6 + j36$ ohms. Thus, a safe Zone-2 setting for AB should be smaller than $(9 + j85) + (3.6 + j36) = 12.6 + j121$ ohms. A good choice would be a setting of $(10 + j110)$ ohms primary. Note that this Zone-2 setting will not overreach all the remote terminals when all infeeds are in service. However, this is the best available Zone-2 setting under the given circumstances.

Zone 3:

With all infeeds in service, Zone 3 should reach beyond the longest of the two lines originating at bus B, that is, line BG. Thus, a desirable Zone-3 setting is $(14.667 + j140) + 1.5 \times (6 + j60) \times [1 + (200 + 300 + 400)/600] = 37.17 + j365 \ \Omega$ primary. This Zone-3 setting must be checked for coordination with whatever protection exists beyond buses G and H. It may be necessary to reduce the setting, or even eliminate Zone 3 if coordination problems could not be resolved.

(5.8) R_d:
Zone 1: Shorter line is DG with no infeed: $(3 + j30)$. Hence, Zone-1 setting should be $0.9 \times (3 + j30) = 2.7 + j27$ ohms primary.
Zone 2: This should be set with infeed. $1.2 \times (2 + j20) + (1 + j10) \times (1 + 0.5) = 9.6 + j96$ ohms primary.
Zone 3: This should be set at $2 + j20 + (1 + 0.5) \times (4 + j40) + 1.5 \times (3 + j30) \times (1 + 0.5) = 14.75 + j147.5$ ohms primary.

Zone 1 of line BA is set at $0.9 \times (3 + j30)$ or $2.7 + j27$ ohms. Without infeed, the Zone-2 setting chosen in the earlier text will reach beyond this, that is, $(2 + j20) + (4 + j40) + 2.7 + j27 = 8.7 + j87$, which is smaller than $9.6 + j96$. The Zone-2 setting chosen in the earlier text is not a good setting. A better setting would be $8 + j80$ ohms primary, which may not be adequate for high infeed, but is probably a good setting.
R_g:
Zone 1 is set on the shorter line without infeed, that is, $0.9 \times (1 + j10 + 2 + j20) = 2.7 + j27$ ohms primary.
Zone 2: This is set at $1.2 \times (1 + j10) + (4 + j40) \times (1 + 1/0.5) = 15.6 + j156$ ohms primary. This clearly is going to be too large, and will interfere with the Zone 2 of line BA. Hence, either Zone 2 of R_g should be shortened, or it must be made slower than Zone 2 of BA.
Zone 3: This is set at $1 + j10 + (1 + 1/0.5) \times (4 + j40) = 1.5 \times (3 + j30) \times (1 + 1/0.5) = 26.5 + j265$ ohms primary. This is probably too large, and may not be a good setting. In that case, Zone 3 may be left out.

(5.9) If we neglect I_{02}, the zero-sequence current in the parallel line, I_a', is in error, given by $I_a' = I_a + mI_{01} = 99.25\angle - 85.23° + 1.25 \times 33.085\angle - 85.23° = 140.60\angle - 85.23°$. The apparent impedance seen by the phase-a distance relay is

$$Z_a = E_a/I_a' = 6649.5\angle -0.95°/140.60\angle -85.23° = 47.29\angle 84.28°$$
$$= 4.71 + j47.05.$$

This is in error above the correct value of $4 + j40$ by about 18%. The zone setting for ground faults should not be changed to accommodate this error. Should it be changed, then when the parallel circuit is out of service, the relay would overreach.

The setting should be chosen as in Example 5.7, in which case the ground distance relays will not reach as far as the phase distance relays, whenever the parallel line is in service.

(5.10) Assume a maximum torque angle of 80°. The two types of relays are illustrated in the following text (see Figure 16.17).
By geometry,

$$98 + (x\cos80)^2 + (x\sin80)^2 = 10^2, \text{ or}$$

$$\begin{aligned} x &= 4.77 \text{ ohms secondary.} \\ &= 4.77 \times 2000/(100 \times 69.3) \\ &= 13.77 \text{ ohms primary.} \end{aligned}$$

By geometry,

$$(3 + x\cos80)^2 + (x\sin80)^2 = 5^2,$$

or

$$\begin{aligned} x &= 3.51 \text{ ohms secondary} \\ &= 3.51 \times 2000/(100 \times 69.3) \\ &= 10.13 \text{ ohms primary.} \end{aligned}$$

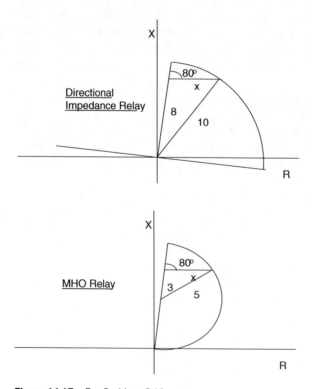

Figure 16.17 For Problem 5.10.

(5.11) This problem involves considerable work. It could be solved on EMTP, and this has been done to confirm the results given here. Bus 1 is the generator terminal, buses 4 and 5 are the terminals of the two lines. Buses 2 and 3 are midpoints of the two lines, where faults will be placed. A b–g fault implies $E_0 + \alpha^2 E_1 + \alpha E_2 = 0$, and $I_0 = \alpha^2 I_1 = \alpha I_2$. A c–g fault implies $E_0 + \alpha E_1 + \alpha^2 E_2 = 0$, and $I_0 = \alpha I_1 = \alpha^2 I_2$. These faults are represented in the symmetrical component circuit by phase shifting transformers in the positive- and negative-sequence circuits, and 1:1 transformers in the zero-sequence circuit. The zero-sequence circuit has mutual coupling between the two lines. This is represented by creating intermediate buses (s1 and s2) so that the common leg of the wye-equivalent is the mutual impedance, and the other two legs of the wye-equivalent are the difference between the zero-sequence mutual and the self-impedances. These ideas are used in developing the symmetrical component circuit for this case. Positive and negative circuits are identical, except for the fact that the positive-sequence circuit has a source in it. The zero-sequence circuit is substantially different. In the final reduced form (see Figure 16.18), only the source and the fault buses are retained, that is, buses

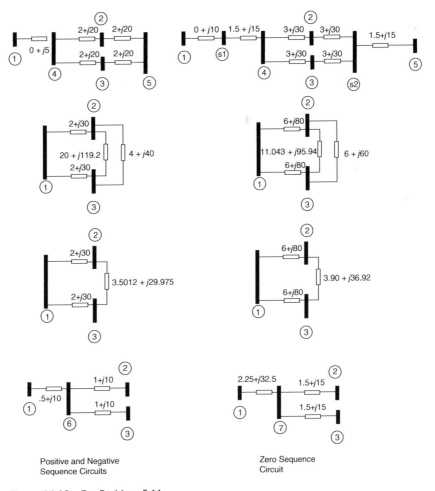

Positive and Negative
Sequence Circuits

Zero Sequence
Circuit

Figure 16.18 For Problem 5.11.

1, 2, and 3. The intermediate steps in the network reduction are shown in the following text, so that after the fault currents have been calculated, the currents in the individual branches could be determined by back-solving the intermediate forms of the network.

The last forms of these networks are now connected in faulted representation as shown in Figure 16.19. Note that the phase-shifting and 1:1 transformers have been added, and the source generator has been connected. The source has been assumed to be solidly grounded. If some other grounding impedance is called for, it would be connected in the zero-sequence circuit.

The values of the impedances are given on the previous page. The next step is to solve this circuit for the currents in the circuit, viz., i_1 and i_2, and then find the voltages in the positive-, negative-, and zero-sequence networks. The loop equations for the two loops in the network are

$$\left\{ e - z_{11}\left(\frac{i_1}{\alpha^2} + \frac{i_2}{\alpha}\right) - z_{12}\left(\frac{i_2}{\alpha}\right) \right\}\alpha + \left\{ -z_{11}\left(\frac{i_1}{\alpha} + \frac{i_2}{\alpha^2}\right) - z_{12}\left(\frac{i_2}{\alpha^2}\right) \right\}\alpha^2$$
$$+ \left\{ -z_{01}(i_1 + i_2) - z_{02}(i_2) \right\} = 0$$

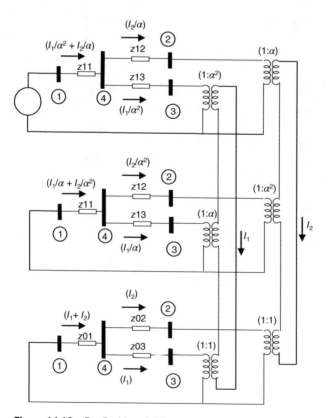

Figure 16.19 For Problem 5.11.

and

$$\left\{ e - z_{11}\left(\frac{i_1}{\alpha^2} + \frac{i_2}{\alpha}\right) - z_{13}\left(\frac{i_1}{\alpha}\right)\right\}\alpha^2 + \left\{ - z_{11}\left(\frac{i_1}{\alpha} + \frac{i_2}{\alpha^2}\right) - z_{13}\left(\frac{i_1}{\alpha^2}\right)\right\}\alpha$$

$$+ \left\{ - z_{01}(i_1 + i_2) - z_{03}(i_1)\right\} = 0.$$

Upon simplifying these two equations, we get

$$z_x i_1 + z_y i_2 = \alpha e,$$

$$z_y i_1 + z_x i_2 = \alpha^2 e,$$

where $z_x = z_{01} - z_{11}$ and $z_y = 2z_{11} + 2z_{12} + z_{01} + z_{02}$. Substituting the values of the impedances,

$$z_x = 1.75 + j22.5 \text{ and } z_y = 6.75 + j87.5.$$

e is set equal to 1.0 and the equations are solved for i_1 and i_2. The result is $i_1 = -0.0136 + j0.0035$ and $i_2 = 0.0129 + j0.0055$. The resulting currents are shown in Figure 16.20.

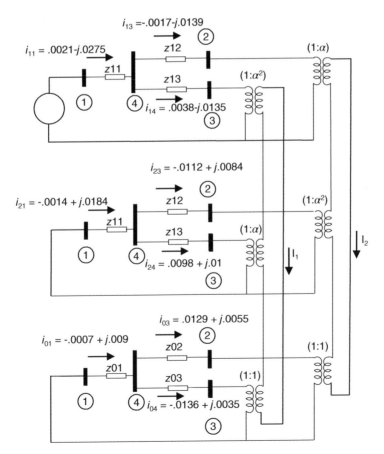

Figure 16.20 For Problem 5.11.

The voltages at the buses in the positive-, negative-, and zero-sequence networks can now be calculated. These can be calculated by using the currents i_{11}, i_{21}, and i_{01}, and calculating the voltage drops in the original impedances up to bus 4—that is, $(0 + j5)$ in the positive- and negative-sequence networks, and $(1.5 + j25)$ in the zero-sequence network. (See earlier steps of network reduction in the positive-, negative-, and zero-sequence networks.) The result is

$$e_{14} = 0.7243 - j0.0074,$$

$$e_{24} = 0.1849 + j0.0049,$$

$$e_{04} = 0.2953 + j0.0024.$$

These voltages are combined to form a, b, c phase voltages, and the delta voltages for the b–c relay.

$$e_a = 1.0452 + j0.0035,$$

$$e_b = -0.4023 - j0.6586,$$

$$e_c = \ = 0.3718 + j0.6761,$$

$$e_{bc} = -0.0305 - j1.3347.$$

The symmetrical components of currents in the two lines are found by determining the distribution of currents found in the earlier text, viz., i_{11}, i_{21}, and i_{01} in the original circuits. From the phase currents, the delta currents and the impedances seen by the b–c relays can be found. The results are:

I7*n* line 1, (with c–g fault):

$$i_{11(1)} = -0.0003 - j0.0138,$$

$$i_{21(1)} = -0.006 + j0.0088,$$

$$i_{01(1)} = 0.0064 + j0.0028,$$

$$i_{b(1)} = -0.01 + j0.0004,$$

$$i_{c(1)} = 0.0292 + j0.0102,$$

$$i_{bc(1)} = -0.0392 - j0.0098,$$

$$Z_{bc(1)} = e_{bc}/i_{bc(1)} = 8.7653 + j31.8396.$$

In line 2, (with b–g fault):

$$i_{11(2)} = 0.0024 - j0.0136,$$

$$i_{11(2)} = 0.0024 - j0.0136,$$

$$i_{01(2)} = -0.0068 + j0.0017,$$

$$i_{b(2)} = -0.0304 + j0.0056,$$

$$i_{c(2)} = 0.0098 + j0.019,$$

$$i_{bc(2)} = -0.0403 + j0.0037,$$

$$Z_{bc(2)} = e_{bc}/i_{bc(2)} = -2.285 + j32.9414.$$

16.6   Chapter 6 problems

(6.1)   Directional Comparison Blocking

Power line carrier (PLC) blocking is the most commonly used pilot scheme in the US. The blocking mode is essential, since the protected line is, itself, the channel and a faulted line would adversely affect a tripping signal. PLC is usually less expensive than microwave unless there is an existing microwave channel between the two or three terminals and the incremental cost of relay channel is low. A pilot wire channel is limited to about 10 miles if metallic cable is used and is generally not applicable for HV or EHV transmission lines.

Permissive Underreaching Transfer Trip

Microwave or pilot wire is necessary, since a fault on the protected line would prevent a tripping signal being received with PLC. P/W with metallic cable is acceptable if the distance is short enough.

Phase Comparison

All three channels are applicable, since phase comparison is essentially a blocking scheme. The criteria would be cost and distance. PLC is usually lowest cost, terminal to terminal when all of the other station requirements such as potential and current transformers are considered.

(6.2) $I_{FL} = \dfrac{120000}{\sqrt{3} \times 138} = 502\,A$—use 600:5 MR CT connect 500/5 from Table 3.1

$$PTR = 138000/120(1150/1)\ V_{ln} = 138000/\sqrt{3} = 79.67\,kV.$$

Carrier Trip Settings:

Line AB @A–B similar

$$175\%x40\ ohms = 70\ ohms\ at\ 85.71°\ primary.$$

Assume line angle is 80°, power factor is 0.8 lagging:

$$70 \times \dfrac{100}{1150} = 6.1\ ohms.$$

Set 6.0 Ω @80°

Check loadability $= \dfrac{3 \times (80)^2 \times 100}{6.0\ \cos(80^0 - 36.9^0) \times 1150} = 381\,MVA$—above maximum load of 120 MVA.

Line BC @ C

$$175\% \times 50\ ohms = 87.5\ ohms\ primary = 7.61\ ohms\ secondary.$$

Set 8.0 Ω @80°

Carrier Start Settings:

Line AB @ B

Reverse MHO relay at B must overreach Trip from A.

$$6.0 \text{ ohms secondary} \times \frac{1150}{100} = 69 \text{ ohms primary.}$$

$$69\text{-}40 = 29 \times 1.25 \text{ ohms primary} \frac{100}{1150} = 3.15 \text{ ohms secondary.}$$

Set 3.5 Ω @80°

Line BC

$$8.0 \times \frac{1150}{100} = 92 \text{ ohms primary} - 50 \text{ ohms} = 42.0 \text{ ohms.}$$

$$42.0 \times 1.25 = 52.5 \times \frac{100}{1150} = 4.6 \text{ ohms primary.}$$

Set 5.0 Ω @80°

See Figure 16.21.

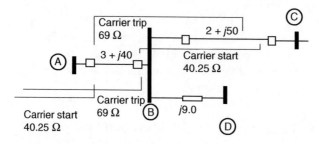

Figure 16.21 For Problem 6.2.

(6.3) A-B $3 + j40 = 40@85.7°$ primary.

$\qquad\qquad = 4.7@85.7°$ secondary.

B-C B $2 + j50 = 50@87.7°$ primary.

$\qquad\qquad = 5.2@87.7°$ secondary.

See Figure 16.22.

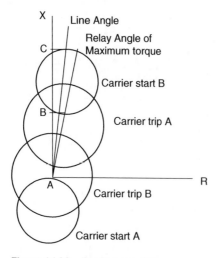

Figure 16.22 For Problem 6.3.

(6.4) Line AB

R_0 at A and B = 150% × (3 + *j*40) = 60.2@85° ohms primary = 5.2 Ω secondary.

Set 5.0 Ω @80°

Line BC

R_0 at B and C = 150% × (2 + *j*25) = 75.0@88° ohms primary = 6.5 Ω secondary.

Set 6.5 Ω @80°

(6.5) Set carrier trip with maximum infeed. Set 175% times the longest impedance (neglect resistance)

From A the longest impedance is A–B.

$$Z_{AB} = 1.75\left[1 + 15 \times \frac{800}{600} + 60.3 \times \frac{1100}{600}\right] = 245.96 \text{ ohms primary} \times \frac{100}{1150} =$$

21.3 ohms secondary.

Set 20 Ω @ 80° (230 Ω primary)

Set carrier start without infeed.

Set 125% of carrier trip setting minus Z_{AB}

A-B@B.1.25 × [230@80°-(j85.3)] = 1.25 × [39.9 + j226-j85.3] = 49.9 + j176.3

= 182.9@74.2° ohms primary = 15.9@74.2° ohms secondary.

Set 16 Ω @75° (184 Ω primary)

Carrier start line BC at C:

Set 125% of carrier trip setting minus Z_{BC} without infeed.

1.25 × [230@80° – (j95.3)] = 1.25 × [39.9 + j226.3-j95.3] = 136.7@73° ohms primary

= 11.9 ohms secondary.

Set 10@75° (115 Ω primary).

(6.6) $F_{3ph}@B = \dfrac{138000}{\sqrt{3} \times 40} = 1991$ A

$F_{ph-ph} = 0.866 \times 1991 = 1724$ amperes

$I_{FL} = \dfrac{120000}{\sqrt{3} \times 138} = 502$ A

Set FD$_1$ = 0.5 × 1724 = 862 amperes.

$\dfrac{862}{502} = 1.72 \times$ maximum load – no distance supervision required.

Set FD = 1.25 × 1724 = 2155 amperes.

16.7 Chapter 7 problems

(7.1) $I_{fl} = \dfrac{975000}{\sqrt{3} \times 22} = 25588\ \text{Aprimary}.$

Select 30 000:5 CT (6000:1)$I_{fl} = 4.26$ A secondary.

$$I_{3ph}@F_1 = I_{3ph}@F_2 = 1.0/0.21 = 4.76\ \text{per unit} \times 4.26$$
$$= 20.3\ \text{amps secondary.}(121799\ \text{amps primary}).$$

Assume no CT error, current through relay is 0 for external fault (F_2) and 20.3 A for internal fault (F_1). Set lightest tap and fastest time:
Pickup = 1.0, Time Dial #1/2

(7.2) Three-phase fault current at $F_2 = 121\ 798$ A

$$I_{1\ sec} = 121798/6000 = 20.3\ \text{amperes},$$
$$I_{2\ sec} = 121798/6000 = 20.3\ \text{amperes},$$
$$\text{Error} = .01 \times 20.3 = 0.2\ \text{ampere}.$$

Set 2–3 times error current, say 0.6 A—minimum tap is 1.0 A. Relay will not operate (a different relay model could be used to achieve greater sensitivity).
From appendix IV b **tap = 1.0,**
Time dial = #1/2.
Check for internal fault (F_1)

$$I_1 = 20.3, I_2 = 0\ I_{rel} = 20.3/1.0 = 20.3 \times \text{pickup}.$$

Relay operates in 0.03 s.

(7.3) $I_{a1} = -I_{a2} = \dfrac{j1.0}{0.21 + 0.21} = j2.3\ \text{per unit}$

$I_b = -I_c = \alpha^2(2.3) + \alpha\,(-2.3) = 2.3 \times (\alpha^2 - \alpha) = -j\sqrt{3} \times j2.3 = 3.98\ \text{per unit}$
$I_b = 3.98 \times 4.26 = {-}I_c = 16.95\ \text{amperes secondary}.$

With no CT error, $I_{rel\ F2} = 0$,

With 1%CT error $I_{rel\ F2} = 0.17$.

Pickup at 1.0 A will be good, that is, relay will not operate. Obviously, if CT has an error of 5–6%, the error will equal relay pickup.

(7.4) See Figures 16.23 and 16.24.

$$\text{Restraint current} = \frac{3.5 + 3.75}{2} = 3.63\ \text{amps}.$$
$$\text{Operating current} = 0.25\ \text{amps}.$$
$$\%\text{operating current}\ \frac{0.25}{3.63} = 0.068\text{-}6.8\%$$

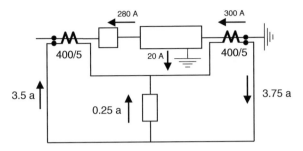

Figure 16.23 For Problem 7.4.

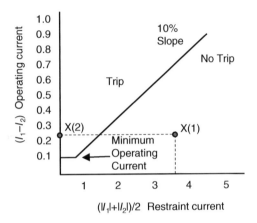

Figure 16.24 For Problem 7.4.

1) The relay will not operate, since the operating current is less than 10% of the restraint current.
2) At no load, the only current is the fault current $= 20/80 = 0.25$ A in the relay. This is above 0.1 A minimum operating current so relay will operate.

(7.5) Convert all impedances to per unit @100 MVA, 22 kV.

$$x_d = x_2 = 0.2 \times \frac{100}{975} = 0.021 \text{ pu}$$

$$x_t = 0.15 \times \frac{100}{1000} = 0.015 \text{ pu}$$

$$x_{sys} = 0.25 \text{ pu}$$

$$I_{fl} = \frac{975000}{\sqrt{3} \times 22} = 25587 \text{ A}$$

Select 30 000:5 (6000:1) CT

$$I_{base} = \frac{100000}{\sqrt{3} \times 22} = 2624 \text{ A},$$

0.021, 0.265 Paralleled: $\dfrac{.021 \times .265}{(.021 + .265)} = .019,$

$I_f(3\,\text{ph}) = 1.0/.019 = 52.63\,\text{pu} = 138105\,\text{amps}$,

$I_{3\text{ph gen}} = .93 \times 138105 = 128438/6000 = 21.4\,\text{amps secondary}$,

$I_{3\text{ph sys}} = .07 \times 138105 = 9667/6000 = 1.6\,\text{amps secondary}$,

$I_{\text{ph-ph gen}} = 0.866 \times 21.4 = 18.53\,\text{amps secondary}$,

$I_{\text{ph-ph sys}} = 0.866 \times 1.6 = 1.38\,\text{amps secondary}$.

Three-phase fault at F_1:

Restraint current $= \frac{21.4 - 1.6}{2} = 9.9\,\text{A}$.

Operating current $= 21.4 + 1.6 = 23\,\text{A}$.

Operating/restraint $= 23/9.9 = 232\%$.

Above 10% – relay will operate.

Phase-to-phase fault at F_1:

Restraint current $= \dfrac{18.53 - 1.38}{2} = 8.575\,\text{A}$.

Operating current $= 18.53 + 1.38 = 19.91\,\text{A}$.

Operating/restraint $= 19.91/8.575 = 232\%$.

Above 10% – relay will operate.

Three-phase fault at F_2:

Restraint current $= \dfrac{21.4 + 21.4}{2} = 21.4\,\text{A}$

Operating current $= 0$, no relay operation.

Phase-to-Phase fault at F_2

Restraint current $= \dfrac{18.53 + 18.53}{2} = 18.53\,\text{A}$.

Operating current $= 0$, no relay operation.

See Figures 16.25 and 16.26.

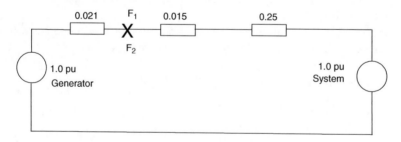

Figure 16.25 For Problem 7.5.

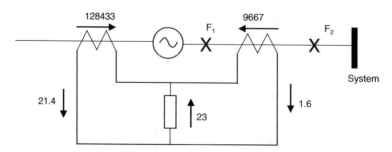

Figure 16.26 For Problem 7.5.

(7.6) I_{fl} @ 975 MVA and 22 kV = 25 617 A.

Select CT = 30 000:5 (6000:1).

Convert impedances to per unit @ 100 MVA, 22 kV.

$$x_d'' = x_2 = 0.21 \times 100/975 = 0.022 \text{ pu,}$$

$$x = .03 \times = 0.0031 \text{ pu,}$$

$$R_{0.5} = 0.5 \times \frac{100}{22^2} = 0.103 \text{ pu,}$$

$$R_{5.0} = 1.03 \text{ pu,}$$

$$R_{50.0} = 10.3 \text{ pu.}$$

For a fault at F_1 and $R_0 = 0.5 \, \Omega$,

$$I_1 = I_2 = I_0 = \frac{1.0}{3(0.103) + j(.022 + .022 + .00311)} = 3.19 \text{ pu,}$$

$$I_{base} = \frac{100000}{\sqrt{3} \times 22} = 2627 \text{ A,}$$

$$I_1 = I_2 = I_0 = 3.19 \times 2627 = 8393 \text{ amperes primary,}$$

$$I_g = I_a = 3 \times I_0 = 25178.9 \text{ amperes primary} = 4.2 \text{ amperes secondary@6000}$$
$$: 1.$$

@ $R_0 = 5.0 \, \Omega$,

$$I_1 = I_2 = I_0 = \frac{1.0}{3.09} = 0.324 \text{ pu} = 850.2 \text{ A primary} = 0.142 \text{ A secondary,}$$

$$I_g = I_a = 2550 \text{ amperes primary} = 0.425 \text{ amperes secondary.}$$

@ $R_0 = 50 \, \Omega$,

$$I_1 = I_2 = I_0 = \frac{1.0}{30.9} = 0.032 \text{ pu} = 84.1 \text{ A primary} = 0.014 \text{ A secondary,}$$

$$I_g = I_a = 252.3 \text{ amperes primary} = 0.042 \text{ amperes secondary.}$$

Set 87G

@$R_a = 0.5 \, \Omega$

$I_s = I_{rel} = 4.2$ A secondary

Relay 87G pickup is 0.2 A. So differential relay will operate.

@$R_a = 5.0 \, \Omega$

$I_s = I_{rel} = 0.425$ A secondary

This is only 2 × pickup of 87G, marginal.

@$R_a = 50 \, \Omega$

$I_s = I_{rel} = 0.042$ A secondary

87G will not pick up.

Set 51N

@$R_a = 0.5 \, \Omega$.

$I_g = 25\,179$ A primary

Select Neutral CT = 25 000/5 (5000:1)
I_{rel} = 5.04 A secondary
Set pick up =1.50 A tap
Set time delay @#1 time dial. This gives 1.3 second @3 × pu, enough to override any transients.
@R_a = 5.0 Ω.
I_g = 2550 A primary
Select Neutral CT = 3000/5 (600:1)
I_{rel} = 4.25 A secondary
Set pick up =1.0 A tap
Set time delay @#1 time dial.
@R_a = 50 Ω.
I_g = 252 A primary
Select Neutral CT = 250/5 (50:1)
I_{rel} = 5.04 A secondary
Set pick up =1.0 A tap
Set time delay @#1 time dial.

Since the generator is not connected to the system when the fault occurs, there is no through restraint current. The area below the percentage differential relay slope characteristic does not apply. 87G will operate as long as the operating current is above 0.2 A. 81N will always operate for phase-to-ground faults, since we set the relays with sufficient margin (see Figure 16.27).

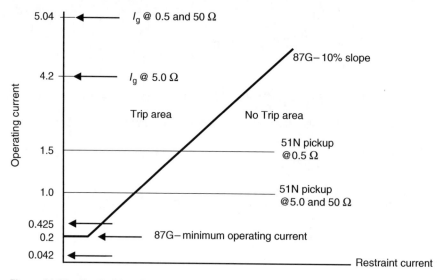

Figure 16.27   For Problem 7.6.

(7.7)   Maximum motor load = 1.15 × 25 = 28.75 A including service factor.
Select 50/5 CT from Table 3.1
Set TDOC (51) at 125% of 25 A. For larger or more critical motors, use the service factor, that is, set at 115% × 125% of I_{fl}.

$$I_{\text{rel}} = \frac{1.25 \times 25}{(50/5)} = 3.125 \text{ A}.$$

For phase faults, set TDOC (51) relays as follows:

CO-6

Use 3.5 A tap for pick up.

$$\text{Multiple of pickup during start-up} = 150/(10 \times 3.5) = 4.3 \times \text{pickup}$$

Must not operate in 1.5 s, #6 TD just operates in 1.5 s. Use #7 TD.

CO-11

Use 3.5 A tap for pickup.

$$\text{Multiple of pickup during start-up} = 150/(10 \times 3.5) = 4.3 \times \text{pickup}$$

Must not operate in 1.5 s, use #4 TD.

IAC-53

Use 3.0 A tap for pickup.

$$\text{Multiple of pickup during start-up} = 150/(10 \times 3) = 5 \times \text{pickup}$$

Must not operate in 1.5 s, use #6 TD.

Set instantaneous relay (50) above locked rotor current – say $2 \times 150 = 300$ A primary to cover offset, low voltage, etc.

@phase–phase fault $15\,000/300 = 50 \times$ pickup, very good.

Set relay $300/10 = 30$ A secondary.

For phase-g fault set TDOC relay (51G) at any available low tap with small time delay.

CO-6 and CO-11 relays use 1.0 A tap and for IAC-53 relay use the 1.5 A tap.

@50:5 CT this is 10 and 15 A secondary, respectively.

@1500 A for phase-g fault this is 150 and 100 × pickup.

Use #1 or #2 TD to avoid trip on false residual current during motor starting. See Figure 16.28.

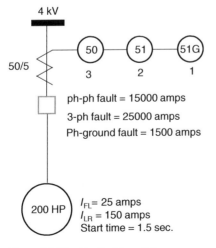

Figure 16.28 For Problem 7.7.

(7.8) For phase faults, set the TDOC relay (51) as follows:

<u>CO-6</u>

Use 3.0 A tap

During start-up $650/(50 \times 3) = 4.33 \times$ pu

Must not pickup in 3 s. This will require a setting above #11 time-dial—do not use this relay.

<u>CO-11</u>

Use 3.0 A tap—#9 time dial.

<u>IAC-53</u>

Use 3.0 A tap—#9 time dial.

<u>Instantaneous relay (50)</u>

Set above locked rotor—say $2 \times 650 = 1300$ A primary.

@phase–phase fault $15\,000/1300 = 11.53 \times$ pickup, good.

Set relay at $1300/50 = 26$ A secondary.

<u>Ground relay (51G)</u>

Set 1.0 or 1.5 A tap pickup. CT is 50/1, so primary pickup current is $1500/(50 \times 1) = 30 \times$ pu, or $1500/(50 \times 1.5) = 2 \times$ pickup. Both are good settings—use #1 or #2 time dial. See Figure 16.29.

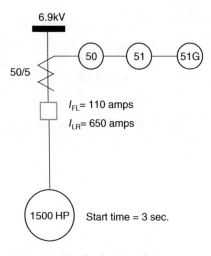

Figure 16.29 For Problem 7.8.

(7.9) To protect 85% of the generator winding, the voltage across the transformer primary equals $0.15 \times (15\,500/\sqrt{3}) = 1342$ V. Maintaining the primary fault current for a full line-to-ground fault at 10 A, the primary resistor remains 895 Ω. The transformer secondary voltage for a fault 85% from the phase end is

$$342 \times (480/14400) = 44.7 \text{ volts.}$$

To protect this much of the winding (i.e., 15% from neutral), the voltage relay must pick up at this value. Changing the secondary resistor will only change the primary current. The pickup of the voltage relay can be anything above 16 V to prevent false operation during normal full-load conditions.

16.8 Chapter 8 problems

(8.1) 100 kVA will produce full-load current of $(100/11) = 9.1$ A on the high side, and a load current of $(100/3.3) = 30.3$ A on the low side. Select CT ratios of 10:5 (i.e., 2:1) and 40:5 (i.e., 8:1) on the two sides. A fault at bus C will produce a current of

$$1/(.01 + j.06 + 0.0 + j0.02) = 12.40\angle\text{-}82.87°\,\text{pu}.$$

R_{bc}:
The pickup setting should be at least 1/3 of this value, and greater than 2 times the full load current of 1.0 pu. A reasonable pickup setting for relay R_{bc} is 3.0 pu. A time dial setting of the fastest available value (1) may be used for this relay. For R_{bc} the pickup setting should be $3 \times 30.3 \times (1/8) = 11.3$ A. Select a value of 10 A as being the nearest relay tap setting available.
Summarizing for R_{bc}, CT ratio $= 40:5$. TD $= 1$, and pickup setting $= 10$ A.
R_{ab}:
This relay should also pickup for a fault current of 12.40 pu. Select a pickup setting which is somewhat higher than that of R_{bc}. A value of 4.0 pu will still be less than 1/3 of the fault current. This value corresponds to $4 \times 9.1 \times (1/2) = 18.2$ A. Select a pickup of 20 A as being the nearest relay setting available.

The fault current at B is $1/(0.0 + j0.02) = 50\angle - 90°$ pu. This current produces in R_{bc} a current of $50 \times 30.3 \times (1/8) \times (1/10) = 19$ times pickup of R_{bc}. With a time-dial setting of 1, this relay will therefore operate in 0.1 s. R_{ab} should therefore operate for this current in $0.1 + 0.3 = 0.4$ s. R_{ab} sees this fault current as $50 \times 9.1 \times (1/2) \times (1/20) = 11.4$ times pickup. From Figure 4.7, the corresponding time dial setting is 3.
Summarizing for R_{ab}, CT ratio $= 10:5$, TD $= 3$, and pickup setting $= 20$ A.

(8.2) Max load $= 120\% \times 5 = 60$ MVA
Error introduced by the LTC $= +10\%$
CT error $= 5\%$

$$I_{high} = \left(60 \times 10^6/110 \times 10^3\right) = 545.45\,\text{amps}$$

Use CT ratio $= 600:5$ (i.e., 120:1)
The secondary current on the high side $I_{high\text{-}s} = 4.545$ A. Use a relay tap of 4.8 A.

$$I_{low} = \left(60 \times 10^6/33 \times 10^3\right) = 1818.18\,\text{amps}$$

Use CT ratio $= 2000:5$ (i.e., 400:1)

The secondary current on the low side $I_{\text{low-s}} = 1818.18/400 = 4.545$ A. Use a relay tap of 4.8 A. With this mismatch, the mismatch current in the differential coil will be zero.

Error due to LTC	= 10%
Error due to CT	= 5%
Safety margin	=5%
Slope	= 20%

Use lowest pickup tap provided on specific relay.

(8.3) Yes. With no tap changer, the only error is due to the CTs. This with the 5% margin would allow the slope to be 10%.

(8.4) Assume a single-phase transformer. Percentage differential relay slope = 20%. And minimum pickup = 0.25 A.

$$\text{Max overload} = 110\% \text{of } 10\,\text{MVA} = 11\text{MVA}$$

$$I_{\text{high}} = \left(11 \times 10^6/69 \times 10^3\right) = 159.42\,\text{amps}$$

Use CT ratio = 200:5 (i.e., 40:1)
The secondary current on the high side $I_{\text{high-s}} = 3.98$ A.

$$I_{\text{low}} = \left(11 \times 10^6/33 \times 10^3\right) 333.33\,\text{amps}$$

Use CT ratio = 400:5 (i.e., 80:1)
The secondary current on the low side $I_{\text{low-s}} = 4.17$ A.
At no load $I_{\text{restr}} = 0$.
$I_{\text{diff}} = (I_f/40)$, which must be greater than the pickup value of 0.25 A. This corresponds to 10 A primary current.
Note: depending upon the construction of the relay, the restraint current could be the average of the absolute values of the currents on the two sides, or the sum of the absolute values of the two currents or the larger of the two currents. The results would correspondingly be different in the various cases. In this solution, we assume that the restraint current is the average of absolute values of the two currents.
At maximum overload

$$I_{\text{restr}} = \frac{1}{2}\left\{\frac{159.42}{40} + \frac{333.33}{80}\right\} = 4.08\,\text{A}.$$

With a slope of 20%, $I_{\text{diff}} = 0.2 \times I_{\text{restr}} = 0.816$ A to operate. For an internal fault, assuming no source on the 33 kV side, $I_f = 0.816 \times 40 = 32.64$ A primary.
At no load, the percentage winding protected is $(1 - 10/159.4) \times 100\% = 93.7\%$. At maximum overload, the percentage winding protected is $(1 - 32.64/159.4) \times 100\% = 79.5\%$.

(8.5) The expression for the current is

$$i(\theta) = I_m \sin\theta; 0 < \theta < \alpha,$$

$$i(\theta) = 0 \;; \alpha < \theta < \pi,$$

$$i(\theta) = I_m \sin\theta; \pi < \theta < \pi + \alpha,$$

$$i(\theta) = 0 \;; \pi + \alpha < \theta < 2\pi.$$

Hence, the Fourier coefficients are

$$a_n = \frac{I_m}{\pi} \int_0^{2\pi} \sin\theta \cos n\theta \, d\theta,$$

$$b_n = \frac{I_m}{\pi} \int_0^{2\pi} \sin\theta \sin n\theta \, d\theta.$$

For even values of n, the two integrands cancel. Hence, there are only odd harmonics, $2m + 1$, where $m = 0,1,2,3, ...$ For odd values of n, the two segments add. Hence,

$$a_{2m + 1} = \frac{2I_m}{\pi} \int_0^{2\pi} \sin\theta \cos (2m + 1)\theta \, d\theta,$$

$$b_{2m + 1} = \frac{2I_m}{\pi} \int_0^{2\pi} \sin\theta \sin (2m + 1)\theta \, d\theta.$$

Using the integration formulas for products of trigonometric functions,

$$a_{2m + 1} = \frac{I_m}{\pi} \left\{ \frac{1 - \cos [2(m + 1)\alpha]}{2(m + 1)} - \frac{1 - \cos 2m\alpha}{2m} \right\},$$

$$b_{2m + 1} = \frac{I_m}{\pi} \left\{ \frac{\sin 2m\alpha}{2m} - \frac{\sin [2(m + 1)\alpha]}{2(m + 1)} \right\}.$$

These formulas have been calculated for $m = 0,2,3,4$ for three values of α, and tabulated below (normalized by i_m):
Cosine coefficients:

Harmonic	$\alpha = 30$	$\alpha = 90$	$\alpha = 120$
3	−0.0796	0.1592	−0.2387
5	−0.0928	0.1061	−0.0398
7	0.0199	0.0796	0.0597
9	0.0040	0.0637	−0.0597
11	−0.0345	0.0531	−0.0080

Sine coefficients:

Harmonic	$\alpha = 30$	$\alpha = 90$	$\alpha = 120$
3	0.1378	0.0000	0.1378
5	−0.0230	0.0000	−0.1149
7	−0.0345	0.0000	0.0345
9	0.0345	0.0000	0.0345
11	−0.0046	0.0000	−0.0505

Magnitude of harmonics:

Harmonic	$\alpha = 30$	$\alpha = 90$	$\alpha = 120$
3	0.1592	0.1592	0.2757
5	0.0956	0.1061	0.1216
7	0.0398	0.0796	0.0689
9	0.0347	0.0637	0.0689
11	0.0348	0.0531	0.0512

If each of the phase currents in the windings of the CTs has the same wave shape, the phase difference between the harmonics of the different phases will be $2n\pi/3$, where n is the order of the harmonics. Thus, all harmonics that are a multiple of 3 will circulate inside the delta. Thus, third and ninth harmonics will not flow in the lines coming out of the delta connected CTs, while fifth and seventh will flow normally. Fifth will be a negative-sequence current, while the seventh will be a positive-sequence current.

(8.6) Three-phase transformer with 500 MVA 34.5/500 kV rating has

$$I_{base} = 8367.4 \text{ amps@34.5 kV}$$
$$= 577.35 \text{ amps@500 kV}$$

The fault current from the 34.5 kV side is

$$|\{1/j(0.3 + 0.08)\}| = 2.632 \text{ pu}$$
$$= 2.632 \times 8367.4 = 22020 \text{ amps primary}$$

The secondary current is $22\,020 \times 5/9000 = 12.235$ A
The fault current on the 500 kV side is

$$|\{1/(j0.2)\}| = 5.0 \text{ pu}$$
$$= 5 \times 577.35 = 2886.75 \text{ amps primary.}$$

This corresponds to $2886.75 \times 5/1000 = 14.434$ A secondary.
The secondaries on the wye side of the main transformer are connected in delta. Hence, on this side, the relay current is $1.73 \times 14.434 = 24.97$ A.

$$I_{restr} = (12.2 + 24.97)/2 = 18.58 \text{ amps,}$$
$$I_{diff} = I_1 - I_2 = (12.2 + 24.97) = 37.2 \text{ amps.}$$

Relay will operate, since $I_{diff} > 0.2 \times I_{restr}$.

(8.7) Fault calculation
For a phase-to-phase fault, the current in the faulted phases is $\sqrt{3}/2$ times the three-phase fault current. Thus, the current in phases b and c on the 500 kV (contribution from the system) are $\sqrt{3}/2 \times 2886.75 = 2499.92$ A.
On the wye side, the current in phases b and c is $\sqrt{3}/2 \times (1/0.38) \times 577.35 = 1315.96$ A. These currents flow in phases b and c in the opposite directions.

The corresponding currents inside the delta windings of these two phases are $1315.96 \times 500/(\sqrt{3} \times 34.5) = 11\,024.22$ A. These currents will flow outside the delta winding as follows:

$\quad I_a = 11024.22$ amps,

$\quad I_b = 2 \times 11024.22 = 22048.43$ amps,

$\quad I_c = 11024.22$ amps.

With the CT ratios on the delta side of 9000:5, the corresponding relay currents are

$\quad i_{a\text{-relay}} = 6.125$ amps,

$\quad i_{b\text{-relay}} = 12.25$ amps,

$\quad i_{c\text{-relay}} = 6.125$ amps.

On the wye side of the main transformer, the CT winding currents are

$\quad 1315.96 \times 5/1000 = 6.58$ amps,

$\quad I_{a\text{-relay}} = 6.58$ amps,

$\quad I_{b\text{-relay}} = 13.18$ amps,

$\quad I_{c\text{-relay}} = 6.58$ amps.

In the a-phase relay, the differential current is $6.58 - 6.125 = 0.455$ A, while the restraint current is $(6.58 + 6.125)/2 = 6.35$ A. Clearly, the relay does not operate. Similar calculations show that the b- and c-phase relays also do not operate (see Figure 16.30).

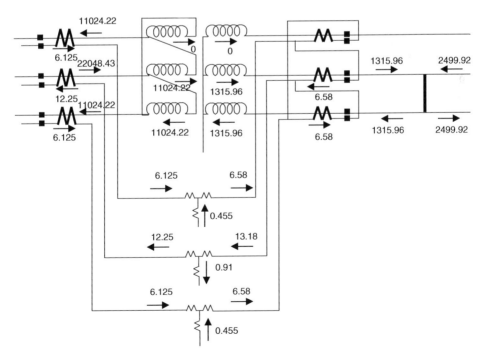

Figure 16.30 For Problem 8.7.

(8.8) A delta zigzag transformer produces no phase shift. Thus, assuming that the CTs on the delta side are connected in wye, and the CTs on the zigzag side are connected in delta (to remove zero sequence currents from the zig-zag side). The CT secondary currents which are fed to the relays will have a phase angle difference of 30°. If we assume that there is no CT ratio mismatch (even taking into account the increase in line current outputs from the delta connected CTs), the current on one side of the relay will be I, while that on the other side will be $I\angle\text{-}30°$. The differential current will then be $(I - I\angle 30°) = 2xI \sin 15° = \text{-}0.52\,I$. The restraint current will be I, thus a slope of 55% will restrain the relay for load or external fault. Any error due to the CT ratio mismatch must be added to determine a new slope (see Figure 16.31).

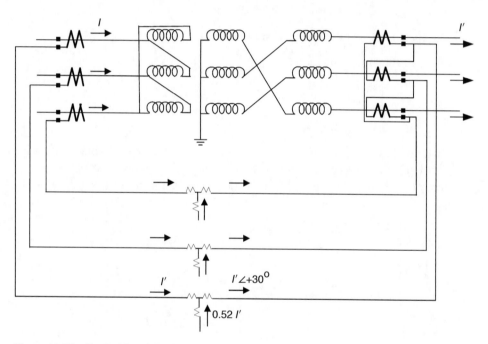

Figure 16.31 For Problem 8.8.

Note that this would result in a very insensitive application, as some error due to CT ratio mismatch or saturation would always be likely. Therefore, a preferred solution would be to use a digital transformer percentage restrained differential relay which is capable of removing the zero sequence current from wye connected CTs. In that case the CTs on the zig-zag side of the transformer could be wye connected with no troublesome phase shift to require an unreasonably high slope setting.

16.9   Chapter 9 problems

(9.1)   Install the CTs on the line side of each circuit breaker so the breaker is included in the differential protective zone. The CTs are connected to their associated bus differential as shown. These connections are permanent regardless of the position of the breaker disconnect switches, that is, open or closed, since the summation around each bus will be correct even if a breaker is removed from service and the line is transferred to the other bus. The relay connection shown in the three-phase diagram is the same whether the relay is an overcurrent, percentage differential, or high or moderately high impedance relay. It is not correct for a linear coupler (see Figures 16.32 and 16.33).

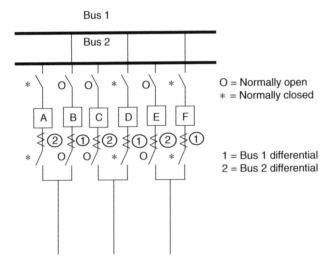

Figure 16.32   For Problem 9.1.

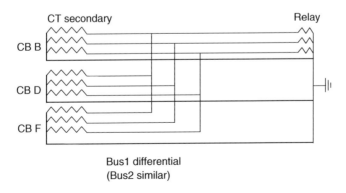

Bus1 differential
(Bus2 similar)

Figure 16.33   For Problem 9.1.

(9.2) The CTs are connected to their associated bus differential for the normal bus configuration. With this bus configuration, there are two options available for maintenance.

1) If a breaker is to be maintained, the bypass disconnect S for the appropriate breaker is closed, bypassing the breaker so it can be removed from service. The bus tie breaker then assumes the function of the line breaker (note: relaying is provided on the bus tie breaker for line protection and must be modified each time the bus tie breaker becomes a different line breaker). All of the other breakers must be transferred to bus 2, the CTs must be reconnected as a bus 2 differential, bus 1 differential is removed from service and bus 1 is protected as part of the bypassed line.

2) If a bus is to be maintained, all of the breakers are transferred to the opposite bus, and CTs are reconnected as a single bus differential and the bus tie breaker is opened (see Figures 16.34 and 16.35).

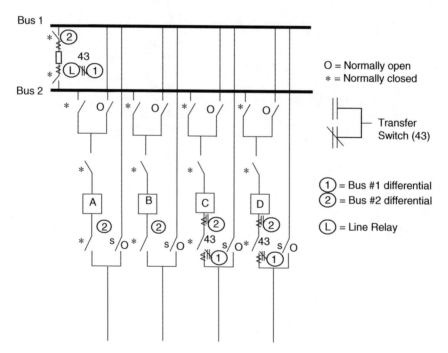

Figure 16.34 For Problem 9.2.

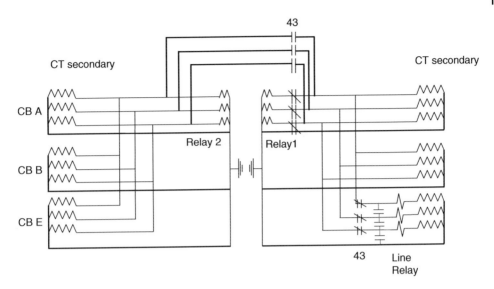

Figure 16.35 For Problem 9.2.

(9.3) The sequence of switching must accommodate several requirements:
1) No load should be lost during the switching process.
2) All elements must be protected at all times.
3) No relays should see an unbalance to the extent that they would misoperate.

Before beginning any primary switching, an overall bus differential should be established. This is accomplished by utilizing transfer switches to reconnect all CT secondaries to form one bus differential (in this case using bus 2 differential relay), and removing bus 1 differential relay from service. Care must be taken during the switching to assure that no CT is open-circuited. This is done by designing the transfer switch to have "make-before-break" contacts so the CT secondaries are always connected to the relay or shorted out. Having a single overall bus differential is, of course, a simple relay application that is always possible, but it negates the purpose of splitting the primary system into two buses to maintain system integrity in the event of a bus fault. During breaker maintenance, however, this is a risk that must be taken. However, it is not taken without due regard to the total system condition, that is, no adverse weather conditions, no other maintenance or construction in the immediate vicinity, and no adverse system conditions. All breakers are now transferred to Bus 2 and the bypass switch around the breaker to be maintained can now be closed. The appropriate relay settings for the line are activated on the tie breaker, either manually or automatically, and the line breaker can be removed from service by opening its disconnect switches. The reverse procedure is used to return the breaker to service, maintaining the overall bus differential until all breakers have been returned to their normal position.

(9.4) $I_{FL} = \dfrac{50\,\text{MVAR}}{\sqrt{3} \times 765} = 37.74\text{ A primary.}$

Select CT ratio 50/5 (10/1), $I_{FL} = 3.77$ A secondary.
<u>Instantaneous relay:</u> use approximately $3 \times I_{FL} = 3 \times 3.77$ A. Set 5 A and check for pickup for a phase-to-phase fault and a phase-to-ground fault at the given station.
<u>TDOC:</u> set $2 \times I_{FL} = 7.5$ A. Set 8.0 A pickup and #1 lever.

(9.5) See Figure 16.36.

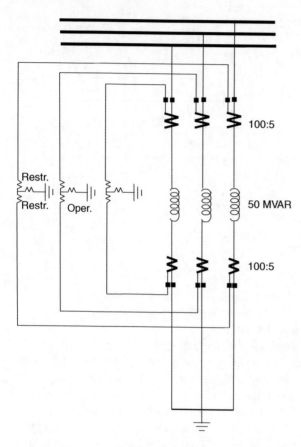

Figure 16.36 For Problem 9.5.

16.10 Chapter 10 problems

(10.1) At the steady-state stability limit, the rotors of the two machines are 90° apart. Let E_2 be the reference phasor. Then E_s is positive imaginary. From the phasor diagram, let the phase angle of E_1 be ϕ.
Then,

$$E_1 = \bar{E}_1 \cos \phi + j\bar{E}_1 \sin \phi.$$

The lengths of E_s and E_2 are divided by the imaginary and real parts of E_1 in the ratio of the impedances of the branches (see the phasor diagram, Figure 16.37):

$$E_2 = \bar{E}_1 \cos \phi[1 + X_t/X_s],$$
$$E_s = \bar{E}_1 \sin \phi[1 + X_s/X_t].$$

The current in the line is given by

$$
\begin{aligned}
I_1 &= [E_s - E_2]/(X_s + X_t) \\
&= [E_s + jE_2]/(X_s + X_t) \\
&= \bar{E}_1 \sin \phi/X_t + j\bar{E}_1 \cos \phi/X_s
\end{aligned}
$$

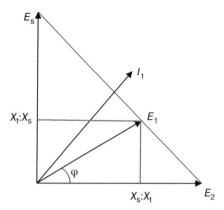

Figure 16.37 For Problem 10.1.

The complex power is given by

$$P_1 + jQ_1 = E_1I_1^* = (\bar{E}_1 \cos \phi + j\bar{E}_1 \sin \phi)(\bar{E}_1 \sin \phi/X_t - \bar{E}_1 \cos \phi/X_s)$$
$$= [E_1^2/X_sX_t][\sin 2\phi(X_s + X_t)/2 + j(X_s \sin^2\phi - X_t \cos^2\phi)]$$

Using trigonometric identities: $1 - \sin^2\phi = \cos^2\phi$, and $\sin^2\phi = (1 - \cos2\phi)/2$, and separating into real and imaginary parts:

$$P_1 = \left[\bar{E}_1^2/2X_sX_t\right](X_s + X_t) \sin 2\phi,$$

$$Q_1 = \left[\bar{E}_1^2/2X_sX_t\right][(X_s-X_t)-(X_s + X_t) \cos 2\phi].$$

The two equations can be solved for $\sin2\phi$ and $\cos2\phi$, and upon squaring and adding, the angle ϕ is eliminated:

$$\cos 2\phi = \left[(X_s-X_t)-2Q_1 X_sX_t/\bar{E}_1^2\right) \cos 2\phi]/(X_s + X_t),$$

$$\sin 2\phi = 2P_1X_sX_t/\left[\bar{E}_1^2(X_s + X_t)\right].$$

When the two expressions are squared and added, the result is

$$4P_1^2\left(X_sX_t/\bar{E}_1^2\right)/(X_s + X_t)^2 + \left[(X_s-X_t)-2Q_1 X_sX_t/\bar{E}_1^2\right)]/(X_s + X_t)^2 = 1.$$

Upon simplifying and rearranging the terms, the desired result is obtained:

$$P_1^2 + \left[Q_1^2 - \frac{E_1^2}{2}\left(\frac{1}{X_t} - \frac{1}{X_s}\right)\right]^2 = \left[\frac{E_1^2}{2}\left(\frac{1}{X_t} + \frac{1}{X_s}\right)\right]^2.$$

(10.2) Substitute $P = E^2R/(R^2 + X^2)$ and $Q = E^2X/(R^2 + X^2)$ in Equation $(P - P_0)^2 + (Q - Q_0)^2 = S_0^2$.

Expanding the squared brackets

$$E^4R^2/\left(R^2 + X^2\right)^2 + P_0^2 - 2E^2RP_0/\left(R^2 + X^2\right) + E^4X^2/\left(R^2 + X^2\right)^2$$
$$+ Q_0^2 - 2E^2XQ_0/\left(R^2 + X^2\right) = S_0^2.$$

Collecting the terms in powers of R and X

$$R^2(P_0{}^2 + Q_0{}^2\text{-}S_0{}^2)\text{-}2E^2RP_0 + X^2(P_0{}^2 + Q_0{}^2\text{-}S_0{}^2) - 2XE^2Q_0 + E^4 = 0.$$

Completing the squares in R and X

$$R^2\text{-}2RE^2P_0/(P_0{}^2 + Q_0{}^2\text{-}S_0{}^2) + \left[E^2P_0/(P_0{}^2 + Q_0{}^2\text{-}S_0{}^2)\right]^2,$$

$$X^2\text{-}2XE^2Q_0/(P_0{}^2 + Q_0{}^2\text{-}S_0{}^2) + \left[E^2Q_0/(P_0{}^2 + Q_0{}^2\text{-}S_0{}^2)\right]^2$$

$$= E^4\left[P_0{}^2 + Q_0{}^2\text{-}P_0{}^2\text{-}Q_0{}^2 + S_0{}^2\right]/(P_0{}^2 + Q_0{}^2\text{-}S_0{}^2) = E^4\left[S_0{}^2\right]/(P_0{}^2 + Q_0{}^2\text{-}S_0{}^2)^2,$$

which leads to the required expression of a circle in the R–X plane:

$$\left[R - \frac{E^2P_0}{P_0^2 + Q_0^2 - S_0^2}\right]^2\left[X - \frac{E^2Q_0}{P_0^2 + Q_0^2 - S_0^2}\right]^2 = \frac{E^4S_0^2}{\left(P_0^2 + Q_0^2 - S_0^2\right)^2}.$$

When $P_0{}^2 + Q_0{}^2 = S_0{}^2$, the circle becomes a straight line through the origin, with the coordinates of the center and the radius becoming infinite.

(10.3) As $t \to \infty$, the RHS of Equation (10.32) tends to infinity. The left-hand side also tends to infinity as the denominator of the log terms tends to 0. Note that the multiplier of the log term in the curly brackets is negative, thus with the minus sign in front of the curly brackets, the entire LHS to $+\infty$ as the log term tends to infinity. Let the limiting frequency as $t \to \infty$ be f_∞. Then,

$$(f_0\text{-}f_\infty)(1 + L_0)d/L_0f_0 = 1,$$

Or, $f_0 - L_0f_0/[d(1 + L_0)] = f_\infty$,

Or, $f_\infty = f_0[1\text{-}L_0/\{(1 + L_0)d\}]$. This is Equation (10.33).

(10.4) The excess load factor L $= (10000\text{-}9000)/9000 = 0.111$.

The exact rate of change of frequency is obtained from the differential equation, Equation (10.26). The initial frequency f_0 is 60 Hz, and with the synchronous frequency f_s is also 60 Hz, a power factor of 0.8, and the inertia constant of 15 s.

$$df/dt = -Lf^2{}_s/(2H_0f_0) = -0.8 \times 0.111 \times 60^2/(2 \times 15 \times 60) = -0.176 \text{ Hz/sec}.$$

Now use the approximate formula, with f_1 of 60 Hz, f_2 equal to 59.9 Hz (a value sufficiently close to 60 Hz, so that these conditions are equivalent to those in the previous calculation):

$$R = 0.8 \times 0.111 \times (60\text{-}59.9)/\left[15 \times \{1\text{-}59.9^2/60^2\}\right] = \text{-}0.177 \text{ Hz/sec}.$$

The two results are practically identical.

(10.5) $f_\infty = f_0[1\text{-}L_0/\{d(1 + L_0)\}]$

$$= 60[1\text{-}0.111/\{2.5(1 + 0.111)\}].$$

$$= 57.60 \text{ Hz}$$

16.11 Chapter 11 problems

(11.1) Consider the impedance triangle shown (Figure 16.38), with the line impedance being Z(line SR), and the impedances seen by distance relays as seen from the two ends being SO and RO. The last two impedances are proportional to the voltages of the two sources. Thus, by making α different from 1, and the length RO α times the length of SO, the two voltages are made unequal. The angle between the two voltage phasors is δ. Let the coordinates of the point O be (x,y). We are to show that the point describes a circle as δ changes, and α remains constant.

The length of RO is $\alpha\sqrt{(x^2+y^2)}$, and the length of SO is $\sqrt{(x^2+y^2)}$. The length of SR is Z, and is made up of SP and PR. Length of SP is y, while the length of PR is $\sqrt{[\alpha^2(x^2+y^2) - x^2]}$. Thus,

$$Z = y + \sqrt{\left[\alpha^2\left(x^2 + y^2\right) - x^2\right]}$$

Simplifying this further,

$$(Z\text{-}y)^2 = \left[\alpha^2\left(x^2 + y^2\right) - x^2\right]$$
$$\text{Or}, x^2\left(\alpha^2\text{-}1\right) + y^2\left(\alpha^2\text{-}1\right) + 2yZ = Z^2$$
$$\text{Or}, x^2 + y^2 + 2yZ/\left(\alpha^2\text{-}1\right) = Z^2/\left(\alpha^2\text{-}1\right)$$
$$\text{Or}, x^2 + \left[y + Z/\left(\alpha^2\text{-}1\right)\right]^2 = Z^2\alpha^2/\left(\alpha^2\text{-}1\right)^2$$

This is the equation of a circle.

Note that when $\alpha = 1$, the above equation degenerates into a straight line with the equation $y=Z/2$.

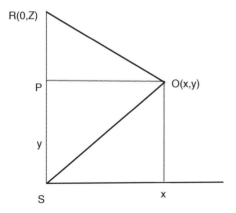

R(0,Z)

P

O(x,y)

y

S

x

Figure 16.38 For Problem 11.1.

(11.2) With the line angle being 80o, the impedance diagram is as shown. The length of line SR is computed as $\sqrt{[SP^2 + PR^2]}$.

$$SR = \sqrt{[(0.11 \cos 10 + 0.06 + 0.1)^2 + (0.11 \sin 10)^2]}$$
$$= 0.2690$$

The angle RSP is $\tan^{-1}[0.11 \sin 10/(0.11 \cos 10 + 0.06 + 0.1)] = 4.06o$.
The angle SOR is 110o.
Hence the length SO is $[0.2690/2]/\sin 55 = 0.1642$.
The length QO is $0.1642\cos51.94 = 0.1012$.
The length QT is $0.1642\sin51.94 - 0.06 = 0.0693$.
Hence the length TO is $\sqrt{[0.1012^2 + 0.0693^2]} = 0.106$.
This is the nearest excursion of the stability swing to the line terminal at T. With a margin of safety, set the inner zone of the out-of-step relay at 0.9 pu. All other swings may remain as in example 10.4. (See Figure 16.39).

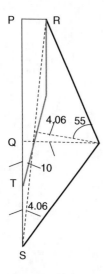

Figure 16.39   For Problem 11.2.

16.12   Chapter 13 problems

(13.1)   Normal voltage = 15 mm × 4 V/mm = 60 V secondary, ph-to-neutral × 3000/1 = 180 kV primary, 180 × $\sqrt{3}$ = 311 kV ph–ph.
Fault voltage = 4 mm × 4 V/mm = 16 V secondary, ph–neutral × 3000/1 = 48 kV primary.
48× $\sqrt{3}$ = 83.14 kV phase-to-phase.

(13.2)   Normal phase current = 4 mm × 1 A/mm = 4 A secondary × 240/1 = 960 A.
Normal ground current = 1 mm ×1 A/mm = 1.0 A secondary × 240/1 = 240 A.
Fault phase current = 16 mm × 1 A/mm = 16.0 A secondary × 240/1 = 3840 A.
Fault ground current = 3840 A primary.

Ground current during normal operation, that is, with no fault, is due to unbalanced phase currents and/or CT errors.

(13.3) Fault is a three-phase fault. All three-phase currents increased. The fault is cleared after 2½ cycles, is de-energized for 6 cycles, and successfully recloses.

(13.4) See Figure 16.40.
   $T = 0$–0.02 s, E_a, E_b, E_c normal
   Current = load current.
   Ground current = 0.
   (No fault)
   $T = 0.02$–0.06 s, E_a (line 1) reduced
   E_b (line 2), E_s (bus) normal
   I_a (line 1) increased
   I_g (line 1) increased
   I_b (line 1) increased
   I_a (line 2) increased
   I_g (line 2) increased
   I_b (line 2) no change
   (phase a to ground fault—could be line 1 or line 2)
   $T = 0.06$–0.08 s, I_a, I_b, I_g (line 1) zero. Breaker A open.

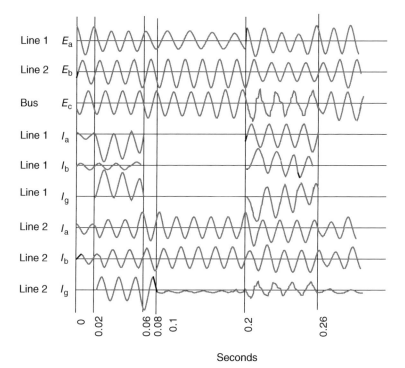

Figure 16.40 For Problem 13.4.

<u>Fault is on line 1.</u>

This could be verified by relay targets.

I_a, I_b, I_g (line 2) increased. This is the contribution of line 2 to the fault on line 1 through breaker B until it opens.

$T = 0.08\text{–}0.2\,\text{s}$: Line 1 is de-energized, line 2 is carrying normal load, voltages on line 2 and bus are normal. E_a (line 1) is showing a trapped charge, slowly decreasing in frequency and magnitude.

$T = 0.2\,\text{s}$ Breaker A recloses

$T = 0.2\text{–}0.26\,\text{s}$ Breaker has reclosed into an a–b–g fault.

$T = 0.26$ Fault is cleared. Line is open.

(13.5)   The symmetrical component connection for a b–c fault is shown in Figure 16.41. At the fault point,

$$E_{1F} = E_{2F} + RI_F,$$

$$\text{and } I_{1F} = -I_{2F}.$$

Note that R in this case and the phase-to-ground fault case may not represent the actual fault resistance, but its symmetrical component representation. This is acceptable, as it does not appear in the final formula.

Using the relationships $E_1 = E_{1F} + kZ_1I_1$ and $E_2 = E_{2F} + kZ_1I_2$, and also $E_b - E_c = (\alpha^2 - \alpha)(I_1 - I_2)$, we get

$E_b - E_c = kZ_1(I_b - I_c) + R(I_{bF} - I_{cF})$. If the positive-sequence distribution factor is defined as

$$d_1 = [Z_{r1} + (1\text{-}k)Z_1]/(Z_{r1} + Z_{s1} + Z_1),$$

the above equation can be written as

$$E_b - E_c = kZ_1(I_b\text{-}I_c) + R[(I_b\text{-}I_c)\text{-}(I_{b0}\text{-}I_{c0})]/d_1,$$

where the subscript "0" indicates the prefault currents. These are the desired equations for the b–c fault.

The symmetrical component connection for a–g fault is shown in Figure 16.42. The fault boundary conditions are given by

$$E_1 = E_{1F} + kZ_1I_1,$$

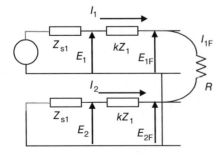

Figure 16.41   For Problem 13.5.

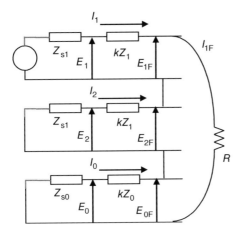

Figure 16.42 For Problem 13.5.

$$E_2 = E_{2F} + kZ_1I_2,$$

$$E_0 = E_{0F} + kZ_0I_0,$$

$$E_{1F} + E_{2F} + E_{0F} = RI_{1F},$$

$$I_{1F} = I_{2F} = I_{0F}.$$

Adding the equations together, and recalling that the phase a voltage and current are the sum of the positive-, negative-, and zero-sequence components, and with the usual definition of compensated phase current for ground faults—see Equation (5.23)—we get

$$E_a = kZ_1I_a' + RI_{1F}.$$

Once again, we may define the positive-sequence distribution factor

$$d_1 = [Z_{r1} + (1\text{-}k)Z_1]/(Z_{r1} + Z_{s1} + Z_1),$$

so that the voltage equation becomes

$$E_a = kZ_1I_a' + R(I_1\text{-}I_{10}).$$

Here also the subscript "0" represents the pre-fault value of the positive-sequence current. In practice, it has been found that it is sufficiently accurate to use the phase a currents instead of the positive-sequence currents in the abovementioned formula.

Since the sequence currents in the fault are all equal to each other, it is not necessary to use negative- or zero-sequence distribution factors. This is indeed a great benefit, as no easy approximations for the zero-sequence distribution factors are available.

(13.6) The exact value of d is $0.6555 - j0.0038$. Changing its angle by $5°$, the distribution factor changes to

$$d' = (\cos 5 + j \sin 5)(0.6555\text{-}j0.0038 = 0.6533 + j0.533.$$

Using this value of d', and the sending-end voltage angle ranging between $-30°$ and $30°$, the calculations performed in Example 11.3 are repeated. These are best done with a computer program, although it is not too time-consuming to do them by hand. The results are as follows:

δ	k
$-30°$	0.4325
$-20°$	0.4085
$-10°$	0.3918
$-0°$	0.3793
$10°$	0.3696
$20°$	0.3618
$30°$	0.3552

If the error in d is $-5°$, $d' = 0.6527 - j0.0609$, and the corresponding values of k are as shown in the following text:

δ	k
$-30°$	0.1536
$-20°$	0.1847
$-10°$	0.2048
$-0°$	0.2190
$10°$	0.2297
$20°$	0.2381
$30°$	0.2449

Appendix A

IEEE Device Numbers and Functions

The devices in switching apparatus are referred to by numbers, with appropriate suffix numbers and letters when necessary. These numbers are based on a system adopted as standard for automatic switchgear by the Institute of Electrical and Electronics Engineers (IEEE) and incorporated in American National Standards Institute (ANSI)/IEEE Standard C37.2. This system is used in electrical elementary and connection diagrams, in instruction books, and in specifications. The list in the following text is a selected group that is commonly used.

1 Master element that operates to place a device in or out of service.

13 Synchronous speed switch that operates at approximately the synchronous speed of the machine.

20 Electrically operated valve.

21 Distance relay which functions when the circuit admittance, impedance, or reactance increases or decreases beyond predetermined limits.

23 Temperature control device.

25 Synchronizing or check-synchronizing device that operates when two AC circuits are within desired limits of frequency, phase angle, or voltage to permit the paralleling of these two circuits.

26 Thermal device that operates when the temperature of the protected apparatus decreases below a predetermined value.

27 Undervoltage relay.

30 Annunciator relay.

32 Directional power relay that operates on a desired value of power in a given direction.

41 Field circuit breaker that applies or removes field excitation to a machine.

42 Running circuit breaker that connects a machine to its running or operating voltage.

43 Manual transfer switch.

46 Reverse-phase or phase-balance relay that operates when the polyphase currents are of reverse-phase sequence, or when the polyphase currents are unbalanced or contain negative-sequence currents of a given amount.

47 Phase-sequence voltage relay that operates upon a predetermined value of polyphase voltage in the desired phase sequence.

48 Incomplete sequence relay that returns equipment to normal if the normal starting or stopping sequence is not completed within a predetermined time.

Power System Relaying, Fifth Edition. Stanley H. Horowitz, Arun G. Phadke and Charles F. Henville.
© 2023 John Wiley & Sons Ltd. Published 2023 by John Wiley & Sons Ltd.

49 Thermal relay that operates when the temperature of a machine exceeds a predetermined value.

50 Instantaneous overcurrent or rate-of-rise relay.

51 AC time-delay overcurrent relay that operates when the current exceeds a predetermined value. The relay operates with either a definite or an inverse time characteristic.

52 AC circuit breaker.

53 Exciter or DC generator relay that forces the DC machine excitation to build up during starting.

55 Power factor relay.

59 Overvoltage relay.

60 Voltage or current balance relay that operates on a given difference in the input or output of two circuits.

62 Time-delay stopping or opening relay.

63 Pressure switch that operates on given values or given rate of change of pressure.

64 Ground protective relay.

65 Governor device used to regulate the flow of water, steam, or other media.

67 AC directional overcurrent relay.

69 Permissive control device.

72 DC circuit breaker.

76 DC overcurrent relay.

78 Phase angle measuring or out-of-step relay that operates at a predetermined phase angle between two currents, two voltages, or between a voltage and a current.

81 Frequency relay.

85 Carrier or pilot wire receiver relay.

86 Lockout relay that is electrically operated and hand or electrically reset to shut down and hold equipment out of service.

87 Differential protective relay that functions on a percentage, or phase angle, or other quantitative difference of two currents or some other electrical quantities.

90 Regulating device that operates to regulate a quantity at a certain value or between certain limits.

91 Voltage directional relay that operates when the voltage across an open circuit beaker or contractor exceeds a given value in a given direction.

94 Trip-free relay that operates to trip a circuit breaker or contractor.

101 Control switch to open and close a circuit breaker or contractor.

If one device performs two relatively important functions so that it is desirable to identify both functions, a double device function number is used, such as 50/51—instantaneous and time-delay overcurrent relay.

Suffix letters are used with device function numbers to further uniquely identify the device. Lowercase suffix letters are used for auxiliary, position, and limit switches. Uppercase letters are used for all other functions.

Suffix numbers are used when two or more devices have the same device function numbers and suffix letters to further differentiate between the devices, such as 52X-1 and 52X-2.

Appendix B

Symmetrical Components

B.1 Definitions

Symmetrical components of voltages or currents are defined through a linear transformation of phase quantities [1, 2]. Let $\mathbf{X_p}$ be the phase quantities and $\mathbf{X_s}$ be their symmetrical components, where X may be voltages or currents. Thus,

$$\mathbf{X_s} = \begin{bmatrix} X_0 \\ X_1 \\ X_2 \end{bmatrix} = \mathbf{SX_p} = \frac{1}{3} \begin{bmatrix} 1 & 1 & 1 \\ 1 & \alpha & \alpha^2 \\ 1 & \alpha^2 & \alpha \end{bmatrix} \begin{bmatrix} X_a \\ X_b \\ X_c \end{bmatrix} \tag{B.1}$$

The elements of $\mathbf{X_s}$ identified by the subscripts 0, 1, and 2 are known as the zero-, positive-, and negative-sequence components, respectively. The inverse of the matrix $\mathbf{S}$ is given by

$$\mathbf{S}^{-1} = 3 \begin{bmatrix} 1 & 1 & 1 \\ 1 & \alpha & \alpha^2 \\ 1 & \alpha^2 & \alpha \end{bmatrix}^{-1} = \begin{bmatrix} 1 & 1 & 1 \\ 1 & \alpha^2 & \alpha \\ 1 & \alpha & \alpha^2 \end{bmatrix}, \tag{B.2}$$

where $\alpha = \left(-\frac{1}{2} + j\frac{\sqrt{3}}{2} \right)$ is a cube root of 1. $\mathbf{S}$ is a similarity transformation on impedance matrices of certain classes of three-phase power apparatus.

B.2 Identities

$$\alpha^3 = 1, \tag{B.3}$$

$$1 + \alpha + \alpha^2 = 0, \tag{B.4}$$

$$1 - \alpha = \sqrt{3}\varepsilon^{-j\pi/6}, \tag{B.5}$$

Power System Relaying, Fifth Edition. Stanley H. Horowitz, Arun G. Phadke and Charles F. Henville.
© 2023 John Wiley & Sons Ltd. Published 2023 by John Wiley & Sons Ltd.

$$1 - \alpha^2 = \sqrt{3}e^{j\pi/6}, \tag{B.6}$$

$$\alpha - \alpha^2 = j\sqrt{3}. \tag{B.7}$$

B.3 Sequence impedances

The impedance matrix (or admittance matrix) of a three-phase element $\mathbf{Z_p}$ transforms into a sequence impedance matrix $\mathbf{Z_s}$ in the symmetrical component frame of reference. The general transformation is

$$\mathbf{Z_s} = \mathbf{S}\mathbf{Z_p}\mathbf{S}^{-1}. \tag{B.8}$$

The elements of $\mathbf{Z_s}$ are known as the sequence impedances of the three-phase element. The three diagonal elements are the zero-sequence, positive-sequence, and negative-sequence impedances, respectively. The off-diagonal elements are zero for all balanced elements. Even in the presence of unbalances, the off-diagonal elements of $\mathbf{Z_s}$ are often neglected. The sequence impedances of some of the more common types of power system elements are given in the following text. For admittances, replace Z's by Y's.

a) Balanced impedances without mutual coupling between phases:

$$\mathbf{Z_p} = \begin{bmatrix} Z_s & 0 & 0 \\ 0 & Z_s & 0 \\ 0 & 0 & Z_s \end{bmatrix}, \tag{B.9}$$

$$\mathbf{Z_s} = \begin{bmatrix} Z_s & 0 & 0 \\ 0 & Z_s & 0 \\ 0 & 0 & Z_s \end{bmatrix}. \tag{B.10}$$

b) Balanced impedances with mutual coupling between phases:

$$\mathbf{Z_p} = \begin{bmatrix} Z_s & Z_m & Z_m \\ Z_m & Z_s & Z_m \\ Z_m & Z_m & Z_s \end{bmatrix}, \tag{B.11}$$

$$\mathbf{Z_s} = \begin{bmatrix} Z_s + 2Z_m & 0 & 0 \\ 0 & Z_s - Z_m & 0 \\ 0 & 0 & Z_s - Z_m \end{bmatrix}. \tag{B.12}$$

c) Balanced rotating machinery:

$$\mathbf{Z_p} = \begin{bmatrix} Z_s & Z_{m1} & Z_{m2} \\ Z_{m2} & Z_s & Z_{m1} \\ Z_{m1} & Z_{m2} & Z_s \end{bmatrix}, \tag{B.13}$$

$$\mathbf{Z_s} = \begin{bmatrix} Z_s + Z_{m1} + Z_{m2} & 0 & 0 \\ 0 & Z_s + \alpha^2 Z_{m1} + \alpha Z_{m2} & 0 \\ 0 & 0 & Z_s + \alpha Z_{m1} + \alpha^2 Z_{m2} \end{bmatrix}. \tag{B.14}$$

d) Flat configuration untransposed transmission line:

$$\mathbf{Z_p} = \begin{bmatrix} Z_{s1} & Z_{m1} & Z_{m2} \\ Z_{m1} & Z_{s2} & Z_{m1} \\ Z_{m2} & Z_{m1} & Z_{s1} \end{bmatrix}, \tag{B.15}$$

$$\mathbf{Z_s} = \frac{1}{3} \begin{bmatrix} 2Z_{s1} + Z_{s2} + 4Z_{m1} + 2Z_{m2} & -\alpha^2(Z_{s1} - Z_{s2} - Z_{m1} + Z_{m2}) \\ -\alpha(Z_{s1} - Z_{s2} - Z_{m1} + Z_{m2}) & 2Z_{s1} + Z_{s2} - 2Z_{m1} - Z_{m2} \\ -\alpha^2(Z_{s1} - Z_{s2} - Z_{m1} + Z_{m2}) & -\alpha(Z_{s1} - Z_{s2} - Z_{m1} + Z_{m2}) \\ & -\alpha(Z_{s1} - Z_{s2} - Z_{m1} + Z_{m2}) \\ & -\alpha^2(Z_{s1} - Z_{s2} + 2Z_{m1} - 2Z_{m2}) \\ & 2Z_{s1} + Z_{s2} - 2Z_{m1} - Z_{m2} \end{bmatrix}. \tag{B.16}$$

Equations (B.15) and (B.16) become Equations (B.11) and (B.12) when $Z_{s1} = Z_{s2}$ and $Z_{m1} = Z_{m2}$. Other types of phase impedance matrices can be transformed into their sequence impedances by using the general formula provided in Equation (B.8).

B.4 Representations of faults

Balanced and unbalanced faults at system buses are represented by appropriate connections of the symmetrical component networks at the fault buses. Figure B.1 shows schematics of various types of faults, followed by the corresponding symmetrical component network connections [3]. All faults are shown to occur through some impedances. If solid short circuits are to be represented, the corresponding impedances must be set equal to 0. In some cases, phase-shifting transformers are shown. These transformers change the phase angles of the currents and voltages in going from the primary to the secondary side.

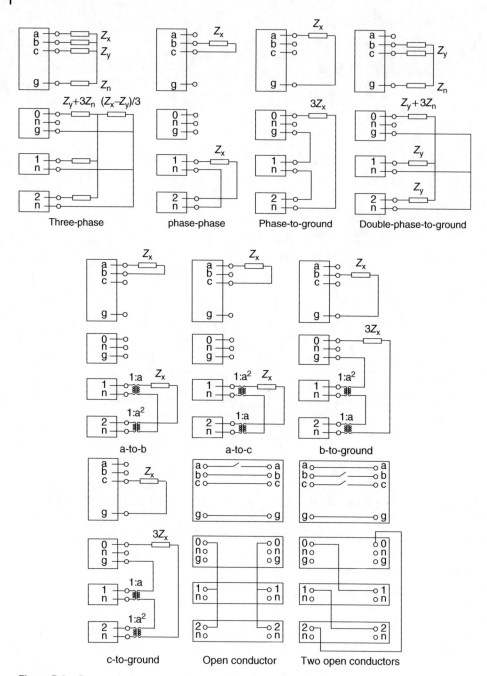

Figure B.1 Symmetrical component connections for various faults.

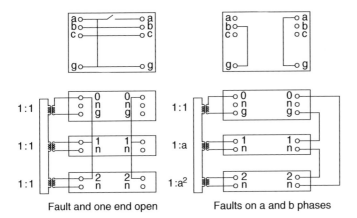

Fault and one end open   Faults on a and b phases

Figure B.1 (Continued)

References

1 Fortescue, C.L. (1918). Method of symmetrical coordinates applied to the solution of polyphase networks. *Trans. AIEE* **37**: 1027–1140.

2 Stevenson, W.D. Jr. (1982). *Elements of Power System Analysis*, 4e. New York: McGraw-Hill.

3 Westinghouse (1976). *Applied Protective Relaying*. Newark, NJ: Westinghouse Electric Corporation.

Appendix C

Power Equipment Parameters

C.1 Typical constants of three-phase synchronous machines

In the following table, the reactances are per unit values on a machine's base [1]. The time constants are in seconds. Average values for a given type of machine are given. Some constants may vary over a wide range, depending upon the design of the machine.

	Two-pole turbine generators	Four-pole turbine generators	Salient-pole machines with dampers	Condensers
X_d	1.10	1.10	1.15	1.80
X_q	1.07	1.08	0.75	1.15
X_d'	0.15	0.23	0.37	0.40
X_d''	0.09	0.14	0.24	0.25
X_2	0.09	0.09	0.24	0.24
X_0	0.0–0.08	0.015–0.14	0.02–0.20	0.02–0.15
T_{do}'	4.4	6.2	5.6	9.0
T_d'	0.6	1.3	1.8	2.0
T_d''	0.035	0.035	0.035	0.035
T_a	0.09	0.2	0.15	0.17

C.2 Typical constants of three-phase transformers

Transformer reactance is given in percent on its own base [2].

BIL of HV winding	BIL of LV winding	Min.	Max.
450	200	10.5	14.5
	350	12.0	17.25

(Continued)

Power System Relaying, Fifth Edition. Stanley H. Horowitz, Arun G. Phadke and Charles F. Henville.
© 2023 John Wiley & Sons Ltd. Published 2023 by John Wiley & Sons Ltd.

(Continued)

BIL of HV winding	BIL of LV winding	Min.	Max.
750	250	12.5	19.25
	650	15.0	24.0
1050	250	14.75	22.0
	825	18.25	27.5
1300	250	16.25	24.0
	1050	20.75	30.5

C.3 Typical constants of three-phase transmission lines

All impedances and susceptances are in ohms and micromhos per mile, respectively, at 60 Hz [3].

a) 362 kV transmission line with flat configuration:

$$\mathbf{Z}_\varphi = \begin{bmatrix} 0.2920 + j\,1.0020 & 0.1727 + j\,0.4345 & 0.1687 + j\,0.3549 \\ 0.1727 + j\,0.4345 & 0.21359 + j\,0.9934 & 0.1727 + j\,0.4345 \\ 0.1687 + j\,0.3549 & 0.1727 + j\,0.4345 & 0.2920 + j\,1.0020 \end{bmatrix},$$

$$\mathbf{Y}_\varphi = \begin{bmatrix} j\,6.341 & -j\,1.115 & -j\,0.333 \\ -j\,1.115 & j\,6.571 & -j1.115 \\ -j\,0.333 & -j\,1.115 & j\,6.341 \end{bmatrix}.$$

b) 800 kV transmission line with flat configuration:

$$\mathbf{Z}_\varphi = \begin{bmatrix} 0.1165 + j\,0.8095 & 0.0097 + j\,0.2691 & 0.0094 + j\,0.2231 \\ 0.0097 + j\,0.2961 & 0.1176 + j\,0.7994 & 0.0097 + j\,0.2691 \\ 0.0094 + j\,0.2213 & 0.0097 + j\,0.2691 & 0.1165 + j\,1.8095 \end{bmatrix},$$

$$\mathbf{Y}_\varphi = \begin{bmatrix} j\,7.125 & -j\,1.133 & -j\,0.284 \\ -j\,1.133 & j\,7.309 & -j\,1.133 \\ -j\,0.284 & -j\,1.133 & j\,7.125 \end{bmatrix}.$$

c) 500 kV parallel transmission lines

$$\mathbf{Z}_\varphi = \begin{bmatrix} Z_{s1} & Z_m \\ Z_m^t & Z_{s2} \end{bmatrix},$$

$$\mathbf{Y}_\varphi = \begin{bmatrix} Y_{s1} & Y_m \\ Y_m^t & Y_{s2} \end{bmatrix},$$

$$
\mathbf{Z}_{\varphi 1} = \begin{bmatrix} 0.1964 + j\,0.9566 & 0.1722 + j\,0.3833 & 0.1712 + j\,0.2976 \\ 0.1722 + j\,0.3833 & 0.2038 + j\,0.9447 & 0.1769 + j\,0.3749 \\ 0.1712 + j\,0.2976 & 0.1769 + j\,0.3749 & 0.2062 + j\,0.9397 \end{bmatrix},
$$

$$
\mathbf{Z}_{\varphi 2} = \begin{bmatrix} 0.2062 + j\,0.9397 & 0.1769 + j\,0.3749 & 0.1712 + j\,0.2976 \\ 0.1769 + j\,0.3749 & 0.2038 + j\,0.9447 & 0.1722 + j\,0.3833 \\ 0.1712 + j\,0.2976 & 0.1722 + j\,0.3833 & 0.1964 + j\,0.9566 \end{bmatrix},
$$

$$
\mathbf{Z}_{\mathrm{m}\varphi}^{\mathrm{t}} = \begin{bmatrix} 0.1642 + j\,0.215 & 0.1701 + j\,0.244 & 0.1742 + j\,0.2863 \\ 0.1600 + j\,0.1922 & 0.167 + j\,0.2120 & 0.1769 + j\,0.3749 \\ 0.1548 + j\,0.1769 & 0.1600 + j\,0.1920 & 0.1642 + j\,0.2150 \end{bmatrix},
$$

$$
\mathbf{Y}_{\varphi 1} = \begin{bmatrix} j\,6.318 & -j\,1.092 & -j\,0.316 \\ -j\,1.092 & j\,6.522 & -j\,1.092 \\ -j\,0.316 & -j\,1.092 & j\,6.388 \end{bmatrix},
$$

$$
\mathbf{Y}_{\varphi 2} = \begin{bmatrix} j6.388 & -j\,1.070 & -j\,0.316 \\ -j\,1.070 & j\,6.522 & -j1.092 \\ -j\,0.316 & -j\,1.092 & j\,6.318 \end{bmatrix},
$$

$$
\mathbf{Y}_{\mathrm{m}\varphi}^{\mathrm{t}} = \begin{bmatrix} -j\,0.069 & -j\,0.143 & -j\,0.434 \\ -j\,0.034 & -j\,0.057 & -j\,0.143 \\ -j\,0.025 & -j\,0.034 & -j\,0.069 \end{bmatrix}.
$$

References

1 Calabrese, G.O. (1969). *Symmetrical Components*. New York: Ronald Press.
2 Weedy, B.M. (1987). *Electric Power Systems*, 3e. New York: Wiley.
3 EPRI (1982). *Transmission Line Reference Book: 345 kV and Above*, 2e. Palo Alto, CA: Electric Power Research Institute.

Appendix D

Inverse Time Overcurrent Relay Characteristics

D.1 Type CO-6 (Courtesy of ABB Power T&D Company)

Definite minimum time relay:
Taps: 1.0, 1.2, 1.5, 2.0, 2.5, 3.0, 3.5, 4.0, 5.0, 6.0, 7.0, 8.0, 10.0, 12.0.

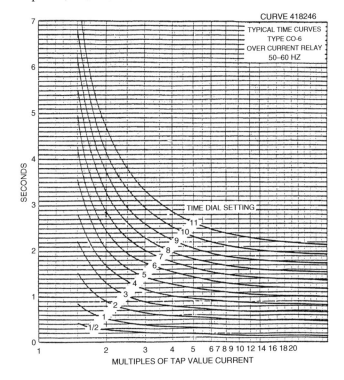

Power System Relaying, Fifth Edition. Stanley H. Horowitz, Arun G. Phadke and Charles F. Henville.
© 2023 John Wiley & Sons Ltd. Published 2023 by John Wiley & Sons Ltd.

D.2   Type CO-11 (Courtesy of ABB Power T&D Company)

Extremely inverse time relay:

Taps: 1.0, 1.2, 1.5, 2.0, 2.5, 3.0, 3.5, 4.0, 5.0, 6.0, 7.0, 8.0, 10.0, 12.0.

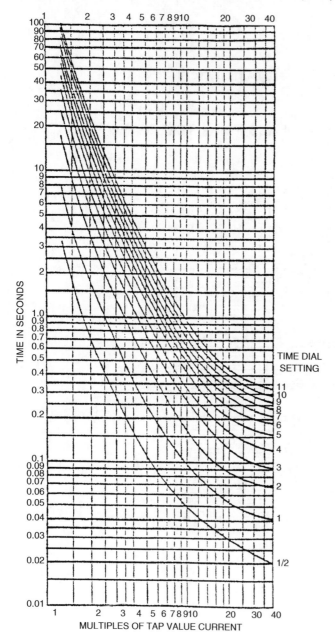

D.3 Type IAC-53 and IAC-54 (Courtesy of General Electric Company)

Very inverse time relay with instantaneous attachment:
Taps: 1.5, 2.0, 2.5, 3.0, 4.0, 5.0, 6.0, 7.0, 8.0, 10.0, 12.0.

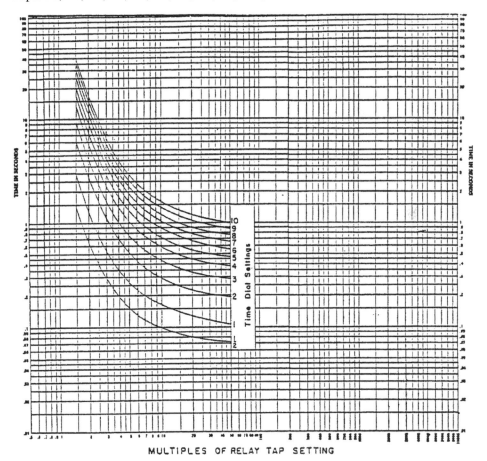

MULTIPLES OF RELAY TAP SETTING

* Figure 9   (0888B0270 [3])   60 Hz Time/Current Characteristics for the Type-IAC53 and IAC54 Relays

Index

Note: **Bold italic** type refers to entries in the Table of Contents and * refers to a Standard Title.

Printed and bound by CPI Group (UK) Ltd, Croydon, CR0 4YY

17/04/2025

14658881-0001